Lecture Notes in Artificial

Edited by R. Goebel, J. Siekmann, a

Subseries of Lecture Notes in Computer Science

Peter Eklund Ollivier Haemmerlé (Eds.)

Conceptual Structures: Knowledge Visualization and Reasoning

16th International Conference
on Conceptual Structures, ICCS 2008
Toulouse, France, July 7-11, 2008
Proceedings

Series Editors

Randy Goebel, University of Alberta, Edmonton, Canada
Jörg Siekmann, University of Saarland, Saarbrücken, Germany
Wolfgang Wahlster, DFKI and University of Saarland, Saarbrücken, Germany

Volume Editors

Peter Eklund
University of Wollongong
School of Information Systems and Technology
Northfields Avenue, Wollongong, NSW 2522 , Australia
E-mail: peklund@uow.edu.au

Ollivier Haemmerlé
Institut de Recherche en Informatique de Toulouse (IRIT)
Université de Toulouse II - Le Mirail
Département de Mathématiques-Informatique
5 allées Antonio Machado, 31058 Toulouse CEDEX, France
E-mail: ollivier.haemmerle@univ-tlse2.fr

Library of Congress Control Number: 2008930415

CR Subject Classification (1998): I.2, G.2.2, F.4.1, F.2.1, H.4

LNCS Sublibrary: SL 7 – Artificial Intelligence

ISSN 0302-9743
ISBN-10 3-540-70595-3 Springer Berlin Heidelberg New York
ISBN-13 978-3-540-70595-6 Springer Berlin Heidelberg New York

Springer is a part of Springer Science+Business Media

springer.com

Typesetting: Camera-ready by author, data conversion by Scientific Publishing Services, Chennai, India
Printed on acid-free paper SPIN: 12322800 06/3180 5 4 3 2 1 0

Preface

This volume contains the proceedings of ICCS 2008, the 16th International Conference on Conceptual Structures (ICCS). The focus of the ICCS conference is the representation and analysis of conceptual knowledge. ICCS brings together researchers to explore novel ways that Conceptual Structures can be used.

Conceptual Structures are motivated by C.S. Peirce's Existential Graphs and were popularized by J.F. Sowa in the 1980s. Over 16 years ICCS has increased its scope to include innovations from a range of theories and related Conceptual Structure practices, among them formal concept analysis and ontologies. Therefore, ICCS presents a family of Conceptual Structure approaches that build on techniques derived from artificial intelligence, knowledge representation, applied mathematics and lattice theory, computational linguistics, conceptual modeling, intelligent systems and knowledge management.

This volume's title – *Knowledge Visualization and Reasoning* – is intended to highlight the shared origins of Conceptual Structures with other visual forms of reasoning. J. Howse's invited survey paper "Diagrammatic Reasoning Systems" sets the scene for this theme, and several other papers in the volume extend and reinforce these connections.

The regular papers in this LNAI volume are split between theoretical and applied contributions. ICCS has traditions in practical systems so the conference includes the one-day Conceptual Structures Tool Interoperability Workshop (CS-TIW 2008) – published as a separate proceedings in the CEUR-WS. Both ICCS 2008 workshop and conference program highlight results achieved with a variety of Conceptual Structures-based software.

The conference also included four invited lectures from distinguished speakers in the field. Three of the four invited speakers submitted accompanying papers which were peer reviewed and are included in this volume. The invited lecture by J.F. Sowa is accompanied by a paper "Pursuing the Goal of Language Understanding," which highlights the computational linguistic aspect of the community and the invited paper "Web, Graphs and Semantics" by O. Corby reports on the implementation of graph theoretical aspects of conceptual graphs, particularly their application to the Semantic Web.

More than 70 papers were submitted to ICCS 2008 for peer review. All papers were assessed by at least three referees one of whom was an Editorial Board member who managed any necessary revisions. The top-ranked 19 papers were selected competitively for this volume. A further 20 papers were published as supplementary proceedings in CEUR-WS.

We wish to thank the Organizing Committee individually: Nathalie Hernandez, Cathy Comparot, Patrice Buche, Lydie Soler, Sophie Ebersold, Jean-Michel

Inglebert, Véronique Debats, Rémi Cavallo, and collectively the Editorial Board and Program Committee members whose input underwrites the scientific quality of these proceedings.

July 2008

Peter Eklund
Ollivier Haemmerlé

Organization

ICCS Executive

General Chair	Ollivier Haemmerlé (Université Toulouse le Mirail)
Program Chair	Peter Eklund (University of Wollongong)
Local Chair	Nathalie Hernandez (Université Toulouse le Mirail)

ICCS Adminstrative

Finance	Cathy Comparot (Université Toulouse le Mirail)
Website	Patrice Buche (INRA Met@risk – Paris)
	Lydie Soler (INRA Met@risk – Paris)
Network and Computers	Jean-Michel Inglebert (Université Toulouse le Mirail)
Logistics	Véronique Debats (IRIT – Toulouse)
Design	Rémi Cavallo (Paris)

ICCS Editorial Board

Galia Angelova (Bulgaria)
Frithjof Dau (Germany)
Aldo de Moor (Netherlands)
Harry Delugach (USA)
Bernhard Ganter (Germany)
Pascal Hitzler (Germany)
Mary Keeler (USA)
Sergei Kuznetsov (Russian Federation)
Guy Mineau (Canada)
Bernard Moulin (Canada)
Marie-Laure Mugnier (France)
Heather D. Pfeiffer (USA)
Simon Polovina (UK)
Uta Priss (UK)
Henrik Schärfe (Denmark)
John F. Sowa (USA)
Rudolf Wille (Germany)
Karl-Erich Wolff (Germany)
Peter Øhrstrøm (Denmark)

ICCS Program Committee

Radim Belohlavek (USA)
Tru Cao (Vietnam)

Dan Corbett (USA)
Madalina Croitoru (UK)
Juliette Dibie-Barthélemy (France)
Pavlin Dobrev (Bulgaria)
David Genest (France)
Udo Hebisch (Germany)
Joachim Hereth Correia (Germany)
Richard Hill (UK)
Adil Kabbaj (Morocco)
Yannis Kalfoglou (UK)
Markus Kroetzsch (Germany)
Leonhard Kwuida (Switzerland)
Sim Kim Lau (Australia)
Robert Levinson (USA)
Michel Liquière (France)
Philippe Martin (France)
Engelbert Mephu Nguifo (France)
Jorgen Fischer Nilsson (Denmark)
Sergei Obiedkov (Russian Federation)
John Old (UK)
Anne-Marie Rassinoux (Switzerland)
Gary Richmond (USA)
Sebastian Rudolph (Germany)
Eric Salvat (France)
Rallou Thomopoulos (France)
William Tepfenhart (USA)
Thomas Tilley (Thailand)
G.Q. Zhang (USA)

Additional Referees

Sebastian Bader
Peter Becker
Patrice Buche
Michel Chein
Vincent Dubois
Maxime Morneau
Amanda Ryan
Bastian Wormuth

Table of Contents

Diagrammatic Reasoning Systems

John Howse

Visual Modelling Group, University of Brighton, Brighton, UK
John.Howse@brighton.ac.uk
http://www.cmis.brighton.ac.uk/research/vmg/

Abstract. Euler diagrams have been used for centuries as a means for conveying logical statements in a simple, intuitive way. They form the basis of many diagrammatic notations used to represent set-theoretic relationships in a wide range of contexts including software modelling, logical reasoning systems, statistical data representation, database search queries and file system management. In this paper we consider some notations based on Euler diagrams, in particular Spider Diagrams and Constraint Diagrams, with particular emphasis on the development of reasoning systems.

Keywords: Visual formalisms, diagrammatic reasoning, automated reasoning, software specification, information visualization.

1 Introduction

Euler diagrams [7] are a simple and familiar visual language for expressing logical or set-theoretic statements. They exploit topological properties of enclosure, exclusion and intersection to represent subset, disjoint sets and set intersection respectively. For example, the Euler diagram d_1 in figure 1 asserts that C is a subset of A and that B and C are disjoint.

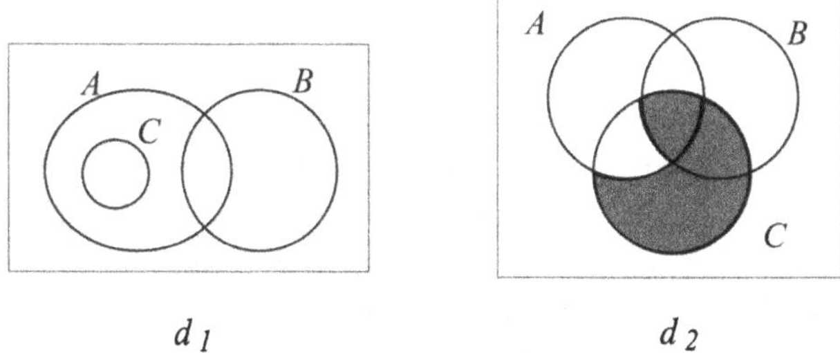

Fig. 1. Euler and Venn diagrams

An Euler diagram consists of a collection of contours (closed curves, usually considered to be simple). A *zone* (sometimes called a *minimal region*) is a set of points in the plane that can be described by a two-way partition of the contour set. For example, in d_1 in figure 1, the set of points in the plane inside A and C

P. Eklund and O. Haemmerlé (Eds.): ICCS 2008, LNAI 5113, pp. 1–20, 2008.

but outside B is a zone. Zones in Euler diagrams represent sets and the union of all the sets represented by the zones in a diagram is the universal set. A "missing" zone represents the empty set. For example, in d_1 in figure 1 the zone that is inside B and C but outside A is missing and hence no element is in B and C but not in A. Some semantic interpretations of Euler diagrams specify that each zone in a diagram represents a non-empty set [38], whereas others do not impose this restriction [24].

Venn diagrams [65] are a special case of Euler diagrams. However, instead of using missing zones to express that a set is empty, shading is used. All possible set intersections are represented in Venn diagrams. The Venn diagram d_2 in figure 1 represents the same information as the Euler diagram d_1 in figure 1. A survey of work on Venn diagrams can be found at [47].

Given certain well-formedness conditions on Euler diagrams (such as contours must be simple), there are statements involving set intersections that Euler diagrams cannot express, identified in [37,66], because there is no drawable diagram with a specified zone set; the proof is based on Kuratowski's theorem for planar graphs [36]. Venn proposed a constructive method for drawing any Venn diagram on n contours which More proved to be valid [39].

Work on reasoning about diagrams expressing logical or set-theoretical properties has a long history, which has been reinvigorated in the last decade or so. In seminal work, Shin [51] demonstrated that diagrammatic reasoning systems could be provided with the logical status of sentential systems. Many other diagrammatic reasoning systems have since been developed and in section 2 we discuss some of them.

Euler diagrams form the basis of many diagrammatic notations used to represent set-theoretic relationships in a wide range of contexts including software modelling, logical reasoning systems, statistical data representation, database search queries, file system management and representing ontologies. They can be used to express logical information. Figure 2 shows a nice example explaining the complexities of the British Isles. This diagram is actually a spider diagram, see section 2. Spider diagrams have also been used in hardware specification [4]; an example related to the safety of power supply components of boiler systems can be seen in figure 3.

Euler diagrams are used in visualizing data of all sorts, particularly data associated with medical experiments or biological processes, for example figure 4 shows output from the GoMiner system visualizing Genetic Set Relations in a Gene Ontology Database [35]. To enable the visualization of statistical data, area-proportional Venn or Euler diagrams [2,3] may be used, where the area of a region is proportional to the size of the set it represents. Figure 5 shows an area-proportional Venn diagram representing heart disease data.

Euler diagrams have been used to represent non-hierarchical directories, replacing the traditional hierarchical structure of file-systems with an Euler diagram based approach [5,6]. An example from the VENNFS system may be seen in figure 6, where the dots placed within a region of overlap of the contours represent files that are in more than one directory.

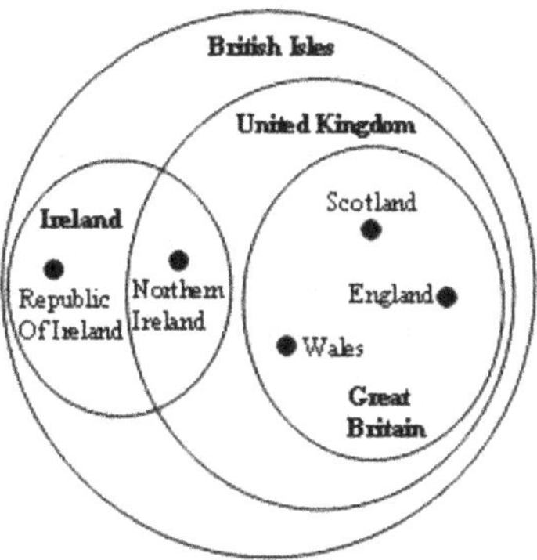

Fig. 2. The British Isles (©Sam Hughes)

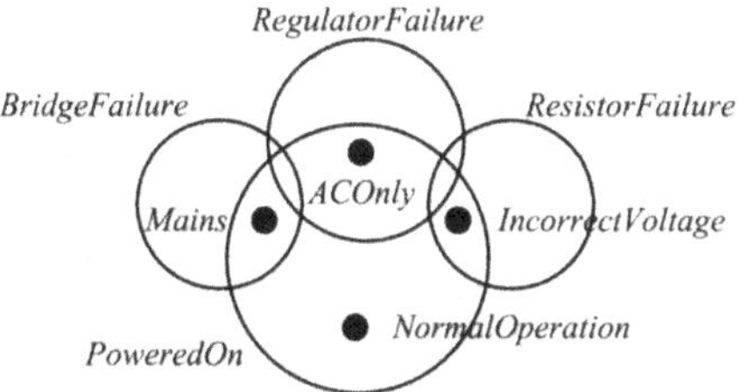

Fig. 3. Safety critical boiler systems

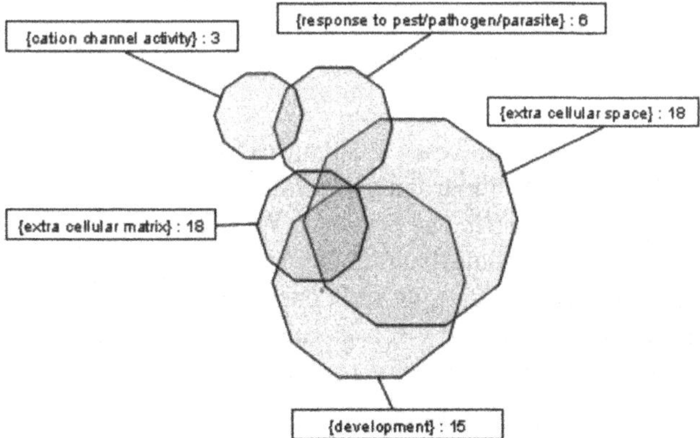

Fig. 4. Visualizing genetic set relations

Euler-based diagrams have been used to represent ontologies in semantic web applications [45,25]. Figure 7 shows an example representing specified values in OWL, the Web Ontology Language. This diagram is a variant of a constraint

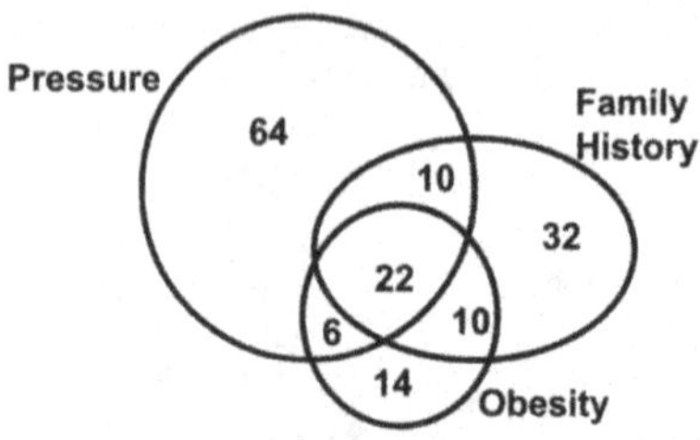

Fig. 5. Heart Disease Data

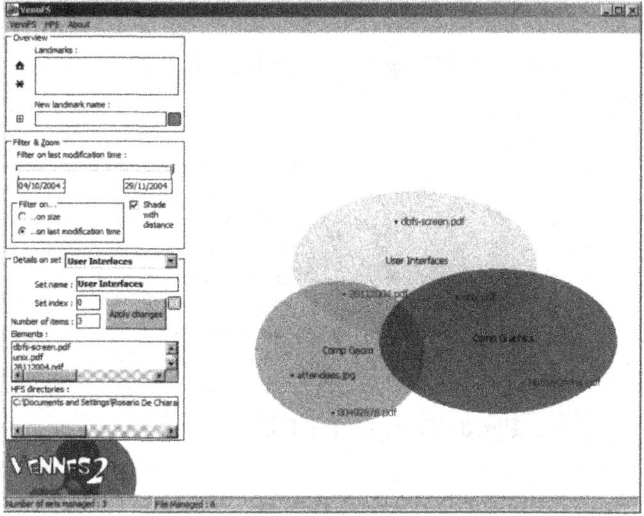

Fig. 6. VennFS2

diagram. Another example can be seen in figure 8 from a new environment COE, the Collaborative Ontology Environment, for capturing and formally representing expert knowledge for use in the Semantic Web [25].

In section 2, we consider some diagrammatic reasoning systems based on Euler diagrams, including spider diagrams and constraint diagrams. In section 3, we discuss diagrammatic reasoning by considering reasoning with spider diagrams. Computer-aided tools are essential for the effective use of diagrammatic notations for practical applications and in section 3 we discuss the tools that have been developed for Euler diagram-based notations and reasoning systems.

2 Reasoning Systems

In this section, we consider some of the notations developed from Euler and Venn diagrams. Reasoning systems have been developed for many of these systems and

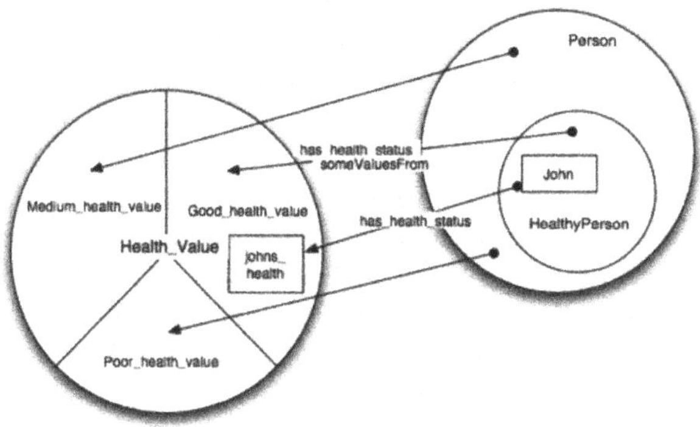

Fig. 7. Web ontology language example

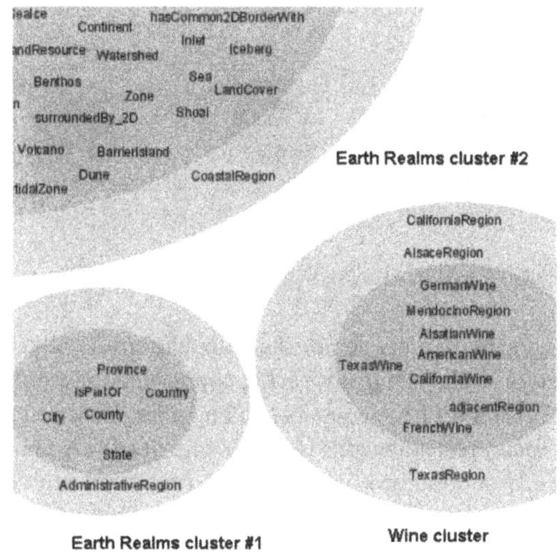

Fig. 8. Collaborative Ontology Environment example

in some cases expressiveness results have been obtained. A survey of reasoning systems based on Euler diagrams can be found at [52].

Basic Euler diagrams. A simple sound, complete and decidable reasoning system based on Euler diagrams is given by Hammer in [24]. The system has just three reasoning rules: *the rule of erasure* (of a contour), *the rule of introduction of a new contour* and *the rule of weakening* (which introduces a zone). A discussion of such reasoning rules is given in section 3

Extensions of Venn diagrams. Venn diagrams cannot assert the existence of elements nor express disjunctive information. To overcome this, Peirce modified

Venn diagrams by introducing the symbol x into the system to represent the existence of an element in a set and o to represent emptiness instead of shading [43]. Peirce also uses lines to connect x's and o's, to represent disjunctive information.

Shin [51] adapted Venn-Peirce diagrams by reverting back to Venn's shading to represent the emptiness of a set rather than using o-sequences, making her Venn-I language less expressive than the Venn-Peirce language. The Venn-I diagram d_1 in figure 9 asserts that there is an element that is a student or a teacher but not both, and the set of *Teachers* is empty. Shin defines six sound reasoning rules for Venn-I and proves completeness. Venn-I cannot express statements of the form $A \subseteq B \vee A \not\subseteq C$, so Shin extends it to a more expressive system, called Venn-II, by allowing Venn-I diagrams to be connected by straight lines to represent disjunction. Shin defines ten reasoning rules for Venn-II and shows that they form a sound and complete set. The Venn-II system is equivalent to monadic first order logic (without equality) [51]. Recently the Venn-II system has been extended to include constants [1].

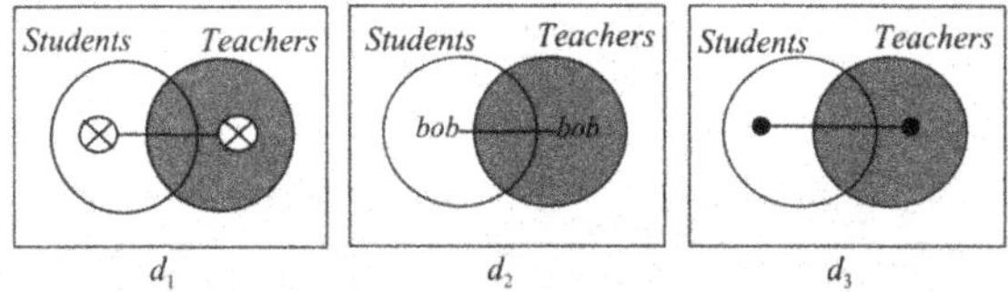

Fig. 9. Venn-I, Euler/Venn and spider diagrams

Euler/Venn Diagrams. Euler/Venn diagrams [60] are similar to Venn-I diagrams but are based on Euler diagrams rather than Venn diagrams, and *constant sequences* are used instead of ⊗-sequences. The Euler/Venn diagram d_2 in figure 9 asserts that *bob* is a student or a teacher but not both and that there are no teachers. In [60], Swoboda gives a set of sound reasoning rules for Euler/Venn diagrams. These rules are extensions of those given by Shin and Hammer [24,51]. In [61] Swoboda and Allwein give an algorithm that determines if a given Euler/Venn monadic first order formula is 'observable' from a given diagram [63]. Information is observable from a diagram if it is explicitly represented in the diagram. Observable formulae are consequence of the information contained in the diagram. Swoboda and Allwein have developed a heterogeneous Euler/Venn diagram and first order logic reasoning system [62].

Spider diagrams. Euler diagrams form the basis of spider diagrams [21,29,30,31,33]. *Spiders* are used to represent the existence of elements and distinct spiders represent the existence of distinct elements. Thus spider diagrams allow finite lower bounds to be placed on the cardinalities of sets. In a shaded region, all of the elements are represented by spiders. So shading, together with spiders, allows finite upper bounds to be placed on the cardinalities of the sets. The spider diagram d_3 in figure 9 asserts that there is an element that is a student or a teacher but not both, and there are no other teachers.

The expressiveness of spider diagrams has been determined to be equivalent to that of monadic first order logic with equality (MFOLe) [55,59]. To show this equivalence in one direction, for each diagram a semantically equivalent MFOLe sentence is constructed. For the significantly more challenging converse it can be shown that for any MFOL sentence S there exists a finite set of models that can be used to classify all the models for S. Using these classifying models, a diagram expressing the same information as S can be constructed. Spider diagrams are, therefore, more expressive than Shin's Venn-II system which is equivalent to monadic first order logic (without equality) [51]. Augmenting the spider diagram language with constants does not increase expressiveness [57].

Several sound, complete and decidable reasoning systems [26,29,27,33] have been developed for spider diagrams. The strategy for proving soundness is straightforward: individual reasoning rules are proved to be valid and then a simple induction argument shows that the application of any finite sequence of rules is valid; however, proving that individual rules are valid is, in some cases, hard. The proof strategy for completeness is to convert the premise and conclusion diagrams to a normal form and then reason about that normal form. This proof strategy can be used for other reasoning systems based on Euler diagrams. An algorithm can be easily extracted from the completeness proof to prove the decidability of the proof.

Constraint diagrams. Constraint diagrams [20,22,34] extend spider diagrams by incorporating additional syntax to represent relations and explicit universal quantification. Figure 10 shows an example of a constraint diagram [32], from which one can infer that "a member cannot both rent and reserve the same title".

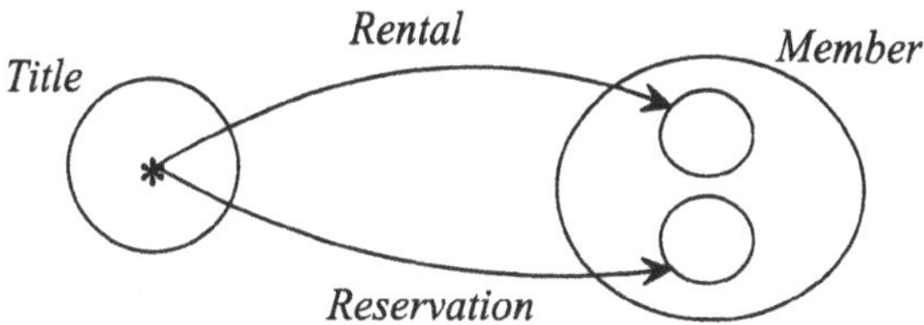

Fig. 10. Modelling with Constraint diagrams

The standard notation for modelling software systems is the Unified Modelling Language (UML) [41]. Diagrammatic notations pervade the UML. Some of these notations are based on Euler diagrams such as Class diagrams and State diagrams. The principal tool for the UML modeller to add constraints to a model is the Object Constraint Language (OCL) [67]. However, OCL is a fusion of navigation expressions and traditional logical notation, rendered in textual form. Constraint diagrams were designed to be a formal diagrammatic alternative to the OCL and can also be used to specify software systems independently of the UML. In [32] a case study is developed which uses a schema notation, developed from a Z-like notation [48,49], to specify operations. Constraint diagrams are used within this schema notation, showing that they can handle dynamic constraints. An event (a state-changing operation) is specified in terms of a pre-condition (above the double line) and a post-condition (below the double line). The following schema specifies the addition of a new member m with associated information i:

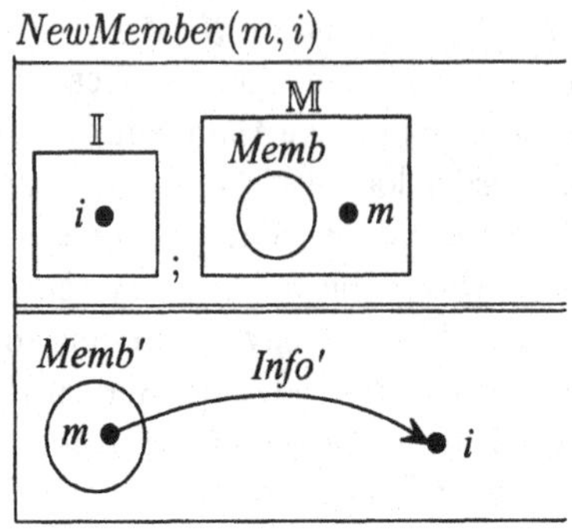

The pre-condition ensures that argument i has type $\mathbb{I}$, and that argument m has type $\mathbb{M}$ and is not in *Memb* (is not already a member). The semi-colon is a separator; the two diagrams in the pre-condition are conjoined. In the post-condition, dashed names denote entities that are changed. The post-condition ensures that m is now in *Memb* and has associated information i.

Constraint diagrams have been formalized [10]. The semantics are defined by translating them into first order predicate logic (FOPL) sentences. Constraint diagrams contain explicit existential quantification and universal quantification; it is not always possible to determine the order in which to read the quantifiers, sometimes rendering a diagram ambiguous. This ordering problem was solved by augmenting the language with *reading trees*, essentially a partial order on the quantifiers, to disambiguate the diagrams [9,10]. The tree provides additional information that is essential for the construction of the FOPL sentence determining where the brackets are placed and, in conjunction with the diagram, the scope of the quantifiers. Figure 11 shows two augmented diagrams with the two interpretations: "For each teacher there is a student who attends only courses taught by that teacher" and "There is a student who attends only courses taught by all teachers", respectively.

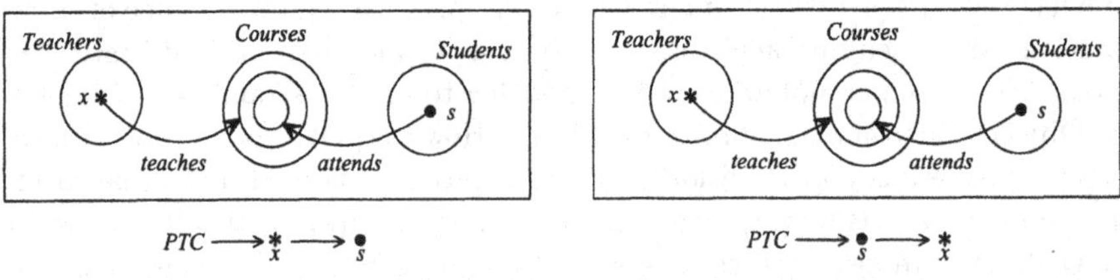

Fig. 11. Augmented constraint diagrams

A constraint diagram may have many reading trees, which can be automatically generated [11]. A set of sound rules for constraint diagrams augmented with reading trees has been developed [8] and a default reading has been proposed [12] and tested [13].

Two sound, complete and decidable fragments of the constraint diagram language have been defined [53,54]. The diagrams in these fragments do not require reading trees, but still include arrows (representing two-place predicates) and one of them includes explicit universal quantification [53]. Some of the reasoning rules for these two systems extend those defined for spider diagrams and additional rules are also defined to give complete systems. The proofs of completeness for these systems are complex.

Abstract syntax. In reasoning systems based on spider diagrams and constraint diagrams a distinction is made between the concrete syntax (the drawn diagrams) and the abstract syntax (a mathematical abstraction of concrete diagrams) [28]; this distinction is not evident in purely textual logics. Reasoning takes place at the abstract level; a motivation for this is that well-formedness conditions may cause problems with the applications of some of the rules at the concrete level; see [50] for an example of this from Shin's Venn II system. Using an abstract syntax brings with it, importantly, a level of precision and rigour that is not always present in diagrammatic systems.

3 Reasoning

In this section we discuss diagrammatic reasoning, by considering reasoning with spider diagrams. There are many ways in which we can reason with diagrams. One way is to interpret a diagram and then make inferences from the information obtained. For example, in figure 12 we can deduce from the diagram that there is an element in A and then infer that there is an element in $A \cup B$.

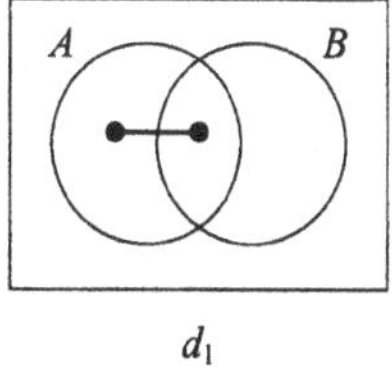

Fig. 12. Deducing information from a diagram

We can also reason about diagrams by comparing them. In figure 13 we can deduce that the two diagrams contain the same information. The diagram d_1 asserts that A and B are disjoint, because these contours do not overlap. The diagram d_2 expresses this same information rather differently; the region inside both contours A and B is shaded, so we can deduce that the sets are disjoint. Thus the two diagrams are semantically equivalent.

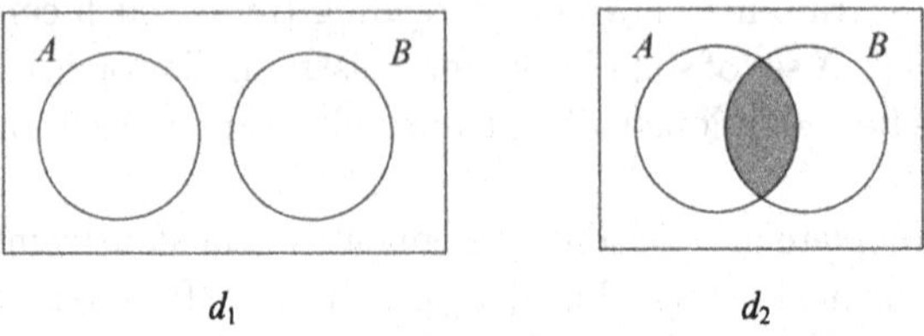

Fig. 13. Equivalent diagrams

In these styles of reasoning, we directly consider the informational content of the diagrams and make inferences about that information. However, we need to be careful when reasoning in this way. Consider the diagram in figure 14. This

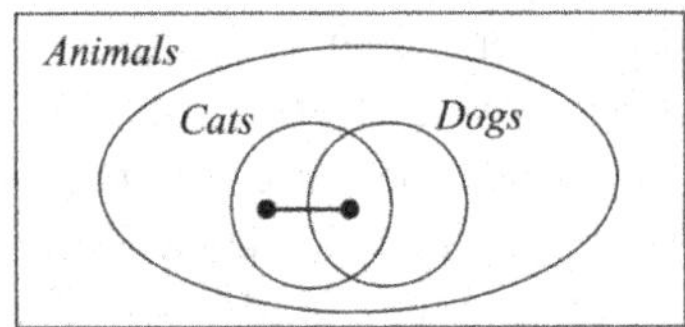

Fig. 14. Cats and dogs

diagram asserts that cats and dogs are animals and that there is an animal that is a cat. From this assertion we might be tempted to infer that there is an animal that is not a dog. However, this is not a valid inference from the diagram; to make this 'inference' we are using knowledge about cats and dogs that is not asserted by the diagram. We are assuming that if an animal is a cat, then it is not a dog; the diagram does not make this assertion.

In order not to make this kind of reasoning error, we can reason with spider diagrams by manipulating the diagrams themselves, using *reasoning rules* which transform one diagram into another. Reasoning rules act on a diagram by introducing, deleting or changing pieces of syntax. For the reasoning rule to be sound the interpretation of the resulting diagram must be derivable from the interpretation of the original diagram. The following examples will illustrate some of the rules, concentrating on those that transform a unitary diagram into another unitary diagram.

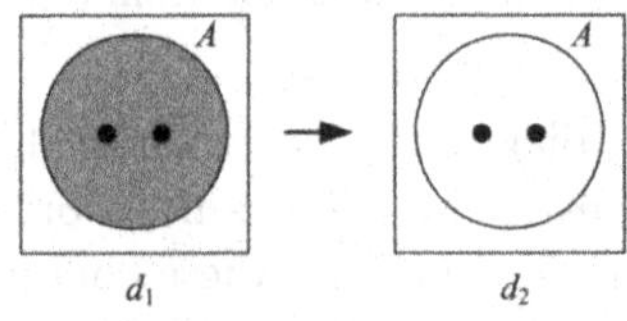

Fig. 15. Removing shading

Example 1. We can remove shading from any zone. In figure 15, d_1 is transformed into d_2 by removing shading. The diagram d_1 expresses that A has exactly two elements, while d_2 expresses that A has at least two elements. The reasoning step allows us to deduce that $|A| \geq 2$ from $|A| = 2$; this deduction is obviously sound. □

Example 2. We can add a foot to a spider. The new foot can be placed in any zone not already containing a foot of the spider. In figure 16, d_1 is transformed

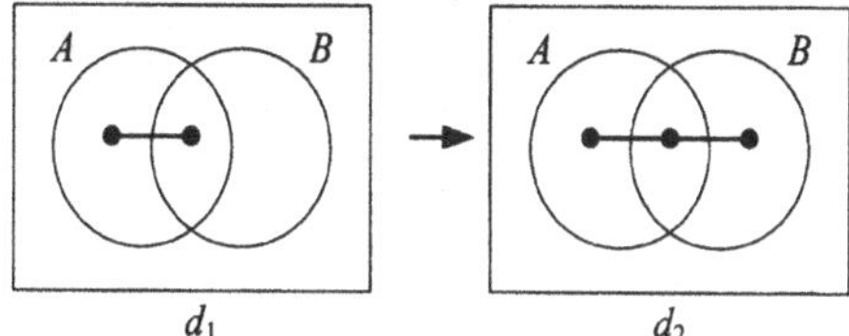

Fig. 16. Adding a foot to a spider

into d_2 by adding a foot, placed in the zone that is inside contour B but outside contour A. The diagram d_1 expresses that there is an element in A, while d_2 expresses that there there is an element in $A \cup B$, providing us with a diagrammatic version of the reasoning step concerning figure 12 with which we began this section. □

Example 3. We can remove a spider. In figure 17, d_1 is transformed into d_2 by

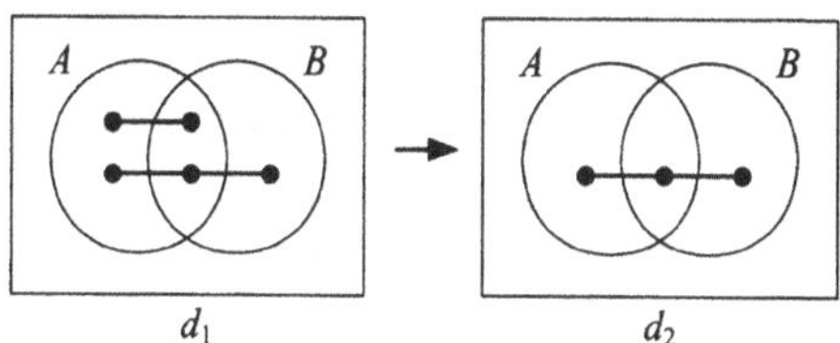

Fig. 17. Removing a spider

removing one of the spiders. The diagram d_1 asserts that there is an element in A and a different element in $A \cup B$, while d_2 asserts that there is an element in $A \cup B$; we have lost information, but the inference is obviously sound. □

However, we have to be careful when applying the remove spider rule. Consider figure 17, in which removing the spider from d_1 would produce d_2. The diagram d_1 asserts that A has exactly one element, while d_2 asserts that A is empty; we obviously cannot infer d_2 from d_1. We cannot remove a spider from a shaded region.

Some rules, such as the three described above, weaken the informational content of a diagram. The next rule does not weaken information, so the two diagrams (the original diagram and the diagram produce as a result of applying the rule) have the same meaning.

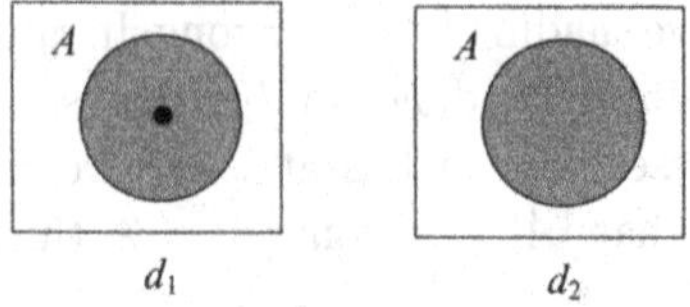

Fig. 18. Removing a spider is sometimes invalid

Example 4. A diagram containing a missing zone can be replaced with a diagram in which that zone is present but shaded. In d_1 in figure 19, the contours A and B do not overlap and hence the zone that is inside both A and B is missing; in d_2 that zone has been added and is shaded. Both diagrams assert that the sets A and B are disjoint. Thus *add shaded zone* is a diagrammatic reasoning rule for the reasoning step considered in figure 13. □

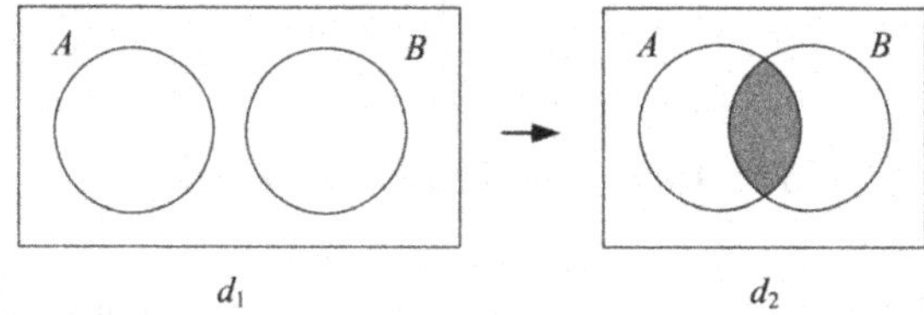

Fig. 19. Adding a shaded zone

As the two diagrams in figure 19 contain the same information, it seems reasonable to suggest a rule that removes a shaded zone, reversing the process of adding a zone. Figure 20 is an example of such a rule; it is the reverse of the application of the rule to add a shaded zone illustrated in figure 19. However, we must be careful when we apply the rule to remove a shaded zone. Consider the diagrams d_1 and d_2 of figure 21. The diagram d_1 asserts that there is at most one element in $A \cap B$, while d_2 asserts that A and B are disjoint and hence that $A \cap B$ is empty; we therefore cannot deduce d_2 from d_1. We have to restrict the conditions under which we can apply a rule to remove a shaded zone.

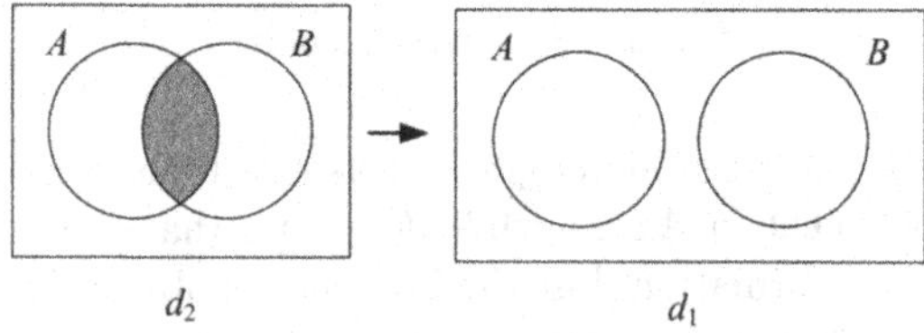

Fig. 20. Removing a shaded zone

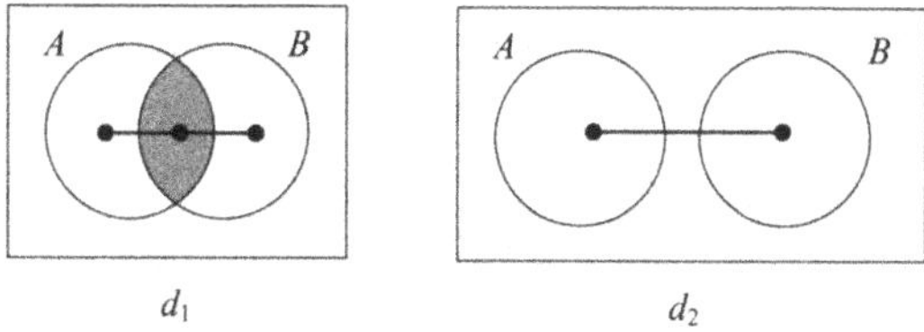

Fig. 21. Removing a shaded zone is not valid in this case

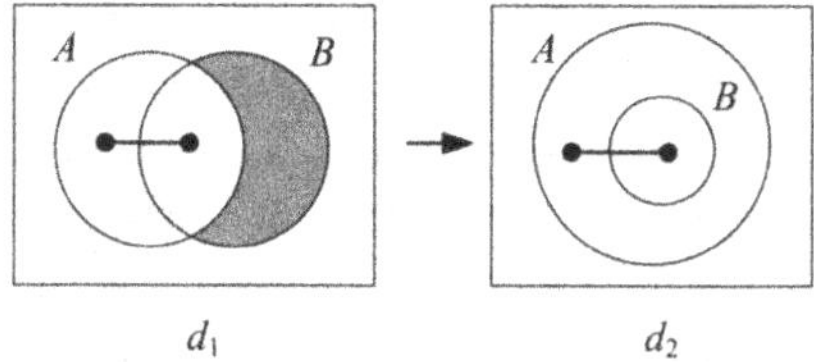

Fig. 22. Removing a shaded zone

Example 5. A shaded zone that is untouched by any spider can be removed. Figure 22 shows another (legitimate) application of this rule. The shaded zone of d_1 has been removed to produce d_2. In d_2, contour B is contained within contour A. The spider is unaffected, in that its habitat remains the same. Both diagrams assert that B is a subset of A and that there is an element in A. This rule does not weaken information. It is interesting to note that d_2 bears little structural similarity to d_1.

The diagram d_1 in figure 23 asserts that A is not empty and there is exactly one element that is in B but not in A. Now consider d_2. It asserts that A is not empty and there is exactly one element that is in B but not in A *and nothing else.* The two diagrams have exactly the same informational content. The extra contour C in d_2 has no effect on the meaning of the diagram. This suggests that there should be a rule to add a contour to a diagram in such a way that it does not affect the information in the diagram.

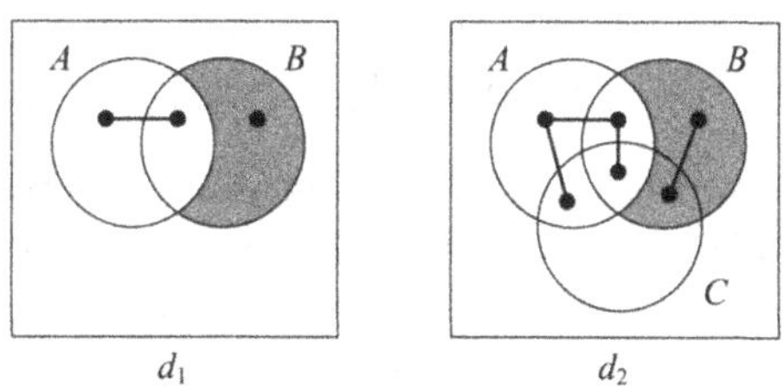

Fig. 23. Adding a contour

Consider now the diagrams in figure 24. A contour has been added to d_1 to produce d_2. However, d_2 asserts that $A - B$ is not empty, while d_1 asserts the weaker statement that A is not empty. Hence, we cannot infer d_2 from d_1.

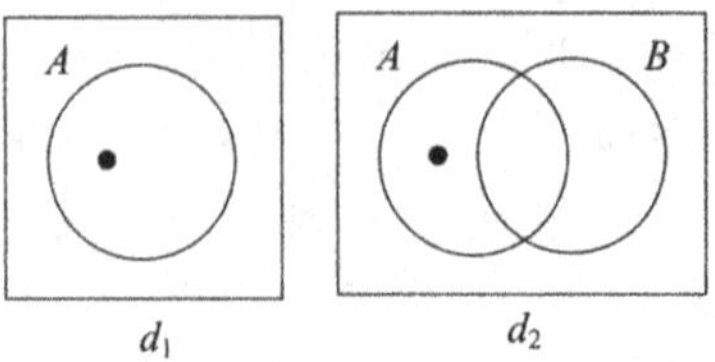

Fig. 24. Adding a contour that introduces new information

Example 6. We can add a contour to a diagram. When we add a contour to a diagram we must do so in a way that does not introduce new information. This is achieved by introducing a contour in such a way that it splits each zone into two zones, each shaded zone into two zones, and each foot of a spider is replaced by a connected pair of feet, one foot in each zone. Figure 25 shows an example of the rule to add a contour; it "fixes" the invalid example given in figure 24. □

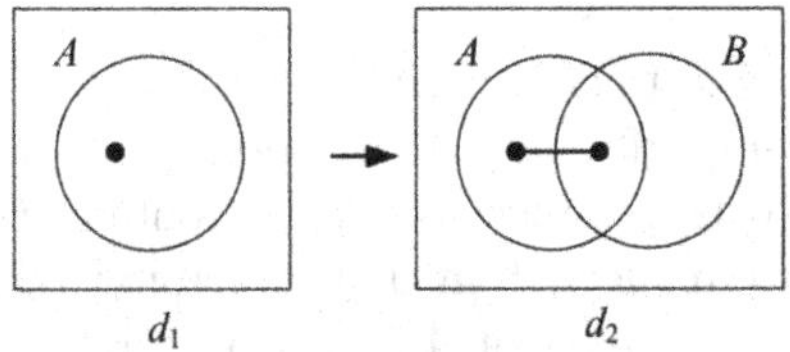

Fig. 25. Adding a contour

Example 7. We can remove a contour from a diagram. When a contour is removed, two feet of the same spider can be placed in the same zone, so we sometimes need to tidy the diagram up a bit, as in figure 26. □

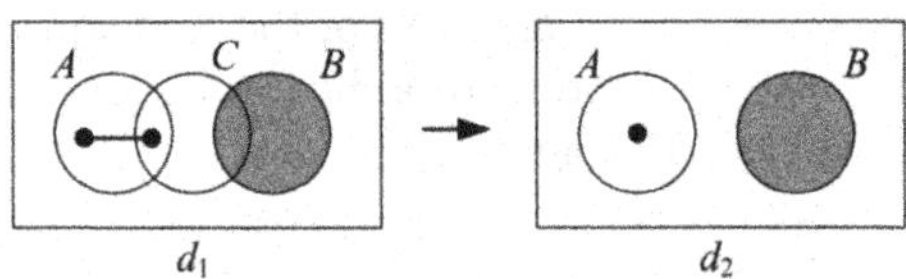

Fig. 26. Removing a contour

We can use sequences of diagrams to obtain further inferences. We conclude this section with an example of a theorem both stated and proved diagrammatically.

Theorem

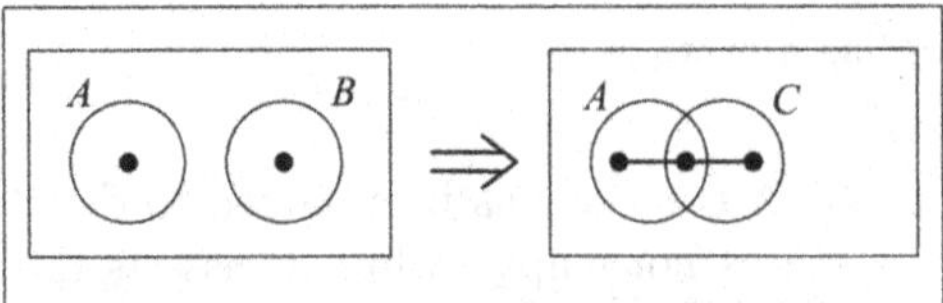

Proof

We assume the premiss diagram:

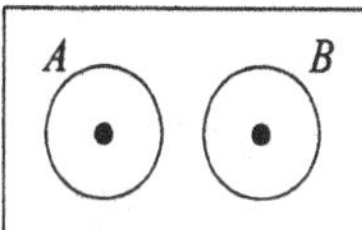

From the premiss diagram we remove the spider in B:

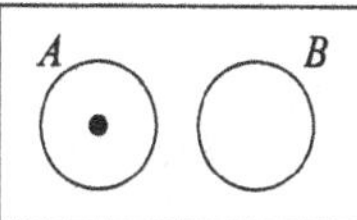

We remove contour B:

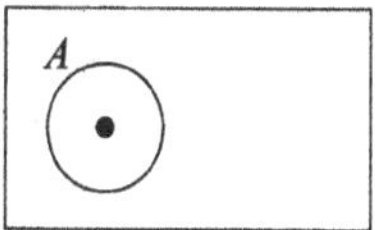

We add a new contour C:

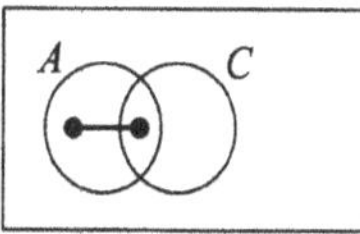

We extend the spider into the zone contained in C but not in A, to obtain the conclusion diagram:

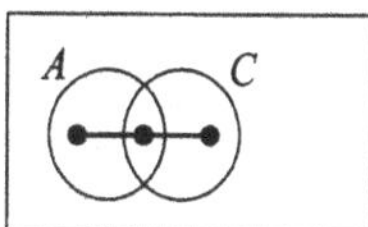

□

4 Software Tools

Work on spider diagrams and constraint diagrams is part of an ongoing project to develop formal visual notations and associated tool support. The effective use of diagrammatic notations for practical applications requires computer-aided support tools. An open source tool for drawing and manipulating Euler diagrams and their extensions can be downloaded from SourceForge.net and at [44]. The editor provides diagram drawing facilities such as editing, cut and paste, and zooming functionality. Diagrams can be laid out automatically and stored in XML format. Associated software tools produced include diagrammatic theorem provers, translators and automatic diagram generators.

Users can access the reasoning functionality from the editor. The interface provides access to theorem provers and allows users to write their own proofs. Tableaux [42] give users a way of visualizing the meaning of a particular diagram, by showing the ways that a diagram can be satisfied. In particular tableaux provide decision procedures for diagram satisfiability and validity. In order to support this varied functionality the software provides sophisticated support for representing and modifying both abstract diagrams (without layout information) and concrete diagrams (with layout information). In addition, translations between diagrammatic and textual representations have been implemented.

The layout of diagrams is of fundamental importance to diagrammatic reasoning systems. Their automatic layout poses several non-trivial challenges. For example, the problem of automatically generating concrete Euler diagrams from abstract descriptions is hard. Algorithm exist to generate concrete diagrams subject to some well-formedness conditions [14,3]. The mechanisms have been integrated to produce an enhanced diagram generation framework [44,64]. The theory developed on nested diagrams [15] has been integrated into this framework.

Automatically generated Euler diagrams are typically not very readable and can be visually unattractive. A function has been implemented to make the diagrams more usable by modifying their layout, whilst maintaining their abstract syntax [18]. Further work introduced a force based method for laying out graphs in Euler diagrams [40], enabling the drawing of spider and constraint diagrams. Furthermore, a key application of the layout work is to visualize sequences of diagrams, such as proofs. For this application (and others), it is desirable to make subsequent diagrams look as similar as possible to previous diagrams. A mechanism to achieve this has been implemented [46].

Gil and Sorkin have also developed an effective but slightly limited editor for drawing and manipulating constraint diagrams [23].

Automated theorem proving. An automated theorem prover has been implemented and evaluated that uses four different Euler diagram reasoning systems [56]. It uses heuristics to guide it through the search space to find shortest proofs. The theorem prover has been empirically evaluated in terms of time taken to find a shortest proof, using each of the four rule sets. The conclusion from this evaluation is that in order to find a shortest proof most quickly, the rule set used is dependent on the proof task [56]. This work on automated reasoning lays the foundations for efficient proof searches to be conducted in many diagrammatic systems.

For spider diagrams, a direct proof writing algorithm can be extracted from the completeness proof strategy given in [33]. An improved version of this algorithm includes functionality to produce counter examples whenever there is no proof [19]. The proofs produced by this algorithm can sometimes be unnecessarily long. In [17] the A^* search algorithm is utilized to produce shortest proofs in a fragment of the spider diagram language and the work has been extended to the full spider diagram language [16].

5 Conclusion

This paper has reviewed and discussed some of the diagrammatic reasoning systems based on Euler diagrams that have been developed recently. Whilst we have not given an exhaustive review, due to space constraints, we have presented an informative overview of current Euler diagrams research. We have concentrated on notations that have been formalized and tried to give a flavour of the applications of these notations which include data representation and database search queries. A growing application area which will become increasingly important is the application of Euler diagrams to visualize information in the context of the semantic web, including OWL and description logic.

Euler diagram based modelling notations have been developed that are sufficiently expressive to be used in software specification on an industrial scale. The development of good software tools, some of which has been described in this paper, is a major advance towards providing sufficient support for the use of these notations in industry.

Research into Euler diagram based notations could be beneficial in other areas. For example, the investigation of decidable fragments of the constraint diagram notation may well deliver previously unknown decidable fragments of first order predicate logic, because “natural” fragments of the diagrammatic notation may not coincide with “natural” fragments of traditional logic.

Acknowledgements. Much of the work described in this paper was developed on the Reasoning with Diagrams and the Visualization with Euler Diagrams projects [64] funded by the UK EPSRC under grants GR/R63509, GR/R63516, EP/E010393 and EP/E011160. We would like to thank Stirling Chow, Robin Clark, Rosario De Chiara, Sam Hughes and Hans Kestler for permission to use figures from their work.

References

1. Choudhury, L., Chakraborty, M.K.: On Extending Venn Diagrams by Augmenting Names of Individuals. In: Blackwell, A.F., Marriott, K., Shimojima, A. (eds.) Diagrams 2004. LNCS (LNAI), vol. 2980, pp. 142–146. Springer, Heidelberg (2004)
2. Chow, S., Ruskey, F.: Drawing Area Proportional Venn and Euler Diagrams. In: Liotta, G. (ed.) GD 2003. LNCS, vol. 2912, pp. 466–477. Springer, Heidelberg (2004)
3. Chow, S., Ruskey, F.: Towards a General Solution to Drawing Area-Proportional Euler Diagrams. In: Proceedings of Euler Diagrams 2004. Electronic Notes in Theoretical Computer Science, vol. 134, pp. 3–18 (2005)
4. Clark, R.P.: Failure mode modular de-composition using spider diagrams. In: Proceedings of Euler Diagrams 2004. Electronic Notes in Theoretical Computer Science, vol. 134, pp. 19–31 (2005)
5. Erra, U., De Chiara, R., Scarano, V.: Vennfs: A venn diagram file manager. In: Proceedings of Information Visualisation, pp. 120–126. IEEE Computer Society, Los Alamitos (2003)
6. Erra, U., De Chiara, R., Scarano, V.: A system for virtual directories using euler diagrams. In: Proceedings of Euler Diagrams 2004. Electronic Notes in Theoretical Computer Science, vol. 134, pp. 33–53 (2005)
7. Euler, L.: Lettres a une princesse dallemagne sur divers sujets de physique et de philosophie. Letters 2, 102–108 (1775) (Berne, Socit Typographique)
8. Fish, A., Flower, J.: Investigating reasoning with constraint diagrams. In: Visual Language and Formal Methods, Rome, Italy. ENTCS, vol. 127, pp. 53–69. Elsevier, Amsterdam (2004)

9. Fish, A., Flower, J., Howse, J.: A reading algorithm for constraint diagrams. In: IEEE Symposium on Human Centric Computing Languages and Environments, Auckland, New Zealand, pp. 161–168. IEEE, Los Alamitos (2003)
10. Fish, A., Flower, J., Howse, J.: The semantics of augmented constraint diagrams. Journal of Visual Languages and Computing 16, 541–573 (2005)
11. Fish, A., Howse, J.: Computing reading trees for constraint diagrams. In: Pfaltz, J.L., Nagl, M., Böhlen, B. (eds.) AGTIVE 2003. LNCS, vol. 3062, pp. 260–274. Springer, Heidelberg (2004)
12. Fish, A., Howse, J.: Towards a default reading for constraint diagrams. In: Blackwell, A.F., Marriott, K., Shimojima, A. (eds.) Diagrams 2004. LNCS (LNAI), vol. 2980, pp. 51–65. Springer, Heidelberg (2004)
13. Fish, A., Masthoff, J.: An Experimental Study into the Default Reading of Constraint Diagrams. Visual Languages and Human Centric Computing, 287–289 (2005)
14. Flower, J., Howse, J.: Generating Euler diagrams. In: Hegarty, M., Meyer, B., Narayanan, N.H. (eds.) Diagrams 2002. LNCS (LNAI), vol. 2317, pp. 61–75. Springer, Heidelberg (2002)
15. Flower, J., Howse, J., Taylor, J.: Nesting in Euler diagrams: syntax, semantics and construction. Software and Systems Modelling 3, 55–67 (2004)
16. Flower, J., Masthoff, J., Stapleton, G.: Generating proofs with spider diagrams using heuristics. In: Proceedings of Distributed Multimedia Systems, International Workshop on Visual Languages and Computing, pp. 279–285. Knowledge Systems Institute (2004)
17. Flower, J., Masthoff, J., Stapleton, G.: Generating readable proofs: A heuristic approach to theorem proving with spider diagrams. In: Blackwell, A.F., Marriott, K., Shimojima, A. (eds.) Diagrams 2004. LNCS (LNAI), vol. 2980, pp. 166–181. Springer, Heidelberg (2004)
18. Flower, J., Rodgers, P., Mutton, P.: Layout metrics for Euler diagrams. In: 7th International Conference on Information Visualisation, pp. 272–280. IEEE Computer Society Press, Los Alamitos (2003)
19. Flower, J., Stapleton, G.: Automated theorem proving with spider diagrams. In: Proceedings of Computing: The Australasian Theory Symposium (CATS 2004), Dunedin, New Zealand. ENTCS, vol. 91, pp. 116–132. Science Direct (January 2004)
20. Gil, J., Howse, J., Kent, S.: Constraint Diagrams: a step beyond UML. In: Proc. TOOLS USA 1999, pp. 453–463. IEEE Computer Society Press, Los Alamitos (1999)
21. Gil, J., Howse, J., Kent, S.: Formalising spider diagrams. In: Proceedings of IEEE Symposium on Visual Languages (VL 1999), Tokyo, pp. 130–137. IEEE Computer Society Press, Los Alamitos (1999)
22. Gil, J., Howse, J., Kent, S.: Towards a Formalization of Constraint Diagrams. In: Proc. IEEE Symposia on Human-Centric Computing (HCC 2001), Stresa, Italy, pp. 72–79. IEEE Press, Los Alamitos (2001)
23. Gil, J., Sorkin, Y.: The constraint diagrams editor, `www.cs.technion.ac.il/Labs/ssdl/research/cdeditor/`
24. Hammer, E.M.: Logic and Visual Information. CSLI Publications (1995)
25. Hayes, P., Eskridge, T., Saavedra, R., Reichherzer, T., Mehrotra, M., Bobrovnikoff, D.: Collaborative Knowledge Capture in Ontologies. In: Proceedings of the 3rd International Conference on Knowledge Capture, pp. 99–106 (2005)
26. Howse, J., Molina, F., Taylor, J.A.: Sound and complete diagrammatic reasoning system. In: Proc. Artificial Intelligence and Soft Computing (ASC 2000), pp. 402–408. IASTED/ACTA Press (2000)

27. Howse, J., Molina, F., Taylor, J.: SD2: A sound and complete diagrammatic reasoning system. In: Proc. IEEE Symp on Visual Languages (VL 2000), pp. 127–136. IEEE Press, Los Alamitos (2000)
28. Howse, J., Molina, F., Shin, S.-J., Taylor, J.: On diagram tokens and types. In: Hegarty, M., Meyer, B., Narayanan, N.H. (eds.) Diagrams 2002. LNCS (LNAI), vol. 2317, pp. 146–160. Springer, Heidelberg (2002)
29. Howse, J., Molina, F., Taylor, J.: On the completeness and expressiveness of spider diagram systems. In: Anderson, M., Cheng, P., Haarslev, V. (eds.) Diagrams 2000. LNCS (LNAI), vol. 1889, pp. 26–41. Springer, Heidelberg (2000)
30. Howse, J., Molina, F., Taylor, J., Kent, S.: Reasoning with spider diagrams. In: Proceedings of IEEE Symposium on Visual Languages (VL 1999), Tokyo, pp. 138–147. IEEE Computer Society Press, Los Alamitos (1999)
31. Howse, J., Molina, F., Taylor, J., Kent, S., Gil, J.: Spider diagrams: A diagrammatic reasoning system. Journal of Visual Languages and Computing 12(3), 299–324 (2001)
32. Howse, J., Schuman, S.: Precise visual modelling. Journal of Software and Systems Modeling 4, 310–325 (2005)
33. Howse, J., Stapleton, G., Taylor, J.: Spider diagrams. LMS J. Computation and Mathematics 8, 145–194 (2005)
34. Kent, S.: Constraint diagrams: Visualizing invariants in object oriented modelling. In: Proceedings of OOPSLA 1997, pp. 327–341. ACM Press, New York (1997)
35. Kestler, H., Müller, A., Gress, T.M., Buchholz, M.: Generalized Venn diagrams: a new method of visualizing complex genetic set relations. Bioinformatics 21(8), 1592–1595 (2005)
36. Kuratowski, K.: Sur le probleme des courbes gauches en topologie. Fundamenta Mathematicae 15, 271–283 (1930)
37. Lemon, O.: Comparing the efficacy of visual languages. In: Barker-Plummer, D., Beaver, D.I., van Benthem, J., Scotto di Luzio, P. (eds.) Words, Proofs and Diagrams, pp. 47–69. CSLI Publications (2002)
38. Lemon, O., Pratt, I.: Spatial logic and the complexity of diagrammatic reasoning. Machine GRAPHICS and VISION 6(1), 89–108 (1997)
39. More, T.: On the construction of Venn diagrams. J. Symb. Logic 24, 303–304 (1959)
40. Mutton, P., Rodgers, P., Flower, J.: Drawing graphs in Euler diagrams. In: Blackwell, A.F., Marriott, K., Shimojima, A. (eds.) Diagrams 2004. LNCS (LNAI), vol. 2980, pp. 66–81. Springer, Heidelberg (2004)
41. OMG. UML 2.0 Specification (2004), `http://www.omg.org`
42. Patrascoiu, O., Thompson, S., Rodgers, P.: Tableaux for diagrammatic reasoning. In: Proceedings of the Eleventh International Conference on Distributed Multimedia Systems. International Workshop on Visual Languages and Computing, pp. 279–286. Knowledge Systems Institute (September 2005)
43. Peirce, C.: Collected Papers, vol. 4. Harvard University Press (1933)
44. Reasoning with Diagrams Project Website (2008), `http://www.cs.kent.ac.uk/projects/rwd/`
45. Rector, A.: Specifying Values in OWL: Value Partitions and Value Sets, W3C Editors Draft 02 (2005)
46. Rodgers, P., Mutton, P., Flower, J.: Dynamic Euler diagram drawing. In: Visual Languages and Human Centric Computing, Rome, Italy, pp. 147–156. IEEE Computer Society Press, Los Alamitos (2004)
47. Ruskey, F.: A Survey of Venn Diagrams. The Electronic Journal of Combinatorics 4 (1997) (update 2001)

48. Schuman, S.A., Pitt, D.H.: Object-oriented subsystem specification. In: Meertens, L.G.L.T. (ed.) Program Specification and Transformation. Proc. IFIP Working Conference, pp. 313–341. North–Holland (1987)
49. Schuman, S.A., Pitt, D.H., Byers, P.J.: Object-oriented process specification. In: Rattray (ed.) Specification and Verification of Concurrent Systems. Proc. BCS FACS Workshop, pp. 21–70. Springer, Heidelberg (1990)
50. Scotto di Luzio, P.: Patching up a logic of Venn diagrams. In: Proceedings 6th CSLI Workshop on Logic, Language and Computation. CSLI Publications (2000)
51. Shin, S.-J.: The Logical Status of Diagrams. Cambridge University Press, Cambridge (1994)
52. Stapleton, G.: A survey of reasoning systems based on Euler diagrams. In: Proceedings of Euler Diagrams 2004, Brighton, UK. ENTCS, pp. 127–151 (2004)
53. Stapleton, G., Howse, J., Taylor, J.: A constraint diagram reasoning system. In: Proceedings of International Conference on Visual Languages and Computing, pp. 263–270. Knowledge Systems Insitute (2003)
54. Stapleton, G., Howse, J., Taylor, J.: A decidable constraint diagram reasoning system. Journal of Logic and Computation (to appear, 2005)
55. Stapleton, G., Howse, J., Taylor, J., Thompson, S.: What can spider diagrams say? In: Blackwell, A.F., Marriott, K., Shimojima, A. (eds.) Diagrams 2004. LNCS (LNAI), vol. 2980, pp. 112–127. Springer, Heidelberg (2004)
56. Stapleton, G., Masthoff, J., Flower, J., Fish, A., Southern, J.: Automated Theorem Proving in Euler Diagram Systems. Journal of Automated Reasoning (to appear, 2007) (accepted)
57. Stapleton, G., Thompson, S., Howse, J., Taylor, J.: The Expressiveness of Spider Diagrams Augmented with Constants. Journal of Visual Languages and Computing (to appear, 2008)
58. Stapleton, G., Thompson, S., Fish, A., Howse, J., Taylor, J.: A new language for the visualization of logic and reasoning. In: International Workshop on Visual Languages and Computing, pp. 263–270. Knowledge Systems Insitute (2005)
59. Stapleton, G., Thompson, S., Howse, J., Taylor, J.: The expressiveness of spider diagrams. Journal of Logic and Computation 14(6), 857–880 (2004)
60. Swoboda, N.: Implementing Euler/Venn Reasoning Systems. In: Anderson, M., Meyer, B., Olivier, P. (eds.) Diagrammatic Representation and Reasoning, pp. 371–386. Springer, Heidelberg (2001)
61. Swoboda, N., Allwein, G.: Using DAG transformations to Verify Euler/Venn Homogeneous and Euler/Venn FOL Heterogeneous Rules of Interence. In: Proceedings of GT-VMT. ENTCS. Elsevier, Amsterdam (2002)
62. Swoboda, N., Allwein, G.: Modeling Heterogeneous Systems. In: Hegarty, M., Meyer, B., Narayanan, N.H. (eds.) Diagrams 2002. LNCS (LNAI), vol. 2317, pp. 131–145. Springer, Heidelberg (2002)
63. Swoboda, N., Barwise, J.: The information content of Euler/Venn diagrams. In: Proceedings LICS workshop on Logic and Diagrammatic Information (1998)
64. Visual Modelling Group, `http://www.cmis.bton.ac.uk/research/vmg`
65. Venn, J.: On the diagrammatic and mechanical representation of propositions and reasonings. The London, Edinburgh and Dublin Philosophical Magazine and Journal of Science 9, 1–18 (1880)
66. Verroust, A., Viaud, M.: Ensuring the Drawability of Extended Euler Diagrams for up to 8 Sets. In: Blackwell, A.F., Marriott, K., Shimojima, A. (eds.) Diagrams 2004. LNCS (LNAI), vol. 2980, pp. 128–141. Springer, Heidelberg (2004)
67. Warmer, J., Kleppe, A.: The Object Constraint Language: Precise Modeling with UML. Addison-Wesley, Reading (1999)

Pursuing the Goal of Language Understanding

Arun Majumdar, John Sowa, and John Stewart

VivoMind Intelligence, Inc.

Abstract. No human being can understand every text or dialog in his or her native language, and no one should expect a computer to do so. However, people have a remarkable ability to learn and to extend their understanding without explicit training. Fundamental to human understanding is the ability to learn and use language in social interactions that Wittgenstein called *language games*. Those language games use and extend prelinguistic knowledge learned through perception, action, and social interactions. This article surveys the technology that has been developed for natural language processing and the successes and failures of various attempts. Although many useful applications have been implemented, the original goal of language understanding seems as remote as ever. Fundamental to understanding is the ability to recognize an utterance as a move in a social game and to respond in terms of a mental model of the game, the players, and the environment. Those models use and extend the prelinguistic models learned through perception, action, and social interactions. Secondary uses of language, such as reading a book, are derivative processes that elaborate and extend the mental models originally acquired by interacting with people and the environment. A computer system that relates language to virtual models might mimic some aspects of understanding, but full understanding requires the ability to learn and use new knowledge in social and sensory-motor interactions. These issues are illustrated with an analysis of some NLP systems and a recommended strategy for the future. None of the systems available today can understand language at the level of a child, but with a shift in strategy there is hope of designing more robust and usable systems in the future.

1 The Goal of Language Understanding

Some early successes of artificial intelligence led to exaggerated expectations. One example was the theorem prover by Hao Wang (1960), which proved the first 378 theorems of the *Principia Mathematica* in 7 minutes — an average of 1.1 seconds per theorem on the IBM 704, a vacuum-tube machine with 144K bytes of storage. Since that speed was much faster than the two brilliant logicians who wrote the book, many pioneers in AI thought that the possibility of exceeding human intelligence was within reach. Good (1965) predicted "It is more probable than not that, within the twentieth century, an ultraintelligent machine will be built and that it will be the last invention that man need make." The movie *2001*, which appeared in 1968, featured the HAL 9000, an intelligent computer that could carry on a conversation in flawless English and even read lips when the humans were trying to communicate in secret. Marvin Minsky, a technical advisor on that movie, claimed it was a "conservative" estimate

P. Eklund and O. Haemmerlé (Eds.): ICCS 2008, LNAI 5113, pp. 21–42, 2008.

of AI technology at the end of the 20th century. Yet intellectual tasks, such as proving theorems or playing chess, turned out to be far easier to process by computer than simulating the language skills of a three-year-old child.

In chess or mathematics, a computer can exceed human abilities without simulating human thought. But language is so intimately tied to thought that a computer probably cannot understand language without simulating human thinking at some level. That point raises many serious questions: At what level? With what theory of thinking? With what kinds of internal mechanisms? And with what theories and mechanisms for relating the internal processes via the sensory-motor systems to other agents and the world? Several kinds of theories have been proposed, analyzed, and discussed since antiquity: thoughts are images, thoughts are feelings, thoughts are propositions, and thoughts are multimodal combinations of images, feelings, and propositions.

The propositional theory has been the most popular in AI, partly because it's compatible with a large body of work in logic and partly because it's the easiest to implement on a digital computer. Figure 1 illustrates the classical paradigm for natural language processing. At the top is a lexicon that maps the vocabulary to speech sounds, word forms, grammar, and word senses. The arrows from left to right link each stage of processing: phonology maps the speech sounds to phonemes; morphology relates the phonemes to meaningful units or morphemes; syntax analyzes a string of morphemes according to grammar rules; and semantics interprets the grammatical patterns to generate propositions stated in some version of logic.

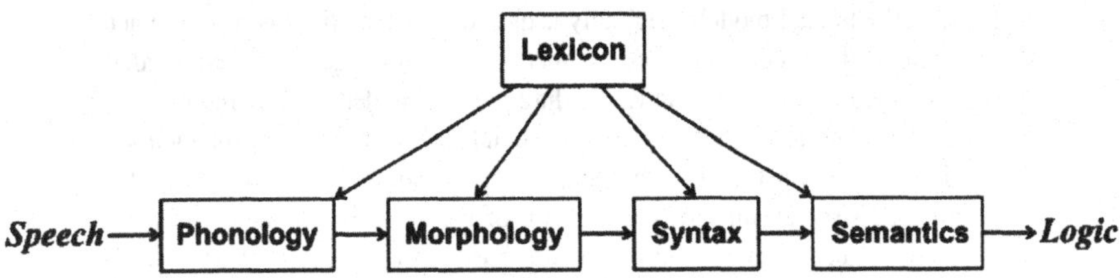

Fig. 1. Classical stages in natural language processing

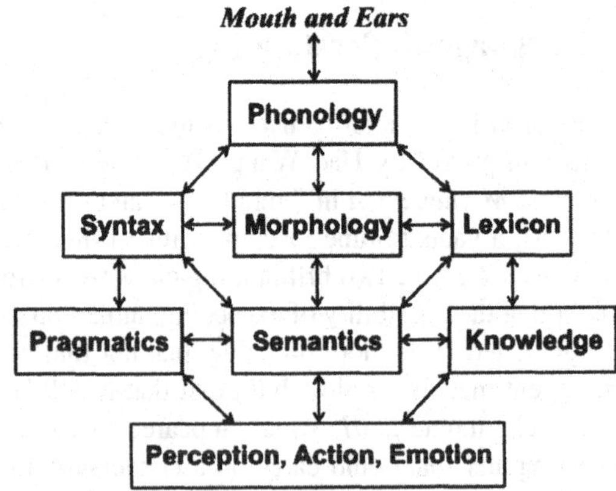

Fig. 2. A more realistic diagram of interconnections

Psycholinguistic evidence since the 1960s has shown that Figure 1 is unrealistic. All the one-way arrows should be double headed, because feedback from later stages has a major influence on processing at earlier stages. Even the arrows from the lexicon should be double headed, because people are constantly learning and coining new words, new word senses, and new variations in syntax and pronunciation. The output labeled *logic* is also unrealistic, because logicians have not reached a consensus on an ideal logical form and many linguists doubt that logic is an ideal representation for semantics. Furthermore, Figure 1 omits everything about how language is used by people who interact with each other and the world. Figure 2 is a more realistic diagram of the interconnections among the modules.

Yet Figure 2 also embodies questionable assumptions. The box labeled *perception, action, and emotion*, for example, blurs all the levels of cognition from fish to chimpanzees. Furthermore, the boxes of Figure 2 correspond to traditional academic fields, but there is no evidence that those fields have a one-to-one mapping to modules for processing language in the brain. In particular, the box labeled *knowledge* should be subdivided in at least three ways: language-independent knowledge stored in image-like form; conceptual knowledge related to language, but independent of any specific language; and knowledge of the phonology, vocabulary, and syntax of specific languages. The box labeled *pragmatics* deals with the use of language in human activities. Wittgenstein (1953) proposed a reorganization in *language games*, according to the open-ended variety of ways language is used in social interactions. That subdivision would cause a similar partitioning of the other boxes, especially semantics, knowledge, and the lexicon. It would also affect the variations of syntax and phonology in casual speech, professional jargon, or "baby talk" with an infant.

In his first book, Wittgenstein (1922) presented a theory of language and logic based on principles proposed by his mentors, Frege and Russell. Believing he had solved all the problems of philosophy, Wittgenstein retired to an Austrian mountain village, where he taught elementary schoolchildren. Unfortunately, the children did not learn, think, or speak according to those principles. In his second book, Wittgenstein (1953) systematically analyzed the "grave errors" (*schwere Irrtümer*) in the framework he had adopted. One of the worst was the view that logic is superior to natural languages and should replace them for scientific purposes. Frege (1879), for example, hoped "to break the domination of the word over the human spirit by laying bare the misconceptions that through the use of language often almost unavoidably arise concerning the relations between concepts." Russell shared Frege's low opinion of natural language, and both of them inspired Carnap, the Vienna Circle, and most of analytic philosophy.

Many linguists and logicians who work within the paradigm of Figure 1 admit that it's oversimplified, but they claim that simplification is necessary to enable researchers to address solvable subproblems. Yet Richard Montague and his followers have spent forty years working in that paradigm, and computational linguists have been working on it for half a century. But the goal of designing a system at the level of HAL 9000 seems more remote today than in 1968. Even pioneers in the logic-based approach have begun to doubt its adequacy. Kamp (2001), for example, claimed "that the basic concepts of linguistics — and especially those of semantics — have to be thought through anew" and "that many more distinctions have to be drawn than are dreamt of in current semantic theory."

This article emphasizes the distinctions that were dreamt of and developed by cognitive scientists who corrected or rejected the assumptions by Frege, Russell, and their followers. Section 2 begins with the semeiotic by Charles Sanders Peirce, who had invented the algebraic notation for logic, but who placed it in a broader framework than the 20th-century logicians who used it. Section 3 discusses the ubiquitous pattern matching in every aspect of cognition and its use in logical and analogical reasoning. Section 4 presents Wittgenstein's language games and the social interactions in which language is learned, used, and understood. Section 5 introduces Minsky's *Society of Mind* as a method of supporting the interactions illustrated in Figure 2. Section 6 summarizes the lessons learned from work with two earlier language processors. The concluding Section 7 outlines a multilevel approach to language processing that can support more robust and flexible systems.

2 Semeiotic and Biosemiotics

Peirce claimed that the primary characteristic of life is the ability to recognize, interpret, and respond to signs. Signs are even more fundamental than neurons because every neuron is itself a semiotic system: it receives signs and interprets them by generating more signs, which it passes to other neurons or muscle cells. Every cell, even an independent bacterium, is a semiotic system that recognizes chemical, electrical, or tactile signs and interprets them by generating other signs. Those signs can cause the walls of a bacterial cell to contract or expand and move the cell toward nutrients and away from toxins. The brain is a large colony of neural cells, whose signs coordinate a symbiotic relationship within an organism of many kinds of cells. The neural system supports rapid, long-distance communication by electrical signs, but all cells can communicate locally by chemical signs. By secreting chemicals into the blood stream, cells can broadcast signs by a slower, but more pervasive method. At every level from a single cell to a multicellular organism to a society of organisms, signs support and direct all vital processes. *Semeiotic* is Peirce's term for the theory of signs. The modern term *biosemiotics* emphasizes Peirce's point that sign processing is more general than human language and cognition.

Deacon (1997), a professional neuroscientist, used Peirce's theories as a guide for relating neurons to language. Figure 3 illustrates his view that the language modules of the brain are a recent addition and extension of a much older ape-like architecture. Deacon used Peirce's categories of *icon*, *index*, and *symbol* to analyze the signs that animals recognize or produce. The calls a hunter utters to control the dogs are indexes, the vocal equivalent of a pointing finger. Vervet monkeys have three types of warning calls: one for eagles, another for snakes, and a third for leopards. Some people suggested that those calls are symbols of different types of animals, but vervets never use them in the absence of the stimulus. More likely, the vervet that sees the stimulus uses the call as an index to tell other vervets to look up, look down, or look around. An early step from index to symbol probably occurred when some hominin proposed a hunt by uttering an index for prey, even before the prey was present. After symbols became common, they would enable planning and organized activities in every aspect of life. The result would be a rapid increase in vocabulary, which would promote the *co-evolution* of language, brain, vocal tract, and culture.

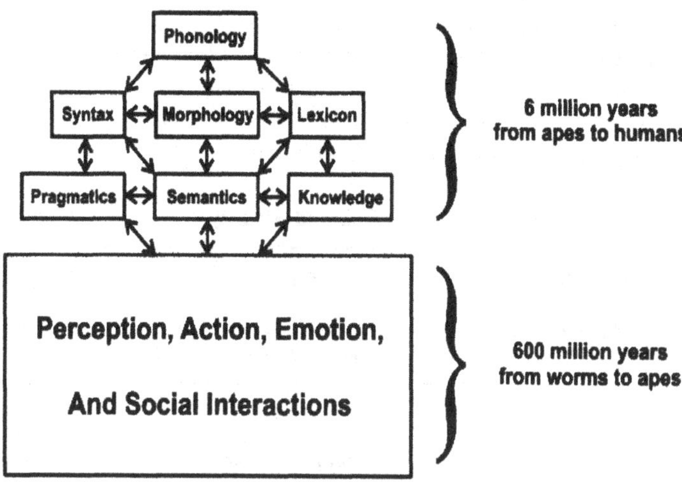

Fig. 3. An evolutionary view of the language modules

Like Frege, Peirce was a logician who independently developed a complete notation for first-order logic. Unlike Frege, Peirce had a high regard for the power and flexibility of language, and he had worked as an associate editor of the *Century Dictionary*, for which he wrote, revised, or reviewed over 16,000 definitions. Peirce never rejected language or logic, but he situated both within the broader theory of signs. In his semeiotic, every sign is a triad that relates a perceptible *mark* (1), to another sign called its *interpretant* (2), which determines an existing or intended *object* (3). Following is one of Peirce's most often quoted definitions:

> A sign, or *representamen*, is something which stands to somebody for something in some respect or capacity. It addresses somebody, that is, creates in the mind of that person an equivalent sign, or perhaps a more developed sign. That sign which it creates I call the *interpretant* of the first sign. The sign stands for something, its *object*. It stands for that object, not in all respects, but in reference to a sort of idea, which I have sometimes called the *ground* of the representamen. (CP 2.228)

A pattern of green and yellow in the lawn, for example, is a mark, and the interpretant is some type, such as Plant, Weed, Flower, SaladGreen, or Dandelion. The guiding idea that determines the interpretant depends on the context and the intentions of the observer. The interpretant determines the word the observer chooses to express the experience. The listener who hears that word uses background knowledge to derive an equivalent interpretant.

As Peirce noted, an expert with a richer background can sometimes derive a more developed interpretant than the original observer. Communication in which both sides have identical interpretants is possible with computer systems. Formal languages are precise, but they are rigid and fragile. The slightest error can and frequently does cause a total breakdown, such as the notorious "blue screen of death."

On the surface, Peirce's triads seem similar to the *meaning triangles* by Aristotle, Frege, or Ogden and Richards (1923). The crucial difference is that Peirce analyzed the underlying relationships among the vertices and sides of the triangle. By analyzing the relation between the mark and its object, Peirce (1867) derived the triad of *icon, index* and *symbol*: an icon refers by some similarity to the object; an index refers by a physical effect or connection; and a symbol refers by a law, habit, or convention. Figure 4 shows this *relational* triad in the middle row.

	1. Quality	2. Indexicality	3. Mediation
1. Material	**Qualisign** *A quality which is a sign.*	**Sinsign** *An actual existent thing or event which is a sign.*	**Legisign** *A law which is a sign.*
2. Relational	**Icon** *Refers by virtue of some similarity to object.*	**Index** *Refers by virtue of being affected by object.*	**Symbol** *Refers by virtue of some law or association.*
3. Formal	**Rheme** *A sign of qualitative possibility.*	**Dicent Sign** *A sign of actual existence.*	**Argument** *A sign of law.*

Fig. 4. Peirce's triple trichotomy

Later, Peirce added the first row or *material triad*, which signifies by the nature of the mark itself. The third row or *formal triad* signifies by a formal rule that relates all three vertices — the mark, interpretant, and object. The basic units of language are characterized by the formal triad: a word serves as a rheme; a sentence, as a dicent sign; and a paragraph or other sequence, as an argument. The labels at the top of Figure 4 indicate how the sign directs attention to the object: by some quality of the mark, by some causal or pointing effect, or by some mediating law, habit, or convention. The following examples illustrate nine types of signs:

1. **Qualisign** (material quality). A ringing sound as an uninterpreted sensation.
2. **Sinsign** (material indexicality). A ringing sound that is recognized as coming from a telephone.
3. **Legisign** (material mediation). The convention that a ringing telephone means someone is trying to call.
4. **Icon** (relational quality). An image that resembles a telephone when used to indicate a telephone.
5. **Index** (relational indexicality). A finger pointing toward a telephone.
6. **Symbol** (relational mediation). A ringing sound on the radio that is used to suggest a telephone call.
7. **Rheme** (formal quality). A word, such as *telephone*, which can represent any telephone, real or imagined.
8. **Dicent Sign** (formal indexicality). A sentence that asserts an actual existence of some object or event: "You have a phone call from your mother."

9. **Argument** (formal mediation). A sequence of dicent signs that expresses a lawlike connection: "It may be an emergency. Therefore, you should answer the phone."

The nine categories in Figure 4 are more finely differentiated than most definitions of signs, and they cover a broader range of phenomena. Anything that exists can be a sign of itself (sinsign), if it is interpreted by an observer. But Peirce (1911:33) did not limit his definition to human minds or even to signs that exist in our universe:

> A sign, then, is anything whatsoever — whether an Actual or a May-be or a Would-be — which affects a mind, its Interpreter, and draws that interpreter's attention to some Object (whether Actual, May-be, or Would-be) which *has already* come within the sphere of his experience.

The mind or quasi-mind that interprets a sign need not be human. In various examples, Peirce mentioned dogs, parrots, and bees. Higher animals typically recognize icons and indexes, and some might recognize symbols. A language of some kind is a prerequisite for signs at the formal level of rhemes, dicent signs, and arguments. In general, Peirce's theory of signs provides a more nuanced basis for analysis than the all-or-nothing question of whether animals have language. Unlike the static meaning triangles of Aristotle or Frege, the most important aspect of Peirce's triangles is their dynamic nature: any vertex can spawn another triad to show three different perspectives on the entity represented by that vertex. During the course of a conversation, the motives of the participants lead the thread of themes and topics from triangle to triangle.

3 Perception, Cognition, and Reasoning

Language affects and is affected by every aspect of cognition. Only one topic is more pervasive than language: signs in general. Every cell of every organism is a semiotic system, which receives signs from the environment, including other cells, and interprets them by generating more signs, both to control its own inner workings and to communicate with other cells of the same organism or different organisms. The brain is a large colony of neural cells, which receives, generates, and transmits signs to other cells of the organism, which is an even larger colony. Every publication in neuroscience describes brains and neurons as systems that receive signs, process signs, and generate signs. Every attempt to understand those signs relates them to other signs from the environment, to signs generated by the organism, and to theories of those signs in other branches of cognitive science. The meaning of the neural signs can only be determined by situating neuroscience within a more complete theory that encompasses every aspect of cognitive science.

By Peirce's definition of *sign*, all life processes, especially cognition, involve receiving, interpreting, generating, storing, and transmitting signs and patterns of signs. Experimental evidence is necessary to determine the nature of the signs and the kinds of patterns generated by the interpretation. Perceptual signs are *icons* derived from sensory stimulation caused by the outside world or caused by internal bodily processes. Recognition consists of interpreting a newly received icon by matching it to previously classified icons called *percepts* and patterns of percepts called *Gestalts*.

The interpretation of an icon is the pattern formed by the percepts, Gestalts, and other associated signs. The interpreting signs may be image-like percepts or imageless *concepts*, which are similar to percepts, but without the sensory connections.

Analogy is a method of reasoning based on pattern matching, and every method of logic is a constrained use of analogy. As an example, consider the rule of deduction called *modus ponens*:

```
Premise:      If P then Q.
Assertion:    P'.
Conclusion:   Q'.
```

This rule depends on the most basic form of pattern matching: a *comparison* of P and P′ to determine whether they are identical. If P in the premise is not identical to P′ in the assertion, then a pattern-matching process called *unification* specializes P by some transformation S that makes S(P) identical to P′. By applying the same specialization S to Q, the conclusion Q′ is derived as S(Q). Each of the following three methods of logic constrain the pattern matching to specialization, generalization, or identity.

1. **Deduction.** Specialize a general principle.
```
Known:   Every bird flies.
Given:   Tweety is a bird.
Infer:   Tweety flies.
```
2. **Induction.** Generalize multiple special cases:
```
Given:   Tweety, Polly, and Hooty are birds.
         Tweety, Polly, and Hooty fly.
Assume: Every bird flies.
```
3. **Abduction.** Given a special case and a known generalization, make a guess that explains the special case.
```
Given:   Tweety flies.
Known:   Every bird flies.
Guess:   Tweety is a bird.
```

These three methods of logic depend on the ability to use symbols. In deduction, the general term *every bird* is replaced by the name of a specific bird *Tweety*. Induction generalizes a property of multiple individuals — Tweety, Polly, and Hooty — to the category Bird, which subsumes all the instances. Abduction guesses the new proposition *Tweety is a bird* to explain one or more observations. According to Deacon's hypothesis that symbols are uniquely human, these three reasoning methods could not be used by nonhuman mammals.

According to Peirce (1902), "Besides these three types of reasoning there is a fourth, analogy, which combines the characters of the three, yet cannot be adequately represented as composite." Its only prerequisite is *stimulus generalization* — the ability to classify similar patterns of stimuli as signs of similar objects or events. Unlike the more constrained operations of generalization and specialization, similarity may involve a generalization of one part and a specialization of another part of the same pattern. Analogy is more primitive than logic because it does not require language or symbols. In Peirce's terms, logic requires symbols, but analogy can also be performed on image-like icons. Case-based reasoning (CBR) is an informal method of reasoning, which uses analogy to find and compare cases that may be relevant to a given problem or question.

Whether the medium consists of discrete words or continuous images, CBR methods start with a question or goal Q about some current problem or situation P. By analogy, previous cases that resemble P are recalled from long-term memory and ranked according to their similarity to P. The case with the greatest similarity (i.e., smallest semantic distance) is the most likely to answer the question Q. When a similar case is found, the part of the case that matches Q is the predicted answer. If two or more cases are equally similar, they may or may not predict the same answer. If they do, that answer can be accepted with a high degree of confidence. If not, the answer is a disjunction of alternatives: Q_1 or Q_2.

For highly regular data, induction can generalize many cases to rules of the form *If P, then Q*. For such data, CBR would derive the same conclusions as a method of deduction called *backward chaining*: a goal Q′ is unified to the conclusion Q of some if-then rule by means of a specialization S; the application of S to P produces the pattern P′, which is a generalization of one or more cases. Formal deduction is best suited to thoroughly analyzed areas of science, for which induction can reduce a large number of cases to a small number of rules. CBR is generally used for subjects with highly varied or frequently changing cases, for which any axioms would have a long list of exceptions. Legal reasoning is a typical example: both the list of laws and the list of cases are enormous, and most generalizations have as many exceptions as applications. Both logic and CBR have a large overlap on which they're compatible: they would generate consistent responses to the same questions.

The world is continuous, all physical motions are continuous, feelings and sensations vary continuously, but every natural language has a discrete, finite set of meaningful units or *morphemes*. No discrete set of symbols can faithfully represent a continuous world, but the two systems must be related by a mapping between language and the world. Wildgen (1982, 1994) maintained that continuous fields are the primary basis for perception and cognition, and he adopted René Thom's *catastrophe-theoretic semantics* for identifying the patterns that map to the discrete units of language. Those ideas are still controversial, but the principle of mapping discrete structures such as *conceptual graphs* (CGs) to continuous fields has proved to be valuable for developing efficient methods for indexing CGs and computing the semantic distance between them (Sowa & Majumdar 2003). Those methods were used for finding analogies in the VivoMind Analogy Engine (VAE), and more precise and flexible mappings have been implemented in a new system called Cognitive Memory™.

4 Language Games

The first language game may have evolved about four million years ago as a system of sounds and gestures for organizing a hunt. At that time, chimpanzees lived in the forests of west Africa, while our ancestor, *Australopithecus*, lived in the grasslands to the east. With fewer trees to climb, the Australopithecines began to walk upright. Chimps supplement their diet by catching and eating small game, but in lands with sparser vegetation, Australopithecines required more meat. Since they weren't as fast as their prey, they had to hunt in organized parties, which require communication. The calls and gestures of the chimps were adequate for occasional hunting, but when hunting became a necessity, any improvement in communication would be an enormous advantage. Vocal calls are convenient because they leave the hands free, and they can be spoken and heard while eyes are focused on the prey. The earliest protowords were probably a few dozen indexical signs, of the sort that modern hunters and shepherds use to control their dogs. The first step from index to symbol likely occurred when some hominin proposed a hunt by uttering the index for prey, even before the prey was present. After symbols were invented, language games could be integrated with every social activity that involved cooperation, negotiation, persuasion, planning, or play.

Wittgenstein's theory of language games has major implications for both semantic theory and computational linguistics. It implies that the ambiguities of natural language are not the result of careless speech by uneducated people. Instead, they result from the fundamental nature of language and the way it relates to the world: each language uses and reuses a finite number of words to represent an unlimited number of topics. A closed semantic basis along classical lines is not possible for any natural language. Instead of assigning a single meaning or even a fixed set of meanings to each word, a theory of semantics must permit an open-ended number of meanings:

- Words are like playing pieces that may be used and reused in different language games.
- Associated with each word is a limited number of lexical patterns that are common to all the language games that use the word.
- Meanings are deeper conceptual patterns that change from one language game to another.
- Metaphor and conceptual refinement are techniques for transferring the lexical patterns of a word to a new language game and thereby creating new conceptual patterns for that game.

Once a lexical pattern is established for a concrete domain, it can be transferred by metaphor to create similar patterns in more abstract domains. By this process, an initial set of lexical patterns can be built up; later, they can be generalized and extended to form new conceptual patterns for more abstract subjects. The possibility of transferring patterns from one domain to another increases flexibility, but it leads to an inevitable increase in ambiguity.

If the world were simpler, less varied, and less changeable, natural languages might be unambiguous. But the complexity of the world causes the meanings of words to shift subtly from one domain to the next. If a word is used in widely different domains, its multiple meanings may have little or nothing in common. As an example, the word *invest,* which originally meant to put on clothing, has come to mean either to surround a fortress or to make a certain kind of financial transaction. In Italian, the related word *investmento* has all the senses of the English *investment,* but with the added sense of traffic accident. As these examples illustrate, the mechanisms of natural languages not only permit, but actually facilitate arbitrarily large shifts in meaning. They have enabled isolated tribes using stone-age tools to adapt to

21st-century cultures within the lifetime of a single generation, while continuing to speak what is called "the same language."

Although Wittgenstein's theory of language games has been highly influential, some linguists and philosophers have raised criticisms and proposed alternative, but related hypotheses. Hattiangadi (1987) proposed that the meaning of a word is the set of all possible *theories* in which it may be used, but that term sounds too formal to cover everyday speech. Kittredge and Lehrberger (1982) used the term *sublanguage* for any specialized language used in any context for any purpose. Whatever it's called, a language game or any related variation must involve an organized system of language patterns and practices that are intimately bound to a system of behavior — or, as Wittgenstein called it, a way of life. Language can only be understood in terms of the social activities of its speakers. Full understanding of the language would require a person or robot to participate in the activity in a way that other participants would consider appropriate. This requirement, which is a variant of the Turing test, is a necessary condition for a single language game. A sufficient condition for general understanding would require the ability to learn, use, and invent a wide range of language games under the same kinds of conditions as native speakers.

By those criteria, the bonobo Kanzi is a nonhuman person who has reached a level of language understanding that is beyond the ability of any computer system yet devised (Savage-Rumbaugh & Lewin 1994). On a test of spoken English with sentences such as "Get the rubber band that's in the bathroom," Kanzi responded with the correct action to 72% of the sentences; Alia, a two-year-old girl, responded correctly to 66% of them. Even more impressive are the reports by Stuart Shanker, a skeptical Wittgensteinian philosopher who became a believer after visiting Kanzi and his teacher, Sue Savage-Rumbaugh. On Shanker's first visit, Sue asked Kanzi, "I'm going to take Stuart around the lab. Could you please water the tomato plants for me while we're doing this?" Following is Shanker's description of what Kanzi did:

> And sure enough, I watched as he trundled over to an outdoor water faucet, picked up a bucket that was lying beside it, turned on the spigot and filled the bucket, turned off the faucet himself, and then walked down to a vegetable patch at the far end of the compound, carrying the bucket in one hand. When he reached the vegetables, I watched as he poured the water on a small patch of tomato plants growing in the corner of the vegetable patch. (Greenbaum & Shanker 2004:107).

There is no evidence of which words of the request Kanzi understood. But he was undoubtedly familiar with the task of watering the tomatoes, and he understood the language games related to that task. Linguists have claimed that the inability to produce detailed syntax indicates that apes have not learned a truly natural language. Yet apes and two-year-old children satisfy the criteria of learning, using, and inventing language games integrated with their behavior. If Kanzi were human, he would be diagnosed as having a language deficit, but no one would doubt his understanding of the language associated with the activities in which he participated.

Current natural language processors have been used in many valuable applications, such as translating languages, finding and extracting information from text, summarizing texts, answering questions, and checking and correcting syntax errors. Some of them have been used to control robots, but none of them have been able to learn, play, and invent language games at the level of Kanzi and other apes. Furthermore, none of them have been able to learn, use, and generate language at the level of a three-year-old child. The following sentences were uttered by a child named Laura at 2 years, 10 months (Limber 1973):

Here's a seat. It must be mine if it's a little one.
I went to the aquarium and saw the fish.
I want this doll because she's big.
When I was a little girl, I could go "geek geek" like that, but now I can go "This is a chair."

Forty years ago, the goal of AI was to meet or exceed all human intellectual abilities. Today, reaching the level of Laura or Kanzi would be a major achievement.

Laura's sentences contain implications (*if*), causal connections (*because*), modal auxiliaries (*can* and *must*), contrast between past and present tenses, metalanguage about her own language at different times, and parallel stylistic structure. Combining modal, temporal, causal, and metalevel logic and reasoning in a single formalism and using it to interpret and generate natural language is still a major research topic. Even though she couldn't prove theorems as fast as Wang's program, Laura used all those operators before the age of three. The assumption that formal logic is the foundation or prerequisite for language understanding seems unlikely.

Some linguists and philosophers have been searching for an elusive "natural logic" that underlies language. Yet there is no sharp boundary between ordinary language and any formal logic. When two logicians talk on the telephone, they can convey the most abstruse ideas with a few words added to ordinary language. A better assumption is that formal logic is a language game played with symbols and patterns abstracted from natural languages. Formal logic may sound unnatural to the uninitiated, but that is true of the language games of any specialized field. Sailors, plumbers, chefs, and computer hackers scorn the "book learning" of novices who try to use their jargon without mastering the associated skills. Book learning is useful, but computer systems must relate it to action in order to demonstrate understanding at the level of Laura or Kanzi.

Some language games involve a disciplined use of syntax, semantics, and vocabulary in a *controlled natural language* that a computer can process without full understanding. An example is the METEO system for translating weather reports to and from English and French (Thouin 1982). For routine reports about the temperature and precipitation, METEO does the translation without human assistance, but for unusual conditions outside the controlled subset, it sends the reports to a human translator. Speech recognition systems for handling telephone calls frustrate people who need to discuss situations that fall outside the controlled subset. More research is necessary to broaden the controlled subsets and determine when to transfer the call to a human being.

5 Society of Mind

In computer systems, the linear flow of Figure 1 is easy to implement: each stage analyzes some data, passes the results to the next stage, and never sees the same data again. But the more complex interconnections of Figure 2 allow other modules, even later stages, to request or propose different interpretations of previously analyzed data. In recordings of the following sentences, for example, Warren (1970) spliced a patch of white noise at each point marked ¿:

The ¿eel is on the shoe.
The ¿eel is on the car.
The ¿eel is on the table.
The ¿eel is on the orange.

Although the sound was identical in each of the four sentences, the listeners who heard the recordings interpreted the four words as *heel*, *wheel*, *meal*, and *peel*, respectively. Apparently, feedback from the semantic stage caused a reinterpretation of the phonology of an earlier word in the sentence. Furthermore, the listeners were not aware of hearing anything unusual. Many similar studies indicate a great deal of parallel processing in the brain with feedback from later stages to earlier stages, usually at a level beneath conscious awareness. To support parallel processing with feedback, a computer system would require a more complicated control structure than a linear flow.

An early AI model of parallel reasoning was Pandemonium (Selfridge 1959), which consisted of a collection of autonomous agents called *demons*. Each demon could observe aspects of the current situation or workspace, perform some computation, and put its results back into the workspace. In effect, Pandemonium was a parallel *forward-chaining* reasoner, whose major drawback was that the demons generated large volumes of mostly useless data that overflowed available storage. For a more disciplined method of passing messages among the linguistic modules, Hahn et al. (1994) designed ParseTalk as a distributed, concurrent, object-oriented parser. In discussing the advantages of ParseTalk, the authors noted that it replaces "the static global-control paradigm" of Figure 1 with "a dynamic, local-control model" that supports "a balanced treatment of both declarative and procedural constructs within a single formal framework." Although ParseTalk is a promising approach, Figure 3 suggests that the language modules use a much older and more pervasive system that supports all aspects of perception, cognition, and action. Therefore, the large box at the bottom of Figure 3 should also be subdivided in modules that operate in parallel and communicate by message passing.

The integration of language games with social activity implies that the language modules should be further subdivided and interconnected with other cognitive modules in dynamically changing ways. The modules for reading, for example, would connect visual perception to the syntactic and semantic mechanisms. Psycholinguistic studies with Japanese syllabic kana symbols and character-based kanji, indicates that they use different neural mechanisms even for reading. Singing integrates language and music in ways that make both the words and the melodies easier to recognize and reproduce. Singing is also connected to dancing, marching, and various kinds of rhythmic work and play. Some linguists claimed that music was based on the syntactic mechanisms of language, but Mithen (2006) presented detailed evidence to show that music is older and independent of language. In fact, syntax may have evolved with or from the music of prosody. Whatever the basis, the number of modules is probably far greater than the eight boxes of Figures 2 and 3. Perhaps there is no limit to the number of modules, and every language game and mode of behavior has its own module or even a collection of interacting modules.

The diversity of mechanisms associated with language is a subset of the even greater diversity involved in all aspects of cognition. In his book *The Society of Mind*, Minsky (1987) surveyed that diversity and proposed an organization of active agents as a computational model that could simulate the complexity:

> What magical trick makes us intelligent? The trick is that there is no trick. The power of intelligence stems from our vast diversity, not from any single, perfect principle. Our species has evolved many effective although imperfect methods, and each of us individually develops more on our own. Eventually, very few of our actions and decisions come to depend on any single mechanism. Instead, they emerge from conflicts and negotiations among societies of processes that constantly challenge one another. (Section 30.8)

This view is radically different from the assumption of a unified formal logic that cannot tolerate a single inconsistency. Unlike the ParseTalk goal of "a single formal framework," Minsky's goal is to build a flexible, fault-tolerant system out of imperfect, possibly fallible components. Such a system can support logic, just as the flexible, fault tolerant, and fallible human brain supports logic. More recently, Minsky (2006) emphasized the role of emotions in driving an engine composed of multiple agents. Without emotions to set the goals, a logic-based theorem prover would have no reason to do anything.

Minsky's proposal for a society of interacting agents could be implemented in a variety of ways. The Flexible Modular Framework™ (FMF) proposed by Sowa (2002, 2004) is an architecture for intelligent systems that was inspired by Minsky's society of agents, by McCarthy's proposal for the logic-based language Elephant 2000, and by Peirce's semeiotic. As in Minsky's society, each module in the FMF is an autonomous agent that communicates with other agents by passing messages. As in McCarthy's Elephant, each message specifies a speech act that indicates its purpose, and the messages may be expressed in logic. As in Peirce's semeiotic, each message is a sign at any level of complexity, each agent is a "quasi-mind" that interprets signs by generating other signs, and many agents use the Peirce-inspired system of logic called *conceptual graphs*. An agent that knows another agent's identity can send it a message directly, but any agent can post a message to a Linda blackboard, and any other agent that can process that type of message can respond to it (Gelernter 1985). Unlike ParseTalk, the FMF does not require a single formal framework, but it can support the *π-calculus*, which is a generalization of Petri nets that allows new agents and communication paths to be created or destroyed dynamically (Milner 1999). Several variations of the FMF have been implemented, and all of them use a lightweight protocol that can be implemented in 8K bytes per agent. Thousands of agents can run simultaneously on a laptop computer, but they can communicate with other agents anywhere across the Internet.

An interactive system of agents that can change their configuration dynamically is strictly more expressive than a conventional Turing machine, and it can compute functions that are not Turing computable (Eberbach et al. 2004). The π-calculus is one such system. Another is the *$-calculus*, which has the same operators as the

π-calculus, but adds a *cost* measure for each computation. A cost measure based on space and time requirements can constrain the excesses of systems like Pandemonium by rewarding agents that produce good results with more resources and reducing the resources of agents that produce useless data.

At VivoMind, the authors have developed a learning method called *Market-Driven Learning* (MDL), which uses a version of $-calculus. The basic idea is that the system of agents is organized in a managerial hierarchy with one agent called the CEO at the top. The CEO is responsible for producing results that earn rewards, measured in units of space and time, in order to keep the society of agents in business. At the bottom of the hierarchy are agents that find data, combine data, or propose hypotheses. Some of them are freelance agents who sell data or hypotheses by advertising them on the Linda blackboards. Other agents are hired by some agent that serves as a manager. Each manager has one or more agents as employees, and every manager except the CEO is allocated resources by a higher-level manager. The managers can use their resources to hire employees, reward employees for good performance, or buy data and hypotheses from freelance agents or from other managers. The managers may combine the data and hypotheses themselves, assign their employees the task of doing the combination, or serve on a committee with other managers to produce a combined report.

The MDL society learns by reorganizing itself to produce consistently good results, which humans are willing to buy. The rewards pass through the hierarchy from manager to employee and create an effect of *backward propagation* similar to the learning methods of a neural network. But unlike the simple switches and numeric functions of a neural network, the MDL agents can be arbitrarily complex programs or reasoning systems, they can hire or fire other agents, and the messages can be propositions or even large documents stated in some version of logic. Also unlike a neural network, the messages that pass through the MDL can be translated to Common Logic Controlled English (CLCE) in order to provide humanly readable explanations about the way any agent derived its data, hypotheses, or reports. By simulating a variety of business methods, the MDL approach has produced good results, and it is used in the VivoMind Language Processor described in Section 7.

6 Experience with Intellitex and CLCE

The theoretical issues discussed in the previous sections influenced the design of two language processors developed and used by the authors: the Intellitex parser, which produced an approximate translation from English to conceptual graphs, and the CLCE parser, which translated the formally defined subset of Common Logic Controlled English to precisely defined conceptual graphs. The two parsers had complementary strengths and weaknesses:

- Intellitex was a fast, but shallow parser that used a version of link grammar (Sleator & Temberley 1993) to translate English sentences to conceptual graphs. Intellitex always generated some conceptual graph as an approximation to the semantics of an English sentence, but its grammar and semantics were not sufficiently detailed to generate an accurate logical form for complex sentences. Its approximations, however, were useful for many applications, such as analogical reasoning and question answering (Sowa & Majumdar 2003). The

VivoMind Analogy Engine (VAE) could enhance and correct the approximate CGs generated by Intellitex, but only if a large knowledge base of precisely defined CGs happened to be available. For an important application, such knowledge enabled Intellitex to perform amazingly well.

- The CLCE parser was a traditional syntax-directed parser, which processed character strings written in Common Logic Controlled English (Sowa 2004). It followed the stages from morphology to semantics in Figure 1 to generate a logical form in the Conceptual Graph Interchange Format (CGIF), as defined by the ISO/IEC 24707 standard for Common Logic. But the CLCE subset of English is a formal language, and the CLCE parser was as rigid and unforgiving as a parser for any formal language. Making it more user friendly or extending it to a broader range of English constructions would require a large number of grammar rules. Furthermore, each grammar rule would require a corresponding semantic rule to generate the correct CG.

Over time, incremental improvements were made to both of these processors, but their methods for parsing English and generating CGs were so different that no synergism between them was possible. Intellitex was more robust and forgiving than the CLCE parser, but it could not detect and correct errors in the input that would cause the CLCE parser to fail. The CLCE parser generated more precise CGs, but it could not improve the output generated by Intellitex. A survey of these two systems can provide some insight into the issues.

The largest and most impressive application combined Intellitex with VAE for a legacy reengineering project that analyzed and related the software and English documentation for a large corporation. The software was written in formal languages: 1.5 million lines of COBOL with embedded SQL statements and several hundred scripts in the IBM Job Control Language (JCL). The documentation consisted of 100 megabytes of English reports, manuals, e-mails, web pages, memos, notes, and comments in the COBOL and JCL code. Some of the documentation and programs were up to 40 years old and still in daily use.

The first goal was to analyze the programs to derive a data dictionary, data flow diagrams, process architecture diagrams, and system context diagrams for all the software. That task could be done with programming-language parsers and conventional methods of analysis. The second and more challenging goal was to analyze the English, detect any errors or discrepancies between the software and the documentation, and generate a humanly readable glossary of terminology for the software and data, including all the variations over the period of 40 years. A major consulting firm estimated that analyzing all the documentation and relating it to the software would require 40 people for 2 years.

By using Intellitex and VAE, two programmers, Arun Majumdar and André LeClerc, accomplished both goals in less than two months, a total of 15 person weeks instead of 80 person years (LeClerc & Majumdar 2002). The results of first analyzing the computer languages and translating them to conceptual graphs were essential for analyzing the English. The names of every program, file, and data element were added to the dictionary used for parsing English. Furthermore, those items were also classified in an ontology of the computer terms that supplemented the ontology derived from WordNet, CoreLex, and other resources. Each term added to the lexicon was associated with one or more conceptual graphs that showed the expected relations: for example, variables occur in programs, programs process files, and files contain data. When parsing English, Intellitex translated every phrase or sentence to a conceptual graph. Those CGs that did not refer to anything in the software were discarded, and the others were used to update a knowledge base of information about the software. The results of the analysis were presented in one CD-ROM: software diagrams, data dictionary, English glossary, and a list of inconsistencies between the software and the documentation.

The reason why Intellitex and VAE succeeded where many natural language processors failed is that it did not attempt to translate informal language to formal logic. Instead, it used the formal CGs derived from COBOL, SQL, and JCL as the background knowledge for interpreting English text and resolving ambiguities. In short, the results were generated by joining formal CGs according to formal rules in order to create a pattern that had a close match to the approximate CGs derived from the English sentences. As an example, the following paragraph is taken from the English documentation:

> The input file that is used to create this piece of the Billing Interface for the General Ledger is an extract from the 61 byte file that is created by the COBOL program BILLCRUA in the Billing History production run. This file is used instead of the history file for time efficiency. This file contains the billing transaction codes (types of records) that are to be interfaced to General Ledger for the given month. For this process the following transaction codes are used: 32 — loss on unbilled, 72 — gain on uncollected, and 85 — loss on uncollected. Any of these records that are actually taxes are bypassed. Only client types 01 — Mar, 05 — Internal Non/Billable, 06 — Internal Billable, and 08 — BAS are selected. This is determined by a GETBDATA call to the client file. The unit that the gain or loss is assigned to is supplied at the time of its creation in EBT.

The common words in this paragraph were found in the dictionary derived from WordNet. Other words such as BILLCRUA, GETBDATA, and EBT were derived from the previous analysis of the software. Those words caused VAE to bring associated CGs from the background knowledge.

The sample paragraph also illustrates how Intellitex can process a wide range of syntactic constructions with a rather simple grammar. A phrase such as "32 — loss on unbilled" is not part of any published grammar of English. When Intellitex found that pattern, it translated it to a rudimentary conceptual graph of the following form:

```
[Number: 32]→(Next)→[Punctuation: "—"]
    →(Next)→[Loss]→(On)→[Unbilled]
```

This graph was stored as a tentative interpretation with a low weight of evidence. But Intellitex found two more graphs, which VAE matched to this graph with a high weight of evidence. Therefore, this syntactic pattern became, in effect, a newly learned grammar rule with a familiar semantic pattern. Although that pattern is not common in the full English language, it is important for the analysis of at least one document. The uninformative relations labeled **Next** were supplemented with background knowledge derived from previously analyzed CGs that formed the best match to those rudimentary graphs.

This discussion illustrates one of the most important lessons learned from Intellitex and VAE: A formal representation is easier to derive by joining conceptual graphs from background knowledge than by limiting the analysis to the details found in the input sentences. In fact, the background knowledge can often correct typos and other careless mistakes in the input text. That process illustrates Peirce's point that the listener may derive "a more developed sign" than the speaker intended. It is colloquially called "reading between the lines." This principle was applied in another application of Intellitex for scoring and correcting student answers to examination questions. Instead of trying to understand every detail of the students' often cryptic and ungrammatical prose, VAE would match the approximate CGs derived from the student answers to previously derived CGs that were known to be correct or incorrect. The results had a high correlation with the scores assigned by experienced teachers.

For applications that require a precise representation in logic, a traditional syntax-directed parser was used to translate Common Logic Controlled English to logic in the Conceptual Graph Interchange Format (ISO/IEC 24707). Following is an example of medical English, as it was written by a physician:

> Exclude patients with a history of Asthma, COPD3, Hypotension, Bradycardia (heart block > 1st degree or sinus bradycardia) or prescription of inhaled corticosteroids.

No system available today can accurately translate this kind of language to any version of logic. But a person with medical expertise and some training in writing controlled English can learn to translate this text to the following CLCE statements:

> Define "x is bradycardia" as "either x is sinus bradycardia or (x is a heart block and x has a degree greater than 1)".
>
> If a patient x has a history of asthma, or x has a history of COPD3, or x has a history of hypotension, or x has a history of bradycardia, or (x is prescribed a drug y, and y is inhaled, and y is a corticosteroid), then x is excluded.

Although these statements can be read as if they were English, CLCE is actually a formal language that has a direct mapping to first-order logic. For somebody who knows the subject matter, reading CLCE requires little or no training. Learning to write CLCE, however, requires training, especially for people who have never taken a course in logic.

To make CLCE more "user friendly," additional grammar rules were added to catch typical errors and to introduce more natural ways of expressing various logical combinations. But as we continued to add rules and inferences, we ran into maintenance problems and interactions between the inferences that resulted in confusing, but consistent readings. Since we were already implementing a new parser to replace Intellitex, we decided to design the new parser to handle CLCE as one kind of language game that could be played with English. The new parser would translate CLCE sentences directly to logic. But instead of rejecting sentences outside the CLCE subset, it would use the methods designed to handle unrestricted English. Then it would translate any CG that was generated back to CLCE as an echo and ask for a confirmation of its accuracy. In effect, CLCE was no longer defined as the language accepted by the parser, but as the language generated as an echo.

7 Designing Robust and Flexible Systems

A computer system that truly understands language would have to address all the issues discussed in this article, perhaps with others that are still unknown. Following is a brief summary:

1. Language learning, by the individual or the species, is grounded in social interactions, and full language understanding must be integrated with social behavior and all the supporting mechanisms of perception, action, knowledge, and reasoning.
2. Wittgenstein was correct in rejecting his early view, as influenced by his mentors Frege and Russell, that logic is the foundation for language. As he said in his notebooks (*Zettel*), language is "an extension of primitive behavior. (For our language game is behavior.)"
3. Instead of being the foundation for language, logic is one among many important games that can be played with the words and syntax of a natural language. Formal logic is an abstraction from those language games, not a prerequisite for them.
4. The elegant syntax of well-edited prose is another important language game, which is used in large libraries of valuable knowledge. But focusing on that game as the prototype of "language competence" is as misguided as privileging any other language game, such as poetry, prayer, casual gossip, technical jargon, or text messaging.
5. Syntactic parsers can be useful for many practical applications, but a rigid linkage of syntactic rules to semantic rules is too inflexible and fragile to support natural languages. The appropriate semantics cannot be determined without knowledge of the context and subject matter.
6. The information needed to understand a sentence can rarely be derived from just its words and syntax. Even when the syntax is unambiguous, background knowledge of the context and subject matter must be added to determine the referents, the exact word senses, and the speaker's intentions.

The new VivoMind Language Processor (VLP) is a modular, open-ended system designed to accommodate the features discussed in this article and others that remain to be invented. The processing is handled by a society of agents, which can dynamically reconfigure their interactions by the market-driven learning methods described in Section 5. New features can be handled by adding new agents to the society, and a failure of one or more agents causes a fail-soft degradation in capability, rather than a hard crash. The syntactic component of VLP generates conceptual graphs as dependency structures by techniques similar to a link-grammar parser (Temperley & Sleator 1993), but with an approach that is similar to the parallel and concurrent ParseTalk (Hahn et al. 1994). Instead of the object-oriented methods of ParseTalk, in which the calling program determines how an object is supposed to respond, the VLP agents have more freedom to make their own decisions. Many of the decisions use a consensus-based approach that combines the results of several agents.

The first major application of VLP was to analyze 79 documents in the geosciences domain. The articles, which ranged in size from 1 to 50 pages, described various sites of interest for oil and gas exploration. The documents were not tagged or annotated in any way, except for the usual formatting tags intended for human readability. The VLP system translated the texts to conceptual graphs, used the new Cognitive Memory system to index and store the graphs, and searched for analogies among the graphs that described various sites. When two sites were found to be similar, the system would state in English which aspects of one site corresponded to which aspects of the other site. A domain expert who examined the output found these side-by-side comparisons to be especially informative and useful.

Several different resources were used to provide lexical knowledge, but no attempt was made to merge all the information in a single lexicon. Such a merger was rejected for several reasons. First, many resources, such as WordNet, CoreLex, and Roget's Thesaurus, are so different in kind, in level of coverage, and in organization that the merger would be difficult to do and difficult to undo, if one resource or the other were inappropriate for a particular application or even a particular sentence. Second, resources are revised and updated at different times, and the constant need to update a large, merged resource would be inconvenient and error prone. Third, there is no need to merge the resources in advance, since the society of agents makes it easy to assign one agent to each resource, and that agent can contribute any relevant information from its resource whenever it seems useful. A voting mechanism among the agents enables them to accept or reject any contribution, depending on the current task and context.

To illustrate the operations of the agents, the following sentence was taken from one of the geoscience articles (Sullivan et al. 2005):

> The Diana field is situated in the western Gulf of Mexico 260 km (160 mi) south of Galveston in approximately 1430 m (4700 ft) of water.

In the first stage, agents for the lexical resources contribute information about the part of speech of each word, its associated concept type, and various formats for measures and other typical qualifiers. Any conflicts among the agents are resolved by voting. The result is

> entity(1, "Diana Field") prep("in the") loc(1, western) loc(2, "Gulf of Mexico") measure(1, "260 kilometers") measure(2, "160 miles") loc(3,south) prep("of") loc(4,"Galveston") prep("in") qualifier(approximately) measure(3,"1430 meters") measure(4,"4700 feet") prep("of") entity(2, water).

Lexical information, context, and heuristics determine that "ft" is "feet". Unknown words or word groups, such as "Diana field" are assumed to be geophysical entities because they are not in the basic lexicons. Other agents specialized for the geoscience domain would later determine that Diana Field is a reservoir and add the information entity([1],reservoir). For this sentence, the syntax is sufficient to determine that Diana field must be in the western part of the Gulf of Mexico, but other attachments are syntactically ambiguous. To prune away unlikely options, some agents use a domain ontology for geoscience, which includes information about reservoirs, bodies of water, and cities. After pruning, the remaining links correctly show that Diana field is south of Galveston and in the water.

The VLP parser is still in an early stage of development, but it has already produced useful results. The market-driven learning methods have proved to be successful on another project, but they haven't yet been extensively tested on VLP. The semantic distance measures of the old VAE were highly efficient for analogy finding, and variations were applied to knowledge capture (Majumdar et al. 2007). More research and testing is necessary, but the new VLP already appears to be more robust and scalable than the old Intellitex.

References

1. Deacon, T.W.: The Symbolic Species: The Co-evolution of Language and the Brain. W.W. Norton, New York (1997)
2. Eberbach, E., Goldin, D., Wegner, P.: Turing's Ideas and Models of Computation. In: Teuscher, C. (ed.) Alan Turing: Life and Legacy of a Great Thinker. Springer, Berlin (2004)
3. Frege, G.: Begriffsschrift (1879); English translation. In: van Heijenoort, J. (ed.) From Frege to Gödel, pp. 1–82. Harvard University Press, Cambridge, MA (1967)
4. Gelernter, D.: Generative communication in Linda. ACM Transactions on Programming Languages and Systems, 80–112 (1985)
5. Good, I.J.: Speculations concerning the first ultraintelligent machine. In: Alt, F.L., Rubinoff, M. (eds.) Advances in Computers, vol. 6, pp. 31–88. Academic Press, New York (1965)
6. Hahn, U., Schacht, S., Broker, N.: Concurrent Natural Language Parsing: The ParseTalk Model. International Journal of Human-Computer Studies 41, 179–222 (1994)
7. Hattiangadi, J.N.: How is Language Possible? Philosophical Reflections on the Evolution of Language and Knowledge, Open Court, La Salle, IL (1987)
8. ISO/IEC, Common Logic (CL) — A Framework for a family of Logic-Based Languages, IS 24707, International Organisation for Standardisation (2007)
9. Kamp, H.: Levels of linguistic meaning and the logic of natural language (2001), http://www.illc.uva.nl/lia/farewell_kamp.html
10. Kittredge, R., Lehrberger, J.(eds.): Sublanguage: Studies of Language in Restricted Semantic Domains, de Gruyter, New York (1982)
11. LeClerc, A., Majumdar, A.: Legacy revaluation and the making of LegacyWorks, Enterprise Architecture 5:9, Cutter Consortium, Arlington, MA (2002)
12. Levin, B.: English Verb Classes and Alternations. University of Chicago Press, Chicago (1993)
13. Limber, J.: The genesis of complex sentences. In: Moore, T. (ed.) Cognitive Development and the Acquisition of Language, pp. 169–186. Academic Press, New York (1973)
14. Majumdar, A., Keeler, M., Sowa, J., Tarau, P.: Semantic distances as knowledge capture constraints. In: Proc. First International Workshop on Knowledge Capture and Constraint Programming (2007)
15. Milner, R.: Communicating and Mobile Systems: the π calculus. Cambridge University Press, Cambridge (1999)
16. Minsky, M.: The Society of Mind. Simon & Schuster, New York (1987)
17. Minsky, M.: The Emotion Machine: Commonsense Thinking. In: Artificial Intelligence, and the Future of the Human Mind, Simon & Schuster, New York (2006)
18. Mithen, S.: The Singing Neanderthals: The Origin of Music, Language, Mind, and Body. Harvard University Press, Cambridge (2006)

19. Ogden, C.K., Richards, I.A.: The Meaning of Meaning, Harcourt, Brace, and World, New York (1923), 8th edn. (1946)
20. Peirce, C.S.: (CP) Collected Papers of C. S. Peirce. In: Hartshorne, C., Weiss, P., Burks, A. (eds.), vol. 8, Harvard University Press, Cambridge, MA (1931-1958)
21. Selfridge, O.G.: Pandemonium: A paradigm for learning. In: The Mechanization of Thought Processes, NPL Symposium No. 10, Her Majesty's Stationery Office, London, pp. 511–526 (1959)
22. Sleator, D., Temperley, D.: Parsing English with a Link Grammar. In: Third International Workshop on Parsing Technologies (1993), http://www.cs.cmu.edu/afs/cs.cmu.edu/project/link/pub/www/papers/ps/LG-IWPT93.pdf
23. Sowa, J.F.: Architectures for intelligent systems. IBM Systems Journal 41(3), 331–349 (2002)
24. Sowa, J.F.: Graphics and languages for the Flexible Modular Framework. In: Wolff, K.E., Pfeiffer, H.D., Delugach, H.S. (eds.) ICCS 2004. LNCS (LNAI), vol. 3127, pp. 31–51. Springer, Heidelberg (2004)
25. Sowa, J.F., Majumdar, A.K.: Analogical reasoning. In: Ganter, B., de Moor, A., Lex, W. (eds.) ICCS 2003. LNCS, vol. 2746, pp. 16–36. Springer, Heidelberg (2003)
26. Sullivan, M.D., Lincoln Foreman, J., Jennette, D.C., Stern, D., Jensen, G.N., Goulding, F.J.: An Integrated Approach to Characterization and Modeling of Deep-water Reservoirs, Diana Field, Western Gulf of Mexico, Search and Discovery, Article #40153 (2005)
27. Thouin, B.: The METEO system. In: Lawson, V. (ed.) Practical Experience of Machine Translation, pp. 39–44. North-Holland, Amsterdam (1982)
28. Wang, H.: Toward mechanical mathematics. IBM Journal of Research and Development 4, 2–22 (1960)
29. Warren, R.M.: Restoration of missing speech sounds. Science 167 (1970)
30. Wildgen, W.: Catastrophe Theoretic Semantics: An Elaboration and Application of René Thom's Theory. John Benjamins Publishing Co., Amsterdam (1982)
31. Wildgen, W.: Process, Image, and Meaning: A Realistic Model of the Meaning of Sentences and Narrative Texts. John Benjamins Publishing Co., Amsterdam (1994)
32. Wittgenstein, L.: Tractatus Logico-Philosophicus. Routledge & Kegan Paul, London (1921)
33. Wittgenstein, L.: Philosophical Investigations. Basil Blackwell, Oxford (1953)
34. Wittgenstein, L.: Zettel. University of California Press, Berkeley (1970)

Web, Graphs and Semantics

Olivier Corby

INRIA Edelweiss Team
2004 route des lucioles - BP 93
FR-06902 Sophia Antipolis cedex
olivier.corby@sophia.inria.fr

Abstract. In this paper we show how Conceptual Graphs (CG) are a powerful metaphor for identifying and understanding the W3C Resource Description Framework. We also presents CG as a target language and graph homomorphism as an abstract machine to interpret/implement RDF/S, SPARQL and Rules. We show that CG components can be used to implement such notions as named graphs and properties as resources.

In brief, we think that CG are an excellent framework to progress in the Semantic Web because the W3C now considers that RDF graphs are– along with XML trees – one of the two standard formats for the Web.

1 Introduction

Conceptual Graphs were introduced by John F. Sowa in 1976 when he was at IBM [32] and were popularized in his foundational book of 1984 [33].

The Semantic Web was introduced in 1998 by W3C along with the Resource Description Framework (RDF) that enables the description of graphs. Recently, the SPARQL Query Language for RDF was published as a *Recommendation* by W3C. Further, Tim Berners-Lee informally reformulated his vision of the Semantic Web as a "Web of Data" and also as a "Giant Global Graph"[1] (GGG). This thinking lead to a dramatic change in the architecture of WWW: *RDF graphs* and XML trees were both considered as data structures for information sharing on the Web according by the W3C[2].

In this paper our presentation follows as an exercise of storytelling about the work done in the Edelweiss/Acacia team from INRIA with CG for RDF Semantic Web. We show how Conceptual Graphs (CG) were a powerful metaphor for identifying, understanding and implementing the W3C Resource Description Framework. We also present CG as a target language and graph homomorphism and as an abstract machine to interpret/implement RDF/S, SPARQL and Rules. In particular, we would like to show that CG components can be used to implement such notions as named graphs and properties as resources.

In short, we believe that CG are a good framework to progress in the Semantic Web because the W3C now considers that RDF graphs are – with XML trees – one of the two standard formats of the Web.

[1] http://dig.csail.mit.edu/breadcrumbs/node/215

[2] http://www.w3.org/Consortium/technology

P. Eklund and O. Haemmerlé (Eds.): ICCS 2008, LNAI 5113, pp. 43–61, 2008.

2 History

The Acacia team previously worked on Information Retrieval through Knowledge Models with CG (PhD Thesis of Philippe Martin [26]) and KADS. We evolved from Knowledge Based Systems to Knowledge Engineering for Corporate Memory Management. We were interested in mixing Knowledge Enginering (KE) and Structured Documents and then KE on the Web.

In 1998 we were interested in and studied XML and XSLT for Structured Web documents. In 1999, RDF was published and, thanks to our CG background, we understood that it could be used to implements graphs (CG/RDF) for documents (XML) and we worked on a first mock-up based on Notio [31,23] in Java. This first mock-up, called Corese for COnceptual REsource Search Engine [10], implemented the first translator from RDF/S to the CG model. In 2000 we had the pleasure to collaborate with Peter Eklund and Philippe Martin on RDF and CG [27].

Then we focussed on Corporate Semantic Web, a mix between Corporate Memory Management and Semantic Web Technologies. We were involved in a European project called Comma for Corporate Memory Management through Agents [18] within which we started to leverage the mock-up into a research prototype.

Ten years later, we have more than 20 running applications using CG/RDF and 5 generic systems based on the technology. We can now acknowledge that CGs were a good metaphor and enabled us to understand and foresee the Semantic Web project and enable us to participate. It was our chance to be members of the CG community and members of INRIA, one of the the founding members of the W3C.

3 CG for RDF

We have proposed a mapping between RDF and CG and extensions to the simple conceptual graph model in order to implement RDF and SPARQL features such as property variables, named graphs, filters and optional parts.

3.1 RDF Schema

We have designed a mapping between RDF and CG, and RDF Schema and CG support. RDF triples are mapped to relations and resources are mapped to concepts. RDFS classes are mapped to concept types, RDF properties are mapped to relation types, domain and range are mapped to relation signature. SubClassOf and subPropertyOf are mapped to concept and property type subsumption respectively. We have designed the type inference algorithm that enables us to create well typed concepts according to their `rdf:type` and to the signatures of their relations. We have implemented some properties of relations such as symmetry, inverse and transitivity.

An interesting feature of RDF Schema is that it follows RDF syntax, i.e. triples made of a resource, a property and a value. Hence, an RDF Schema statement can be understood as a relation in a graph. For example the RDFS triples below:

```
Human subClassOf Primate
Human label 'human'@en
```

can be translated into the graph relations:

```
[Class:Human]
-(subClassOf)-[Class:Primate]
-(label)-[Literal:'human'@en]
```

RDF Schema statements loaded in the graph can be seen as *annotations*. They are related to the instances via the `rdf:type` relation.

```
[Human:Jules]-(rdf:type)-[Class:Human]-(subClassOf)-[Class:Primate]
```

Once present in the graph, the RDFS statements represent (reify) the real types that are present in the support. There is no semantics attached to these relations, the semantics comes from the CG support as usual. They are used as proxies for querying purpose. Two occurrences of the same identifier may represent two different entities according to the context, e.g. `Human` identifies a class and an instance.

For example, the query below retrieves instances of classes whose English label contains the string 'human' and hence finds Jules:

```
?x r ?y . ?y rdf:type ?class .
?class rdfs:label ?l
filter(regex(str(?l), 'human' ) && lang(?l) = 'en' )
```

This feature happens to be extremely useful in real applications where we can query the graph and its schema within the same formalism. Once again, the operational semantics w.r.t. graph projection is carried out by the support.

3.2 Type Intersection

One main difference between RDF and CG is that in RDF a resource may have several types whereas in CG a given concept has but one type. We solved this problem by assigning as a concept type the intersection of the types. Hence, we had to design an algorithm that computes, on the fly, the intersection of two types.

```
x rdf:type T1        p1 domain T1
x rdf:type T2        p2 domain T2
=>                   x p1 y
[T1 AND T2 : x]      x p2 z
                   =>
                    [T1 AND T2 : x]
```

The algorithm maintains the consistency in the type hierarchy. Which means that subtypes of types for which we compute an intersection must then be subtypes of this intersection. In the example below, Aircraft must be a subclass of the intersection of Mobile and Object:

```
Flying  subClassOf Mobile     Aircraft subClassOf Flying
Artefact subClassOf Object   Aircraft subClassOf Artefact

Mobile_AND_Object subClassOf Mobile
Mobile_AND_Object subClassOf Object
=>
Aircraft subClassOf Mobile_AND_Object
```

The intersection algorithm also takes into account disjoint types that cannot generate intersections in their descendants.

3.3 Datatype Values

In order to implement RDF we had to design a datatype extension. Some nodes in the graph carry datatype values such as (integer, 45) or (string, 'Garfield'). Datatype values are implemented as Java objects whose classes implement operators, such as *equal, greater than*, etc., through method overloading. Markers of literal nodes contain such Java objects.

Two input strings may lead to the same datatype value:
`'01'xsd:integer` and `'1'xsd:integer` represent the same value. Hence they must be mapped to the same marker containing the same value.

Operators are implemented through method overloading that realizes type checking, i.e. numbers can compare with numbers, strings with strings, etc. We decided for efficiency reasons to rely on Java polymorphism to tackle type checking.

3.4 Property Concept

In a standard query graph, there may be generic markers associated to concepts but not with relations. Property variables enable the use a variable in a query in place of a property (relation). For example in the query below we search two concepts, ?x and ?y, related by any property, denoted by variable `?p`.

```
?x ?p ?y
```

The advantage of using a variable is that we can retrieve the property in the result by getting the value of variable `?p` just as any other variable (e.g. `?x`). In addition, we can search for concepts that are related by the *same* property by using the same variable.

```
?x ?p ?y . ?y ?p ?z
```

Eventually, we can express constraints on the property by means of the variable. For example, we can look for transitive properties:

```
?x ?p ?y . ?p rdf:type owl:TransitiveProperty
```

Or we can search for properties from a specific ontology:

```
?x ?p ?y
filter(regex(str(?p), 'http://www.inria.fr/edelweiss/schema#'))
```

In order to implement the processing of property variable within standard graph projection, we have proposed to reify the property by an additional concept. This concept is of type `rdf:Property` and its marker is the name of property. Each occurrence of relation in a graph contains the additional concept that represents (reify) the property.

```
x1 r y1 -> r(x1, y1, r)
x2 q y2 -> q(x2, y2, q)
```

Hence we manage hyperarcs, i.e. arcs that relate more that two nodes. It is remarkable that several authors [3,14,22] propose the same extension from a theoretical point of view.

In our extension, a query relation may or may not use a property variable. If not, the property concept is invisible and is not processed during graph projection.

3.5 Named Graph

Following the same design pattern, we have implemented a second extension for named graph. A named graph is a graph which is associated a name by means of a URI. This URI is a standard resource that can itself be annotated by means of properties.

In the example below, `g1` is the name of a graph:

```
g1 { cat on mat . cat name 'Garfield' }
g1 author James
```

The name of the graph (the URI) is reified as an additional concept and each relation of a given graph contains this additional concept. With the example above, and with `std` as the URI of the standard graph (the graph with no name):

```
on(cat, mat, on, g1)
name(cat, 'Garfield', name, g1)
author(g1, James, author, std)
```

Note that `name` (resp. `author`) appears once as the name of the relation and once as the concept that reifies the relation, according to the hyperarc point of view explained above. Hence, the same name is used for different entities.

We are then able to process queries with graph patterns by matching the graph URI with the additional argument carried by the hyperarcs:

```
select * where {
graph ?g { cat on ?place }
}
```

This query is translated into the following hyperarc where `_:b` represents a query blank which means that we don't care about the property concept:

```
on(cat, ?place, _:b, ?g)
```

We obtain as result:

```
?g = g1 ; ?place = mat
```

It is remarkable that this very simple idea, implementing named graphs with an additional argument, solves the problem of representing *and* querying named graphs. This is what we mean by considering CG as a valuable target abstract machine to implement RDF processing. The SPARQL *from* and *from named* clauses are implemented by adding appropriate filters on the graph variables.

```
select *
from <g1>                    on(?cat, ?place, on, ?g)
where { ?cat on ?place }     filter(?g = <g1>)
```

3.6 Inference Rules

We have designed a forward chaining graph rule language with an RDF/SPARQL syntax. This language is inspired by Salvat and Mugnier [30]. The syntax of the rule condition and conclusion patterns is that of SPARQL patterns (i.e. collections of triples). We have included the graph pattern in the syntax, hence it is possible to take named graphs into account.

```
graph ?g { ?x ?p ?y . ?y rdf:type owl:SymmetricProperty }
=>
graph ?g { ?y ?p ?x }
```

3.7 Projection

We have designed and implemented an hypergraph homomorphism algorithm based on relation enumerations following heuristics to optimize the search. The order in which the query relations are considered is compiled according to heuristics such as the relation's cardinality (number of occurrences), connexity, presence of filters, etc. In addition to compiling the order of query relations, the algorithm is able to backjump in case of a failure due to the absence of a target relation or due to the failure of a constraint. By backjump we mean that it is able to backtrack – not systematically to the preceding query relation – but to a preceding query relation that may solve the failure. The index of where to backjump is determined statically and compiled.

In addition to property variables and graph patterns for named graphs, the algorithm is able to process optional query parts. If an optional part fails, the query does not fail. If it succeeds, the answer contains additional information.

Example: retrieve resources which have a name (mandatory) and which may have an age (optional).

```
[?x]-(name)-[?name]
optional { [?x]-(age)-[?age] }
```

An optional part may contain several relations, in which case it succeeds if *all* relations succeed. It may contain filters in which case it succeeds if the filters evaluate to true. It may contain nested optional parts which are processed only if the current optional part succeeds. Hence, the processing of queries with optional parts imply the introduction of scopes surrounding the optional parts.

Eventually, the algorithm has been adapted to interpret SPARQL queries with *select, distinct, order by* and *limit* operations. The *distinct* operation is an interesting constraint that ensures that two answers do not contain the same variable bindings, e.g. `select distinct ?x ?y` ensures that the bindings of *?x, ?y* differ in all answers. Hence, we need to manage a list of current answers to the homomorphism and check that the current answer that is computed is distinct from all previous answers. An optimization computes the distinct set as soon as all variables are bound in the partial result. If it is not the case, the graph homomorphism backtracks and searches for other bindings. In practice, the algorithm *backjumps* to a new binding.

In addition, we have added a *group by* operation that enables us to group results that share same variable binding for some variables and *count()* that enables to count the number of values of a variable after grouping. We have also added the possibility of returning the result of an expression in the result (in the select clause). For example, the query below retrieves persons that are the authors of documents, groups the results by person, counts the documents of each author and returns the counter in the result.

```
select ?person count(?doc) as ?count
where { ?person :author ?doc }
group by ?person
```

These operations fits smoothly within graph homomorphism but the SPARQL union operation does not fit well into this paradigm. It needs to be implemented as an operator of an interpreter that would implement AND, UNION and OPTIONAL operations applied to elementary graph homomorphisms.

3.8 Constraints

Another originality of our homomorphism algorithm is that it is able to take additional constraints on node values into account. Example of constraints are: `?x != ?y`, `?date <= '2008-01-01'` and `fun:foo(?x, ?y)` where `?x, ?y, ?date` represent the value of the target nodes associated by homomorphism to the query nodes denoted by the variables.

A query graph with constraint matches a target subgraph found by homomorphism if the constraint evaluates to true when applied to the appropriate nodes of the target graph. Constraints are prefixed by the `filter` keyword.

Examples:

```
[?x]-(r)-[?z]-(p)-[?y]   filter(?x != ?y)

[?x]-(birth)-[?date]     filter(?date <= '2008-01-01')
```

We have designed a constraint language that has been extended to process SPARQL filters. The language enables us to define simple operations such as comparisons between node values: `?x != ?y`, boolean expressions such as: `?x != ?y && ?z <= '2004-01-01'` and function calls such as: `xsd:datatype(?x)`.

The atomic entities of the language are constants and variables. Constants are values carried by the nodes of the target graph. They may be URIs of resources or literal values such as strings, integers, booleans and dates. Variables represent the values of target nodes found by graph homomorphism. Values of target nodes are datatype objects, similar to the constants, that implement polymorphic operators according to type checking rules (integers do not compare with strings, floats compare with doubles, etc.).

Constraint expressions (`EXP`) are built on top of the atomic entities (`CST`, `VAR`) with function calls (`FUN`) and terms (`TERM`). Terms are recursively build with expressions related by operators. An abstract syntax of the constraint language is given below:

```
EXP  ::= CST | VAR | FUN | TERM
FUN  ::= NAME ( EXP* )
TERM ::= EXP and EXP | EXP or EXP | not EXP |
         ( EXP )     | EXP OPER EXP
OPER ::= < <= =  != >= > + - * /
```

The projection algorithm cooperates with a constraint evaluator that is able to evaluate partial constraints according to a current partial binding. As soon as the variables of a constraint are bound by target nodes, the constraint is evaluated. If the expression evaluates to true, the projection continues (the current partial projection is successful). If it fails, the projection algorithm backtracks in order to find another binding for the variables. In fact, the algorithm *backjumps* in order to effectively change the binding.

The evaluator is a recursive function that has two arguments: an expression of the constraint language and an environment that contains variable bindings. Variable bindings are computed by the projection and are the values of the target nodes corresponding to the query node variables, e.g. `?x = 12 ; ?y = '2007-01-01' ; ?z = URI`. The evaluator returns values of the same domains as the constants. The final result of a constraint evaluation must evaluate to true.

A scheme of the constraint evaluator is given below where `exp` is the constraint expression and `env` is the variable binding environment.

```
eval(exp, env){
switch(exp){
case constant : return exp;
case variable : return env.get(exp);
case funcall  : values = for all arg(exp) : eval(arg, env);
                return apply(fun(exp), values);
case not       :return ! eval(arg(exp), env);
default : return apply(operator(exp), eval(arg1(exp), env),
                                      eval(arg2(exp), env));}}
```

Complex constraint expressions are decomposed into smaller ones which are associated to subpart of the query where their variables are bound and they are evaluated as soon as possible in order to cut the search tree.

It must be noted that – as in SPARQL – it is possible to test a *negation as failure* query using an optional pattern and a `! bound()` constraint. As an example, the query below searches persons that are *not* author of a document. The query search for an optional `author` relation. If it is not found, the query succeeds; if it is found, the constraint fails because the `?doc` variable is bound and hence the query fails.

```
select * where {
?x rdf:type :Person
optional { ?x :author ?doc }
filter(! bound(?doc))
}
```

3.9 Type Relaxation

Our projection algorithm is able to perform approximate search wrt types. It is possible to relax type checking according to subsumption. For example, when searching for a person author of an article, we may return a research team author of a report. We relax the type Person by Team and the type Article by Report. We compute a semantic distance between concept types which decreases with depth like in [35] and try to minimize the sum of the distances.

This idea happens to be quite interesting and we have generalized this relaxation process. It is now possible to design and program a new distance algorithm and specify such a user defined algorithm in a query. Hence, the user can try different relaxation algorithms according to the domain and/or the query. The syntax is the following where the `more` keyword authorizes relaxation, the prefix specifies where to fin the Java package of the user defined distance algorithm and the `relax by` statement requires the user defined distance.

```
prefix dd: <fun://fr.inria.edelweiss.Distance>
select more * where { PATTERN }
relax by dd:distance
```

3.10 Graph Path

We implemented an extension to SPARQL to process path queries, inspired by [25,1]. The path algorithm avoids cycles. Using a path variable in place of the property is done by introducing a $ prefixed variable, which means find a path of one or more relations that links `a` and `b`:

```
a $path b
```

It is possible to test the length of the target path:

```
pathLength($path) >= 2 && pathLength($path) <= 8
```

It is possible to associate a regular expression that must be matched by the types of the relations of the target path. In the following case, we want the properties to be either p1 or p2. By default, we also accept subproperties.

```
match($path, star(p1 || p2))
```

We have designed and implemented the following original extension in order to match the target relations that have been found in the path. The target relations of the path are grouped in a transient *named graph* whose name is given by the path variable. Hence this named graph is accessible by means of a graph pattern on the path variable. It has for effect to enumerate the target path relations as shown below, where **$path** is the path variable:

```
graph $path { ?x ?p ?y }
```

The purpose of this pattern, in addition to enumerate the path relations in the result, is to enable us to specify additional constraints such as in the examples shown below. For instance, to go through a specific resource within the path:

```
graph $path { ?x ?p ?y filter(?x = a || ?y = a) }
```

Or not to go through a specific resource:

```
graph $path {
optional { ?x ?p ?y filter(?x = a || ?y = a) }
filter (! bound(?p))
}
```

Or to find a specific pattern within the path:

```
graph $path { ?x p a . a q ?z }
```

This path algorithm has been applied to the Insee RDF base that describes French territory[3] with 500,000 relations and a version which computes shortest path was able to find a shortest path between Nice and Grenoble in 0.3 sec.

Another extension of the path algorithm for navigating through recursively nested contexts is explained below.

4 Context

Recently, we have been working on contexts using named graphs. A named graph is a graph which has a name given by a URI and which is accessible by means of a graph pattern in a query. In addition, this URI is itself a resource that can be part of graphs.

[3] http://rdf.insee.fr/geo/

Two named graphs, **g1** and **g2**, are shown below:

```
g1 { a p b . b q c}    g2 { a r d }
```

A query with a graph pattern to retrieve relations in named graphs:

```
select *
from named <g1>
from named <g2>
where {
graph ?g { ?x ?p ?y }
}
```

A special case of named graphs enables us to describe nested graphs such as nested Conceptual Graphs.

```
:Alice c:tell :story
:story { :Cat :on :Mat }
```

Named graphs and graph patterns are simple but powerful notions that enable us to model contextual metadata where a context is a named graph and is denoted by its name.

4.1 Hierarchy of Type of Context

It is possible to model a hierarchy of class of context and to type the URI of the named graphs. Hence we can retrieve contextual metadata according to context types and exploit subsumption.

```
Past subClassOf Context
Prehistory subClassOf Past
 Paleolithic subClassOf Prehistory
 Neolithic subClassOf Prehistory
Present subClassOf Context
Future subClassOf Context

g1 rdf:type :Paleolithic
g1 { :man :practice :hunting }

g2 rdf:type :Neolithic
g2 { :man :practice :agriculture }
```

A query that retrieves activities in contexts of type Prehistory, i.e. Paleolithic and Neolithic:

```
graph ?g { :man :practice ?activity }
?g rdf:type :Prehistory
```

4.2 Annotation of Context

Thanks to its uniform nature, it is possible to annotate context by means of its name which is a URI.

```
g1 rdf:type :Paleolithic   g2 rdf:type :Neolithic
g1 :start -2500000         g2 :start -10000
g1 :location :Europe       g2 :location :MiddleEast
```

Note that it is possible to have several contexts of type Neolithic that start at different dates according to the location. We can then query contextual metadata:

```
graph ?g { ?x :practice ?activity }
?g :start ?date filter(?date <= -10000)
```

4.3 Contextual Relations

We can now model semantic relations between contexts such as temporal relations. It is possible to define spatio/temporal relations, linguistic relations such as those used in rhetorical structure theory (RST), logical relations, etc. Note that contextual relations can themselves be contextualized. For example, `g1 sequence g2` is true in context `state1`:

```
g1 { ... }   state1 { g1 sequence g2 }
g2 { ... }   state2 { g3 sequence g4 }
g3 { ... }   state3 { state1 parallel state2 }
g4 { ... }
```

We can then query what happens in a context `?g2` after a given context `?g1`:

```
graph ?g1 { ?x ?p ?y }
graph ?g2 { ?z ?q ?t }
?g1 sequence ?g2
```

It is of course desirable to specify the algebraic properties of the contextual relations, e.g. parallel is symmetric and transitive, sequence is transitive, etc. This can be done using OWL light statements that are interpreted in Corese. More complex algebraic properties of relations can be modeled by rules.

```
parallel rdf:type owl:SymmetricProperty
parallel rdf:type owl:TransitiveProperty
sequence rdf:type owl:TransitiveProperty
```

4.4 Rec Graph Pattern

In order to enable querying contextual relations, we have designed a generalized version of the path algorithm dedicated to nested contexts.

In the example above, suppose that we want to search/retrieve triples recursively nested within the state3 context, i.e. the triples in state1, state2, g1 and g2. We need to know the exact relations between the nested context to retrieve these triples. It may be impossible to be aware of the whole structure. To solve this, we propose a new query pattern called `rec graph` (recursive graph) as follows:

```
rec graph state3 { ?x ?p ?y }
```

The result of the query will be the triples from the state3 graph and the triples from the recursively nested graphs, e.g. state1 and g1. This is computed by the path algorithm described above. Instead of searching for path from ?x to ?y (first and second arguments), the algorithm searches for path from state3 to ?x, (i.e. from graph name argument to first argument). An example of such a path of length 3 is: `(state3, state1), (state1, g1), (g1, a)` as shown below:

```
(1) state3 { state1 parallel state2 }
(2) state1 { g1 sequence g2 }
(3) g1 { a p b }
```

Another path would be: `(state3, state2), (state2, g4), (g4, b)`. This query pattern also enables to search if several triples are recursively related by an embedding context:

```
rec graph ?g { a p b . c q d }
```

It is also possible to specify a regular expression on the relations that link the nested contexts as shown below:

```
rec graph ?g {
?x ?p ?y
filter(match(?p, star(log:property)))
}
```

4.5 Defining a Resource Using a Named Graph

The named graph statement enables to assign a name (a URI) to a graph. We propose to use this statement in a slightly different way in order to assign a definition (a graph) to a URI (it's name). This enables to define composite objects made of atomic objects where none of the sub objects plays a special role. Hence we assign a URI to a composite structure made of several related objects. The URI can then be used in other composite structures.

For example, we define the H_2O molecule as a named graph containing a description of two hydrogens related to one oxygen. The `cos:graph` attribute is a syntactic extension to RDF/XML, (W3C member submission [19]), that enables to define the URI of a named graph. Note that in the example below, there are two different Hydrogens (two blank nodes) related to the same Oxygen (one blank node with 'o' ID).

```
<c:Hydrogen cos:graph='&c;H2O' >
 <c:related><c:Oxygen rdf:nodeID='o'/></c:related>
</c:Hydrogen>
<c:Hydrogen cos:graph='&c;H2O' >
 <c:related><c:Oxygen rdf:nodeID='o'/></c:related>
</c:Hydrogen>
```

This RDF description is equivalent to the named graph:

```
H2O { [H]-(r)->[O]<-(r)-[H] }
```

Then we define the CH_4 molecule as a named graph containing a description of one carbon related to four hydrogens.

```
<c:Carbon cos:graph='&c;CH4'>
 <c:related><c:Hydrogen/></c:related>
 <c:related><c:Hydrogen/></c:related>
 <c:related><c:Hydrogen/></c:related>
 <c:related><c:Hydrogen/></c:related>
</c:Carbon>
```

We can now query the structure of a molecule using a named graph pattern.

```
select ?atom countItem(?at) as ?count where {
  graph c:H2O { ?at rdf:type ?atom }
}
group by ?atom

atom = H ; count = 2
atom = O ; count = 1
```

We can then define a product *Prod* as a named graph containing two molecules, one instance of H_2O and one instance of CH_4. Note that molecules are now considered as classes that are instantiated. We could also use a property to relate a molecule to its definition as H_2O, e.g. `_:b :definition c:H2O`.

```
<c:H2O cos:graph='&c;Prod' />
<c:CH4 cos:graph='&c;Prod' />
```

Then we define an instance of *Prod* that will hence contain one H_2O and one CH_4.

```
<c:Prod cos:graph='&c;exp' />
```

Queries

We can now write queries to check the structure of the product and of the molecules. The graph pattern query below destructures the named graph in order to retrieve the molecules that compose the product `c:Prod`. The `c:isTypeOf` property is the inverse of `rdf:type`.

```
select ?part countItem(?p) as ?count where {
 graph c:Prod { ?part c:isTypeOf ?p }
}
group by ?part
```

The result is:

```
part = H2O ; count = 1
part = CH4 ; count = 1
```

The recursive graph pattern query below (**rec graph**) recursively destructures the named graphs in order to retrieve the molecules and the atoms that compose the product `c:Prod`. Note that the query is the same as the one above except that we have added the keyword `rec`.

```
select ?part countItem(?p) as ?count where {
 rec graph c:Prod { ?part c:isTypeOf ?p }
}
group by ?part
```

The result is:

```
part = H2O ; count = 1        part = H  ; count = 6
part = CH4 ; count = 1        part = O  ; count = 1
                              part = C  ; count = 1
```

This example shows the power of representing a composite object through a URI of named graphs as we have the inverse operation that enables us to walk through the internal structure recursively by means of the **rec graph** pattern.

5 Applications

In this section we show that the couple CG/RDF has proved to be a very fruitful idea in term of systems and applications.

5.1 Generic Systems

There are now several generic CG/RDF based systems that have been designed and developed in the Edelweiss team:

Corese[4] is a generic RDF/S, SPARQL & Rules Semantic Factory that is entirely based on CGs and where CGs are the abstract machine which implements RDF graph operations by means of graph homomorphism.

Sewese[5] is a Semantic Web Server Platform based on Tomcat and Java Taglib [17]. Sewese is built on the Corese engine and provides a set of primitives to build interfaces for queries, edition and navigation, and for the management of the transverse functions of a portal (presentation, internationalization, security, etc.). An ontology editor, a generic annotation editor and a basic rule editor are parts of the Sewese platform. The main purpose of Sewese is to integrate recurrent semantic web operations (e.g. perform a SPARQL Query, transform a result binding in a given view) within a classic web technology framework (e.g. JSP pages, servlet calls).

[4] http://www-sop.inria.fr/edelweiss/wiki/wakka.php?wiki=Corese

[5] http://www-sop.inria.fr/edelweiss/wiki/wakka.php?wiki=Sewese

SweetWiki[6] is a wiki built around a semantic web server that uses semantic web technologies to support and ease the life cycle of wikis [4]. It implements folksonomy based navigation into the wiki pages.

Ecco is a Cooperative Ontology Editor dedicated to support end-users with different profiles (domain expert, engineer, ontologist, ...) in a cooperative process of ontology construction and evolution. The Ontology is managed by the Corese Factory.

SemAnnot is Generic platform for annotation extraction from text using NLP parsers and ontologies [24], the ontology and annotation processor is also Corese.

5.2 Applications

We have been involved in more than 20 applications that use the RDF/CG mapping.

SevenPro[7] is an European project on Semantic Virtual Engineering Environment for Product Design. Corese is used as Semantic Engine for Text mining and Virtual Reality annotation.

e-WOK[8] is a french ANR project that aims at designing a Semantic Web Platform for Geo Sciences. It aims at building a set of communicating portals (called e-WOK Hubs), offering both: (a) web applications accessible to end-users through online interfaces, and (b) web services accessible to applications through programmatic interfaces As applicative objectives, e-WOK aims at enabling the management of the memory of several projects on CO_2 capture and storage, with use of results of technological watch on the domain.

Two projects focus on semantic text mining of scientific literature in biology. SeaLife[9] is an European project on "A Semantic Grid Browser for the Life Sciences Applied to the Study of Infectious Diseases". ImmunoSearch[10] is a French project on searching biomarkers for controlling and maintaining the harmlessness of molecules used in perfumes, aromatics and cosmetics.

The Palette[11] European project is about "Pedagogically sustained Adaptative LEarning Through the exploitation of Tacit and Explicit Knowledge". It aims at designing semantic web services to help communities of practice communicate and share knowledge.

In the past we have also worked on Knowledge Management Platforms[21] (KMP and KM2) and on Corporate Memory Management through Agents (Comma). There are also projects that we are not members of and that make use of Corese. For example, Neurolog[12] is an ANR Funded project on Medical Imaging with Software technologies for integration of process, data and knowledge in medical imaging.

[6] http://www-sop.inria.fr/edelweiss/wiki/wakka.php?wiki=SweetWiki
[7] http://www.sevenpro.org
[8] http://www-sop.inria.fr/edelweiss/projects/ewok
[9] http://www.biotec.tu-dresden.de/sealife
[10] http://www-sop.inria.fr/edelweiss/wiki/wakka.php?wiki=Projects
[11] http://palette.ercim.org
[12] http://neurolog.polytech.unice.fr/doku.php

6 Conclusion

Semantic Web and RDF provide a unique opportunity to use CGs in large scale applications. The basic idea is to consider RDF/S as an input format to build CGs and hence test CG algorithms on large scale real applications. In effect, it is now possible to load schema and data from all over the world as more and more RDF Schema and RDF metadata are available online.

We have shown that it is possible to mix several languages – among which are RDF/S, SPARQL and its XML Result format, RDF Rule language and CGs. In addition, XSLT can be used for interoperation and presentation. Our work demonstrates that CG technology can be integrated into a complex software system. The system itself can then be used to build various applications such as Semantic Wikis, Semantic engine for an ontology based natural language processing platform or a Virtual Reality semantic engine. The point is to rely on standard languages for input/output, to focus on information and knowledge retrieval and not on presentation or editing issues within the semantic engine itself. Presentation is delegated to external processors such as XSLT engines and web servers. CG were highly successful for understanding and implementing RDF and SPARQL. The only hard problem that we encountered was the SPARQL UNION operator. We have also shown that interesting performance can be obtained – we answer in less than half a second to queries to a graph with 500,000 relations (the insee RDF base[13]).

Further, we identify open problems that would be interesting to tackle within a mix CG/RDF viewpoint: library of semantic distances, scaling to graphs with some giga-relations, indexing such giant graphs, processing queries by distributed graph homomorphism.

To finish, it has always been a great surprise that so little work on the Semantic Web makes use of CGs.

Acknowledgement

This work was performed in the Edelweiss (formerly Acacia) team at the INRIA Sophia Antipolis - Méditerranée. We would like to thank all our colleagues, among whom: Rose Dieng-Kuntz, Alain Giboin, Fabien Gandon, Jean-François Baget, Priscille Durville, Khaled Khelif, Hacène Cherfi, Virginie Bottollier, Olivier Savoie, Francis Avnaim as well as Catherine Faron-Zucker and Michel Buffa from University of Nice-Sophia Antipolis and Philippe Martin.

References

1. Alkhateeb, F., Baget, J.-F., Euzenat, J.: RDF with Regular xpressions. Technical Report RR-6191, `http://hal.inria.fr/inria-00144922/en`
2. Anyanwu, M., Maduko, A., Sheth, A.: SPARQL2L: Towards Support for Subgraph Extraction Queries in RDF Databases. In: Proc. WWW 2007, Banff, Alberta, Canada (May 2007)

[13] http://rdf.insee.fr/geo/

3. Baget, J.-F.: RDF Entailment as a Graph Homomorphism. In: Gil, Y., Motta, E., Benjamins, V.R., Musen, M.A. (eds.) ISWC 2005. LNCS, vol. 3729, pp. 82–96. Springer, Heidelberg (2005)
4. Buffa, M., Gandon, F., Erto, G.: A Wiki on the Semantic Web. In: Rech, J., Decker, B., Ras, E. (eds.) Emerging Technologies for Semantic Web Environments: Techniques, Methods and Applications, Fraunhofer Institute for Experimental Software Engineering (IESE), Germany (July 2007)
5. Carroll, J.J., Bizer, C., Hayes, P., Stickler, P.: Named Graphs, Provenance and Trust. In: Proc. of WWW 2005, Chiba, Japan (2005)
6. Chein, M., Mugnier, M.-L.: Conceptual Graphs: Fundamental Notions. Revue d'Intelligence Artificielle 6(4), 365–406 (1992)
7. Corby, O., Dieng-Kuntz, R., Faron-Zucker, C.: Querying the semantic web with corese search engine. In: Lopez de Mantaras, R., Saitta, L. (eds.) Proc. of the 16th European Conference on Artificial Intelligence (ECAI 2004), Prestigious Applications of Intelligent Systems, Valencia, Spain, August 22-27, pp. 705–709 (2004)
8. Corby, O., Dieng-Kuntz, R., Faron-Zucker, C., Gandon, F.: Ontology-based Approximate Query Processing for Searching the Semantic Web with Corese. INRIA Research Report RR-5621, INRIA (July 2005)
9. Corby, O., Dieng-Kuntz, R., Faron-Zucker, C., Gandon, F.: Searching the Semantic Web: Approximate Query Processing based on Ontologies. IEEE Intelligent Systems & their Applications 21(1), 20–27 (2006)
10. Corby, O., Dieng-Kuntz, R., Hebert, C.: A Conceptual Graph Model for W3C Resource Description Framework. In: Ganter, B., Mineau, G.W. (eds.) ICCS 2000. LNCS (LNAI), vol. 1867, pp. 468–482. Springer, Heidelberg (2000)
11. Corby, O., Faron-Zucker, C.: Corese: A corporate semantic web engine. In: Proceedings of the International Workshop on Real World RDF and Semantic Web Applications, 11th International World Wide Web Conference, Hawai, USA, May 7 (2002)
12. Corby, O., Faron-Zucker, C.: Implementation of SPARQL Query Language based on Graph Homomorphism. In: Proc. of the 15th ICCS, Sheffield, UK, pp. 472–475 (July 2007)
13. Corby, O., Faron-Zucker, C.: RDF/SPARQL Design Pattern for Contextual Metadata. In: Proc. of IEEE/WIC/ACM International Conference on Web Intelligence, Silicon Valley, USA (November 2007)
14. Dau, F.: RDF as Graph-based, Diagrammatic Logic. In: Esposito, F., Raś, Z.W., Malerba, D., Semeraro, G. (eds.) ISMIS 2006. LNCS (LNAI), vol. 4203, pp. 332–337. Springer, Heidelberg (2006)
15. Detwiler, T.: GLEEN: Regular Paths for ARQ SparQL. Technical report, University of Washington, `http://sig.biostr.washington.edu/projects/ontviews/gleen`
16. Dieng-Kuntz, R., Corby, O.: Conceptual Graphs for Semantic Web Applications. In: Dau, F., Mugnier, M.-L., Stumme, G. (eds.) ICCS 2005. LNCS (LNAI), vol. 3596, pp. 19–50. Springer, Heidelberg (2005)
17. Durville, P., Gandon, F.: Sewese: Semantic Web Server. In: WWW 2007 Developers track, Banff, Canada (2007)
18. Gandon, F.: Distributed Artificial Intelligence And Knowledge Management: Ontologies and Multi-Agents Systems for a Corporate Semantic Web. PhD thesis, university of Nice-Sophia Antipolis (November 2002)
19. Gandon, F., Bottollier, V., Corby, O., Durville, P.: RDF/XML Source Declaration, W3C Member Submission (September 2007), `http://www.w3.org/Submission/rdfsource`

20. Gandon, F., Bottollier, V., Corby, O., Durville, P.: RDF/XML Source Declaration. In: Proc. of IADIS International Conference WWW/Internet, Vila Real, Portugal (October 2007)
21. Giboin, A., Gandon, F., Gronnier, N., Guigard, C., Corby, O.: Comment ne pas perdre de vue les usage(r)s dans la construction d'une application à base d'ontologies ? retour d'expérience sur le projet KmP. In: Jaulent, M.C. (ed.) Actes des 16e Journées francophones d'Ingénierie des connaissances (IC 2005), Grenoble. France, pp. 133–144 (2005)
22. Hayes, J., Gutierrez, C.: Bipartite Graphs as Intermediate Model for RDF. In: Proc. International Semantic Web Conference, ISWC (2004)
23. Hebert, C.: Modèle de traitement de RDF basé sur les graphes conceptuels. Master thesis, I3S, University of Nice-Sophia Antipolis (1999)
24. Khelif, K., Dieng-Kuntz, R., Barbry, P.: An Ontology-based Approach to Support Text Mining and Information Retrieval in the Biological Domain. Journal of Universal Computer Science (JUCS), Special Issue on Ontologies and their Applications (2007)
25. Kochut, K.J., Janik, M.: SPARQLeR: Extended SPARQL for Semantic Association Discovery. In: Franconi, E., Kifer, M., May, W. (eds.) ESWC 2007. LNCS, vol. 4519, pp. 145–159. Springer, Heidelberg (2007)
26. Martin, P.: Exploitation de graphes conceptuels et de documents structurs et hypertextes pour l'acquisition de connaissances et la recherche d'informations. PhD thesis, University of Nice-Sophia Antipolis (October 1996)
27. Martin, P., Eklund, P.: Conventions for Knowledge Representation via RDF. In: Proc. of WebNet 2000, San Antonio, Texas, USA (November 2000)
28. Mugnier, M.-L., Chein, M.: Représenter des connaissances et raisonner avec des graphes. Revue d'IA 10(1), 7–56 (1996)
29. Rudolph, S., Krtzsch, M., Hitzler, P.: Quo Vadis, CS? On the (non)-Impact of Conceptual Structures on the Semantic Web. In: Conceptual Structures: Knowledge Architectures for Smart Applications, Proc. of the 15th ICCS, Sheffield, UK. LNCS. Springer, Heidelberg (2007)
30. Salvat, E., Mugnier, M.-L.: Sound and Complete Forward and Backward Chainings of Graph Rules. In: Proc. of the 4th ICCS, Sydney, Australia. LNCS (LNAI), vol. 1115, pp. 248–262. Springer, Heidelberg (1996)
31. Southey, F., Linders, J.G.: Notio - a java API for developing CG tools. In: Proc. ICCS, pp. 262–271 (1999)
32. Sowa, J.: Conceptual Graphs for a Database Interface. IBM Journal of Research and Development (4), 336–357 (1976)
33. Sowa, J.: Conceptual Structures - Information Processing in Mind and Machine. Addison-Wesley, Reading (1984)
34. Stoermer, H., Palmisano, I., Redavid, D., Iannone, L., Bouquet, P., Semeraro, G.: RDF and Contexts: Use of SPARQL and Named Graphs to Achieve Contextualization. In: Proc. of the 1st Jena User Conference, Bristol, UK (2006)
35. Zhong, J., Zhu, H., Li, J., Yu, Y.: Conceptual graph matching for semantic search. In: Priss, U., Corbett, D.R., Angelova, G. (eds.) ICCS 2002. LNCS (LNAI), vol. 2393, pp. 92–106. Springer, Heidelberg (2002)

Transdisciplinarity and Generalistic Sciences and Humanities

Rudolf Wille

Technische Universität Darmstadt,
Fachbereich Mathematik, D–64289 Darmstadt
wille@mathematik.tu-darmstadt.de*

Abstract. The term *"transdisciplinarity"* applies to forms of research which are used by disciplines with the effect that their ways of thinking are rationally understandable, available, and can be activated beyond their boundaries for being able especially to contribute to solutions of problems which cannot be mastered purely disciplinary. This paper elaborates the thesis that the disciplines can fulfill best the request for transdisciplinarity if they develop, maintain, and activate their part of generalistic sciences and humanities in the largest possible breath. For the transdisciplinary border crossing in the sense of generalistic sciences and humanities, it has been proven efficient to search for general interpretations of disciplinary concepts, propositions, and theories which can be made understandable (possibly in standard language). Examples are given by applications of Formal Concept Analysis.

Contents

1 Multi-, Inter-, and Transdisciplinarity

Since more than two hundred years the science develops "into the form of an increasing and differentiating system of sciences and humanities, of scientific departments and disciplines" ([Ko87]; p.7). *Disciplinarity* therefore constitutes the kernel of university science today. However, the increasing specialization and division of scientific disciplines cause that it becomes more and more difficult to say what at all a discipline is. In his article "unity of the world - multitude in the science" [Kr87], L. Krüger examines by the four aspects "objects", "methods", "interest in discovery", and "theories and their systematic and historic connections" to what extent disciplines can be characterized. In spite of quite considerable difficulties of demarcation, Krüger views the identity of disciplines founded by their paradigms in the sense of Th. Kuhn (cf. [Ku73]), which comprise the four named aspects, and derives from that:

* This paper is an English version of the German publication [Wi02].

P. Eklund and O. Haemmerlé (Eds.): ICCS 2008, LNAI 5113, pp. 62–73, 2008.

> "Disciplines are historic units; they are neither in their internal subdisciplinary structure nor in their inter- and supra-disciplinary external relationships to be determined once and for all. They are individuals which
> ◇ arise in the history of the sciences and humanities,
> ◇ stand to each other in genealogical relationships,
> ◇ form families or drift apart, and
> ◇ can enter into new connections with different luck." ([Kr87]; p.116f.)

The great significance of disciplines, which make possible substance, connection, and continuity, has as reverse side that disciplines have to limit themselves in their objects, methods, interest in discovery, and theories. These limitations cause that many important problems cannot be disciplinarily mastered (for example in areas of environment and consequences of technics). For enabling the scientific treatment of such problems, the opening to *multi-, inter-, and transdisciplinarity* becomes more and more demanded (see, for instance, [Mi87], [Mi89]).

Since the terms "multi"-, "inter"-, and "trans"-disciplinarity are not always used with the same meanings, we have to clarify first how these terms should be understood in the scope of this paper.

A form of research shall be called *"multidisciplinary"* if several disciplines combine additively in such a form and if each discipline contributes with its own way of thinking. Multidisciplinarity which does not aim to aggregate forms of thinking of different disciplines are often criticized without good reason (cf. [Mi98], p.32). Many problems are quite solved multidisciplinarily without coming to ways of aggregate thinking. As an example we have first of all the successful collaboration of physics and mathematics; for instance, in the second half of the 20th century, physicists discovered and explored the so-called quasi-crystals whose quasi-symmetries could later be made understandable by the mathematician N. G. de Bruijn in a way that it was not necessary to integrate the used mathematical patterns of thinking into the physical theory of quasi-crystals (cf. [Br81]). The multidisciplinary cooperation of physics and mathematics usually succeeds on the reason that many physical connections are described by mathematical concepts which can be used as bridges to different theories of mathematics. In many cases of multidisciplinarity the successful connection between disciplines is already produced by the practice of the considered field of problems. As an example for this, the research about "pure intonation" in the scope of the Darmstadt research project "Mathematical Music Theory" shall be mentioned: For this research the experimental instrument MUTABOR was developed [GHW85] for which primarily music-theoretical and electrotechnical competences were necessary which could be successfully combined on the basis of the existing practice of constructing musical instruments.

A form of research shall be called *"interdisciplinary"* if, in such a form, several disciplines combine integratively , if each discipline contributes with its way of thinking, and if it aims together with the other disciplines at ways of problem-related, aggregated research which still satisfies general claims of scientific rationality. Interdisciplinarity therefore represents a form of research which demands from the participating disciplines integration efforts to a high degree for which most of the scientists today are not well prepared. Nevertheless, convincing interdisciplinary research projects have already been performed in great plenty which merely prove the research projects supported by the Darmstadt *"Zentrum für Interdisziplinäre Technikforschung"* (ZIT). A concrete example shall briefly illustrate how it may come in an interdisciplinary research cooperation to ways of problem-related, aggregated thinking: For a text of introduction to a volume of case studies about regimes in international relations [Ko89], the politician B. Kohler-Koch (one of the first members of the ZIT-board) wanted a comprehensive analysis of a data-table in which 18 regimes were classified by a larger number of attributes; she therefore asked for support the "Research Group on Formal Concept Analysis" in the department of mathematics of the TU Darmstadt. The principal item of the initiated interdisciplinary cooperation was the aggregation of mathematical and political ways of thinking in more than 100 concept structures (derived from the data-table) in which interesting connections could be uncovered and examined (cf. [Ko89], [VWW91], [KV00]). The success of this cooperation had rececently in its train a further interdisciplinary cooperation project (supported by ZIT) in which the data of more than 90 regimes were analyzed (see [WW01]).

A form of research shall be called *"transdisciplinary"* if, in such a form, disciplines have an effect thereupon that their ways of thinking are rationally understandable, available, and can be activated beyond their boundaries for being able especially to contribute to solutions of problems which cannot be mastered purely disciplinary. Such forms of research allow to exceed disciplinary and specializing limitations to the advantage of an - as J. Mittelstraß formulates it in [Mi96] - "extension of scientific perception abilities and problem solving competences". Mittelstraß however goes on in his determination of transdisciplinarity by seeing in the transdisciplinarity the "real interdisciplinarity" which set aside disciplinary and specializing parcellations and views the original unity of science and humanities as the unity of scientific rationality in the practical-operational sense (see [Mi98], p.44f.). This extensive concept of transdisciplinarity shall not be taken over because this concept would finally expire the already introduced concepts of multidisciplinarity and interdisciplinarity which have proven useful in a manifold and convincing practice. The more limited concept of transdisciplinarity, proposed in this paper, seems in addition to be more fruitful with respect to research methodology. For instance, with this concept, one can clarify that each increase of transdisciplinarity produces better presumptions not only for transdisiplinarity, but in general for a more practical relationship of disciplinarity to reality. How such an increase of transdisciplinarity can be reached methodologically shall be explained in the following in the frame work of "generalistic sciences and humanities".

2 Generalistic Sciences and Humanities, and "Good" Disciplinarity

The conception of "generalistic sciences and humanities" arose in the last twenty-five years out of efforts for a "good" disciplinarity. "Good" disciplinarity can be adjudicated to a science and a humanity, respectively, if they are conscious of their social tasks, productively acting for that, and making their actions accessable for general critics by suitable imparting. How "good" disciplinarity can be realized, this was described by H. von Hentig in his book "Magician or master? On the unity of sciences and humanities in the process of understanding" as follows: The individual sciences and humanities must examine

> "their disciplinarity, and this means: to uncover their unconscious purposes, to declare their conscious purposes, to select and adjust their means, to explain publicly and understandably their justification and their possible consequences and, for that, to make accessible their path of findings and their results via the common language." ([He74], p.136f.)

The question is: by which way a science or humanity may reach "good" disciplinarity. In any case, it has to examine its conception of oneself, its relationship to the world, and questions about sense, meaning, and connection of its actions. In particular, sciences and humanities must comprehend the general imparting of their scientific ideas as their task, which is still not enough understood. To that, Hentig shall be quoted again:

> "The restructuring of sciences and humanities in themselves - to make them better learnable, mutually accessible, and more generally (i.e. also beyond the disciplinary competence) criticizable - may and must be undertaken by patterns which are drawn from the general forms of perception, thought, and actions." ([He74], p.33f.)

About thirty years ago professors, assistants, and students of the Darmstadt department of mathematics tried to seriously fulfill Hentig's charges and to reach a more general understanding of mathematics [DW82]. They have approached the required *restructuring of sciences and humanities* for different parts of mathematics which has been proven eminently fruitful for teaching as well as for research (cf. [Wi01]). However, critical reactions and discussions showed that the restructuring should be seen in the frame work of the larger conception of generalistic sciences and humanities. What is understood by *"generalistic sciences and humanities"* has been first carried out 1987 by a contribution to a lecture series on *"responsibility in sciences and humanities"* at Darmstadt University of Technology (see [Wi88]). According to this lecture, the conception of generalistic sciences and humanities cover all efforts to disclose and to make accessible sciences and humanities so that the general public may, in particular, discuss possible consequences and results of scientific doings. Thus, generalistic science and generalistic humanity are not understood as autonomous scientific discipline, respectively, but as part of each scientific discipline and subdiscipline too. What does now characterize such part of a scientific discipline? According to [Wi88], that is

- the *attitude* to open the scientific discipline for the general public and to principally make the discipline learnable and criticizable,
- the *presentation* of scientific developments in their sense, meanings, and conditions,

- the *imparting* of the scientific discipline in its lifeworld connections even across the borders of subjects,
- the *discussion* about goals, procedures, value representations, and validity claims of the discipline.

What causes often difficulties is to bring such an understanding of generalistic sciences and humanities into correspondence with usual standards of "good" disciplinarity. A central reason for this lies in the often dominating instrumental understanding of rationality. J. Habermas therefore pleads in his "theory of communicative action" for subordinating the cognitive-instrumental rationality under a concept of *"communicative rationality"* which refers to more older logos ideas:

> "This concept of communicative rationality carries with it connotations which finally goes back to the central experience of the unconstrainedly unifying, argumentative discource, in which the different participants overcome their first only subjective views and ascertain, thanks to the common ground of reasonably motivated convictions, simultaneously the unity of the objective world and the intersubjectivity of their continuity of life." ([Ha81]; p.28)

As K.-O. Apel made clear in [Ap76], not only the subjective and the objective, but also the intersubjective is - in the transcendental-philosophical sense - constitutive for the human thinking and acting. Therefore the validity of propositions about the world can only be assured by rational argumentation in the scope of the intersubjective community of communication. Therefore "good" disciplinarity basically depends on understanding and communicative rationality, what also takes in account the mentioned historical conditionality of disciplines. Hence "good" disciplinarity for the sciences and humanities means that they are capable and willing to impart their discipline and its impacts to the general public and, therewith, to take on the social discussion, or differently phrased: that they develop, maintain, and activate their part in the generalistic sciences and humanities in the largest possible breath (cf. [Wi97]).

3 Transdisciplinarity by Generalistic Sciences and Humanities

My thesis is:

> *The disciplines can fulfill best the request for transdisciplinarity if they develop, maintain, and activate their part of generalistic sciences and humanities in the largest possible breath.*

According to the expositions in Section 2, this thesis can also be brought into the short form: *transdisciplinarity by "good" disciplinarity.* Our thesis obtains support by H. von Hentig's critical evaluation of *interdisciplinarity* which he presented to the programmatic discussion about the foundation of the University of Bielefeld (Germany):

> "Interdisciplinarity has become a problem because disciplinarity is misunderstood and misused. One has to discipline the disciplines and not to create new interdisciplinary institutes. The communication and cooperation, the convertibility and the mobility between sciences and humanities, respectively, do not need as much interdisciplinary bridging as rather the development of certain pre-disciplinary learning processes and continued trans-disciplinary problem tasks. The real problem is: how do basic sciences and humanities, which are always specialized in some form, come into a practicable relationship to a practice which is almost never restricted disciplinarily and to a preparation which must be general by many reasons, but above all for the sake of sciences and humanities themselves." ([He74], p.33)

For Hentig, real interdisciplinarity cannot be realized without transdisciplinary competence of the participating disciplines which can be activated because, without those competences, it cannot come to interdisciplinarily aggregated ways of thinking. For successful transdisciplinarity it is decisive that disciplinary thinking and results are made accessable and understandable across the borders of the disciplines (even by the standard language). This opens the possibility for an integration below or before the step to disciplinary specialization. Interdisciplinarity is therefore realizable, as a rule, on a pre-disciplinary step which allows so much common understanding that the necessary aggregations of disciplinary ways of thinking and results can be achieved.

But how does one come to the evidently necessary *transdisciplinary competence* of the participating disciplines? As the Darmstadt efforts about generalistic sciences and humanities have shown, such efforts can lead to the development of transdisciplinary competence of considerable extent. An example concerning transdisciplinary border crossing of mathematics shall be described in more detail:

Hentig's required restructuring in patternsof general perception, thought and actions, to make disciplines more learnable, accessible, and criticizable, is anything but self-evident in mathematics ([Wi01]). A fruitful question, which decisively accelerated the process of restructuring, was the question concerning the "interfaces" between mathematics and reality. To our surprise we found out that there is obviously one dominating type of interface between mathematics and reality: the *data table*! In its most simple form: the *cross table*, in which a cross indicates that an object has an attribute, respectively, the data table corresponds to what is called an *incidence structure* in mathematics and is defined as binary set structure (G, M, I) with $I \subseteq G \times M$; the set G corresponds to the whole of objects listed in the data table, the set M corresponds to the whole of attributes listed in the data table, and the binary relation I corresponds to the whole of object-attribute-pairs indicated by the crosses in the data table. If a data table indicates more generally that an object has an attribute value with respect to a (many-valued) attribute, respectively, then such a data table corresponds to a set structure (G, M, W, I) with $I \subseteq G \times M \times W$. This more general case can be led back to the more elementary case of a cross table by methods which are adapted to the contents, respectively (see [GW99], chapter 1).

For the process of restructuring the effort for suitable mathematizations of basic forms of rational thinking was particularly supporting. In this case the decisive breakthrough was the mathematization of *concept* and *concept hierarchy* in the sense of a philosophical understanding which underlies the standards DIN2330 "Begriffe und Benennungen" and DIN2331 "Begriffssysteme und ihre Darstellungen". It has been approved that this mathematization was founded on the set structure (G, M, I) which corresponds to a cross table and is named *"formal context"*. Concepts have been mathematically modeled in a formal context (G, M, I) as pairs (A, B) of sets $A \subseteq G$ and $B \subseteq M$; the "concept extent" A contains exactly those objects of G which have all attributes of B, and the "concept intent" B contains exactly those attributes of M which apply to all objects of A. The "formal concepts" (A, B) of (G, M, I) always form with respect to the ordering "subconcept-superconcept" the mathematical structure of a complete lattice which is called the *concept lattice* of the formal context (G, M, I) [Wi82]. What makes concept lattices in the sense of transdisciplinarity so valuable lies, according to many years of experiences, in the fact that the conceptual relationships inherent in data tables become transparent and can be activated; in particular, suitable representations of concept lattices contribute to that by labelled line diagrams (see for more details in [GW99]).

For the transdisciplinary border crossing in the sense of generalistic sciences and humanities, it has been proven efficient to search for *general interpretations* of disciplinary concepts, propositions, and theories which can be made understandable (possibly in standard language). With the approach of a restructuring of mathematics as described up to now, we obtain, for example, for the mathematical term "set" the general interpretation "concept extent" and "concept intent", respectively, and for "lattice" the interpretation "concept hierarchy". The mathematical lattice theory could even be interpreted altogether as formal theory of concept hierarchies. This has not only stimulated new applications of lattice theory to a surprising extent, but also inspired considerably the further elaboration of lattice theory. All four aspects of generalistic sciences and humanities and with them the transdisciplinary competence are gaining by the general interpretations and by the extensions of research animated by those interpretations: The general interpretations are the expression of a disciplinary opening and support in this way a positive *attitude* to border crossing, they deliver a basis for the *presentation* of disciplinary developments in their sense and meanings, they facilitate the *imparting* across the borders of subjects, and they strengthen convictions which enable the *discussion* about goals, value representations, and validity claims of the discipline.

Transdisciplinary competence can also be increased by working out for the discipline concerned in which way and with which effect *general actions of thinking* can be supported by the discipline. For the described approach of restructuring, this is carried out in the article "Begriffliche Wissensverarbeitung: Theorie und Praxis" [Wi00]: There it is explained by examples of application how actions of thinking are supported mathematically and concept-analytically, in particular:

1. *Exploring* by concept lattices constructed by experts for orientating processes of learning,
2. *Searching* by conceptual retrieval structures in the form of combinative concept lattices,
3. *Recognizing* by transparence of conceptual connections in line diagrams of concept lattices,
4. *Identifying* by gradual specialization in classifying concept lattices,
5. *Investigating* by systematic elaboration of relationships in data contexts,
6. *Analyzing* by purpose- and theory-accompanying examination of data contexts and their concept lattices,
7. *Making aware* by interpreting communication guided by concept lattices,
8. *Deciding* by transparence of the concept-analytically represented alternatives,
9. *Improving* by making understandable structural connection concept-analytically,
10. *Restructuring* by disclosing substructures and subsystems concept-analytically,
11. *Memorizing* by representations in data contexts and conceptual retrieval structures,
12. *Informing* by concept-analytically conceived knowledge and information systems.

Transdisciplinary competence obtains promotion naturally by concrete project work in non-disciplinary problem fields which however presupposes certain capabilities concerning transdisciplinarity. Thus, it has needed altogether years of efforts toward generalistic sciences and humanities and the restructuring of mathematics before the Darmstadt group came to serious transdisciplinary project work. But then a development started which became more and more intensive so that finally the Darmstadt group alone were involved in more than 200 application projects. This development even led in 1994/95 to a the firm *"Navicon Gesellschaft für Begriffliche Wissensverarbeitung"*, founded by four members of the Darmstadt group, which offers its clients systems of exploring and analyzing data and knowledge based on concept lattices. The application projects of the Darmstadt group are concerned with an extremely wide spectrum of special subjects and problem fields which shall here become distinct only by the following twelve projects, by which, in [Wi00], the above described general actions of thinking are explained (same sequence!):

1. TOSCANA-exploration-system for the search of literature of the library of the "Center of Interdisciplinary Technology Research" of the TU Darmstadt [RW00],
2. TOSCANA-retrieval-system about the laws and technics concerning buildings developed for the ministry of building and residing in Nordrhein-Westfalen (Germany) [EKSW00],
3. Application of the program TOSCANA in the data warehouse domain for supporting database marketing activities of a combine for retail trade and real estate in Swizzerland [Hr00],

4. Computer program for identifying symmetry types of plane ornaments and their presentation at the symmetry exhibition in Darmstadt 1986 [Ga86],
5. Investigation of a data context about norm- and rule-guided international cooperations (regimes) [KV00],
6. TOSCANA-system for analyzing speech act verbs of German performed at the "Institut für deutsche Sprache" in Mannheim [GH00],
7. Concept-analytical evaluation of repertory grid interviews of anorectic patients at the psychosomatic clinic of the University Gießen [Sp90],
8. Analysis of data about the pollution of Lake Ontario from the National Research Water Institute in Burlington (Ontario) [SW92],
9. Optimization of a chip-production with methods of formal concept analysis [WS93],
10. Formal concept analysis in software engineering: restructuring and reengineering of software [LS00],
11. TOSCANA-system applied in a PhD-thesis about "Simplicity. Reconstruction of a conceptual landscape in the music esthetics of the 18th century" [MW99], [Ma00],
12. Interactive information maps about flight connections in Austria and Australia [EGSW00].

Among the application projects, there were quite a larger number of projects for which no multi- or inter-disciplinary cooperation were necessary. For instance, the already mentioned project for the symmetry exhibition needed transdisciplinarity with regard to the general understandability for the visitors of the exhibition, but otherwise the mathematical competence together with elementary knowledge about programming were sufficient. For the most application projects the existing transdisciplinary competence were however the decisive presumption for an effective interdisciplinary cooperation. An impressive example for it is the mentioned project of the concept-analytical evaluation of repertory grid interviews of anorectic patients at which the physician N. Spangenberg and the mathematician K. E. Wolff productively cooperated over many years; during this time, each one learned from year to year more from the other to reach ways of aggregated thinking and by that to be successful. Especially, they benefited from the transdisciplinary competence which continuously grew during the project.

References

[Ap76] Apel, K.-O.: Transformation der Philosophie. 2 Bände. Surkamp-Taschenbuch Wissenschaft 164/165, Frankfurt (1976)

[Br81] de Bruijn, N.G.: Algebraic theory of Penrose's non-periodic tilings plane. Proc. Kon. Ned. Acad. Wetensch. Ser. A 84(1) (1981)

[DW82] Dorn, G., Frank, R., Ganter, B., Kipke, U., Poguntke, W., Wille, R.: Forschung und Mathematisierung - Suche nach Wegen aus dem Elfenbeinturm. In: Berichte der AG Mathematisierung, Bd. 3, GH Kassel, pp. 228–240 (1982); also in: Wechselwirkung 15, 20–23 (1982)

[EGSW00] Eklund, P., Groh, B., Stumme, G., Wille, R.: A contextual-logic extension of TOSCANA. In: Ganter, B., Mineau, G. (eds.) ICCS 2000. LNCS, vol. 1867, pp. 453–467. Springer, Heidelberg (2000)

[EKSW00] Eschenfelder, D., Kollewe, W., Skorky, M., Wille, R.: Ein Erkundungssystem zum Baurecht: Methoden der Entwicklung eines TOSCANA-Systems. In: Stumme, G., Wille, R. (eds.) Begriffliche Wissensverarbeitung: Methoden und Anwendungen, pp. 254–272. Springer, Heidelberg (2000)

[Ga86] Ganter, B.: Der Ornamentcomputer. In: Mazzola, G.(Hrsg.) Symmetrie in Kunst, Natur und Wissenschaft. Band 3 - Spiel, Natur und Wissenschaft, p. 130. Roether Verlag, Darmstadt (1998)

[GHW85] Ganter, B., Hempel, G., Wille, R.: MUTABOR - ein rechnergesteuertes Musikinstrument zur Untersuchung von Stimmungen. ACUSTICA 57, 284–289 (1985)

[GW99] Ganter, B., Wille, R.: Formal Concept Analysis: mathematical foundations. Springer, Heidelberg (1996)

[GH00] Großkopf, A., Harras, G.: Begriffliche Erkundung semantischer Strukturen von Sprechaktverben. In: Stumme, G., Wille, R. (Hrsg.) Begriffliche Wissensverarbeitung: Methoden und Anwendungen, pp. 325–340. Springer, Heidelberg (2000)

[Ha81] Habermas, J.: Theorie des kommunikativen Handelns. 2 Bände. Suhrkamp, Frankfurt (1981)

[He74] von Hentig, H.: Magier oder Magister? Über die Einheit der Wissenschaft im Verständigungsprozeß. Suhrkamp-Taschenbuch 207, Frankfurt (1974)

[Hr00] Hereth, J.: Formale Begriffsanalyse im Data Warehousing. Diplomarbeit, FB Mathematik, TU Darmstadt (2000)

[Ko87] Kocka, J., zu, E.: Kocka, J. (Hrsg.): Interdisziplinarität: Praxis - Herausforderung - Ideologie. Suhrkamp-Taschenbuch Wissenschaft 671, Frankfurt, pp. 7–14 (1987)

[Ko89] Kohler-Koch, B.: Zur Empirie und Theorie internationaler Regime. In: Kohler-Koch, B. (Hrsg.) Regime in den internationalen Beziehungen. Nomos Verlagsgesellschaft, Baden-Baden, pp. 17–85 (1989)

[KV00] Kohler-Koch, B., Vogt, F.: Normen und regelgeleitete internationale Kooperationen. In: Stumme, G., Wille, R. (Hrsg.) Begriffliche Wissensverarbeitung: Methoden und Anwendungen, pp. 325–340. Springer, Heidelberg (2000)

[Kr87] Krüger, L.: Einheit der Welt - Vielheit der Wissenschaft. In: Kocka, J. (Hrsg.) Interdisziplinarität: Praxis - Herausforderung - Ideologie. Suhrkamp-Taschenbuch Wissenschaft 671, Frankfurt 1987, pp. 106–125 (1987)

[Ku73] Kuhn, T.S.: Die Struktur wissenschaftlicher Revolutionen. Suhrkamp-Taschenbuch Wissenschaft 25, Frankfurt (1973)

[LS00] Lindig, C., Snelting, G.: Formale Begriffsanalyse im Software Engineering. In: Stumme, G., Wille, R. (Hrsg.) Begriffliche Wissensverarbeitung: Methoden und Anwendungen, pp. 151–175. Springer, Heidelberg (2000)

[Ma00] Mackensen, K.: Genese und Wandel einer musikästhetischen Kategorie des 18. Jahrhunderts. Bärenreiter, Kassel 200

[MW99] Mackensen, K., Wille, U.: Qualitative text analysis supported by conceptual data systems. Quality & Quantity 33, 135–156 (1999)

[Mi87] Mittelstraß, J.: Die Stunde der Interdisziplinarität? In: Kocka, J. (Hrsg.) Interdisziplinarität: Praxis - Herausforderung - Ideologie. Suhrkamp-Taschenbuch Wissenschaft 671, Frankfurt 1987, pp. 152–158 (1987)

[Mi89] Mittelstraß, J.: Wohin geht die Wissenschaft? Über Disziplinarität, Transdisziplinarität und Wissen in einer Leibniz-Welt. Konstanzer Blätter für Hochschulfragen 26, 97–115 (1989)

[Mi96] Mittelstraß, J.: Transdisziplinarität. In: Mittelstra, J. (Hrsg.) Enzyklopädie Philosophie und Wissenschaftstheorie. Bd. 4. Metzler, Stuttgart 1996, p. 329 (1996)

[Mi98] Mittelstraß, J.: Interdisziplinarität oder Transdisziplinarität? In: Mittelstra, J.(ed.) Die Häuser des Wissens: wissenschaftstheoretische Studien. Suhrkamp-Taschenbuch Wissenschaft 1390, Frankfurt 1998, pp. 29–48 (1998)

[RW00] Rock, T., Wille, R.: Ein TOSCANA-Erkundungssystem zur Literatursuche. In: Stumme, G., Wille, R. (Hrsg.) Begriffliche Wissensverarbeitung: Methoden und Anwendungen, pp. 239–253. Springer, Heidelberg (2000)

[Sp90] Spangenberg, N.: Familienkonflikte eßgestörter Patientinnen. Eine empirische Untersuchung mit der Repertory Grid Technik. Habilitationsschrift, Universität Gießen (1990)

[SW93] Spangenberg, N., Wolff, K.E.: Datenreduktion durch die Formale Begriffsanalyse von Repertory Grids. In: Scheer, J.W., Catina, A. (Hrsg.) Einführung in die Repertory Grid Technik. Bd.2: Klinische Forschung und Praxis. Huber, Bern 1993, pp. 38–54 (1993)

[SW92] Strahringer, S., Wille, R.: Towards a structure theory for ordinal data. In: Schader, M. (ed.) Analyzing and modeling data and knowledge, pp. 129–139. Springer, Heidelberg (1992)

[VWW91] Vogt, F., Wachter, C., Wille, R.: Data analysis based on a conceptual file. In: Bock, H.-H., Ihm, P. (eds.) Classification, data analysis, and knowledge organization, pp. 131–140. Springer, Heidelberg (1991)

[Wi82] Wille, R.: Restructuring lattice theory: an approach based on hierarchies of concepts. In: Rival, I. (ed.) Ordered sets, pp. 445–470. Reidel, Dordrecht-Boston (1982)

[Wi88] Wille, R.: Allgemeine Wissenschaft als Wissenschaft für die Allgemeinheit. In: Böhme, H., Gamm, H.-J. (Hrsg.) Verantwortung in der Wissenschaft. TH Darmstadt, pp. 159-176 (1988); (Gekürzter) Nachdruck in: Conceptus - Zeitschrift für Philosophie 60, 117–128 (1989)

[Wi97] Wille, R.: Verständigungsorientierte Wissenschaft und Didaktik. In: TUD (Hrsg.) Für eine neue Lernkultur - Martin Wagenschein zum 100. Geburtstag. TU Darmstadt 1997, pp. 93–100 (1997)

[Wi00] Wille, R.: Begriffliche Wissensverarbeitung: Theorie und Praxis. Informatik Spektrum 23, 357–369 (2000); gekürzte Version In: Schmitz, B. (Hrsg.) Thema Forschung: Information, Wissen, Kompetenz (TU Darmstadt), Heft 2/2000, pp. 128–140 (2000)

[Wi01] Wille, R.: Allgemeine Mathematik - Mathematik für die Allgemeinheit. In: Lengnink, K., Prediger, S., Siebel, F. (Hrsg.) Mathematik und Mensch: Sichtweisen der Allgemeinen Mathematik. Verlag Allgemeine Wissenschaft, Mühltal, pp. 3–19 (2001)

[Wi02] Wille, R.: Transdisziplinarität und Allgemeine Wissenschaft. In: Krebs, H., Gehrlein, U., Pfeifer, J., Schmidt, J.C. (Hrsg.) Perspektiven Interdisziplinärer Technikforschung: Konzepte, Analysen, Erfahrungen, pp. 73–84. Agenda-Verlag, Münster (2002)

[WW01] Wolf, K.D., Wille, R.: Möglichkeiten zur Unterstützung der Theoriebildung in der Politikwissenschaft durch ein TOSCANA-System: Politische Steuerung technikinduzierter Problemstellungen. Projektantrag an das ZIT. TU Darmstadt (2001)

[WS93] Wolff, K.E., Stellwagen, M.: Conceptual optimization in the production of chips. In: Janssen, J., Skiadas, C.H. (eds.) Applied stochastic models and data analysis, vol. II, pp. 154–164. World Scientific Pupl.Comp, Singapore (1993)

Jacob Lorhard's Ontology: A 17th Century Hypertext on the Reality and Temporality of the World of Intelligibles

Peter Øhrstrøm[1], Henrik Schärfe[1], and Sara L. Uckelman[2]

[1]Department of Communication and Psychology
Aalborg University
[2]Institute for Logic, Language, and Computation
Universiteit van Amsterdam
poe@hum.aau.dk, scharfe@hum.aau.dk, suckelma@illc.uva.nl

Abstract. Jacob Lorhard published his ontology in 1606. In this work the term *ontologia* 'ontology' was used for the first time ever. In this paper, it is argued that Lorhard's ontology provides a useful key to the understanding of the early 17th-century world view in Protestant Europe. Among other things, Lorhard's ontology reflects how the relations between scientific investigation and religious belief were seen. It is also argued that several of the conceptual choices which Lorhard made in order to establish his ontology may still be relevant for modern makers of ontological systems. In particular, Lorhard's considerations on the notions of reality and time deserve modern reflections. Also his assumption of the educational value of diagrammatical ontology deserves a modern discussion. Along with this paper an online hypertext version of Lorhard's ontology has been presented in order to create a useful tool for historical research in early 17th-century thought and in order to illustrate the problems, which characterized the early attempt at establishing a diagrammatical approach to ontology.

Note: References with just a page number (e.g. [p.17]) refer to the English translation of Lorhard's *Ontology* [7].

1 Introduction

The term *ontologia* ('ontology') was coined by Jacob Lorhard (1561--1609), who used this new term for the first time in his volume of eight books *Ogdoas Scholastica* (1606), in which he demonstrated how ontology could be presented in a diagrammatical manner (see [8] and [7, 9]). Lorhard's way of presenting his ontology makes it natural, from a modern perspective, to read it as a hypertext. This makes it straightforward to implement Lorhard's ontology as a modern hypertext (see [7]).

Lorhard's ontology is relevant for contemporary historians as a semi-formal key to a 17th-century world-view. The creation of such world-views was typical within academic circles of early 17th century Protestantism, so Lorhard's work provides us with a snapshot of the framework within which scientifc study was carried out in the 17th century. Moreover, we argue that several of Lorhard's considerations regarding

P. Eklund and O. Haemmerlé (Eds.): ICCS 2008, LNAI 5113, pp. 74–87, 2008.

the organization of his ontological system also warrant consideration by modern makers of ontological systems. The two considerations that we focus on in most detail are the concepts of reality and time. The first concept, in particular, is of central importance, both in Lorhard's ontology and in modern ontologies.

We begin in section 2 with some historical and conceptual background to Lorhard's ontology, providing a cultural and philosophical setting against which his views on time and reality should be evaluated. In section 3 we discuss Lorhard's concept of reality as it is manifested both in beings and attributes, using the reality of moral qualities and of structures or orders in the world as specific examples. In section 4 we consider Lorhard's treatment of the temporal aspects of beings. In section 5 we comment on Lorhard's presentation of his ontology. We show that an implementation of the system provides a useful tool for research in the history of science and philosophy. It turns out that Lorhard's ontology can be represented as a hypertext dealing with aspects of reality and time which are essential for the understanding of the world. We make some concluding remarks in section 6.

2 Some Comments on the Historical Setting

In order to understand Lorhard's approach to ontology, and the effect that this had on his approach to doing science, we must understand the philosophical and social milieu within which he worked. Lorhard's *Ogdoas Scholastica* was designed as a grammar school text-book. As such, it was meant to introduce essential parts of the scientific and religious aspects of the world to the students. Lorhard was influenced by the tradition from Peter Ramus (1515--72), who wanted to transform dialectical reasoning into a single method of pedagogical logic partly by using diagrammatical tools (see [9]). Lorhard sees diagrammatical ontology in this context, believing that the students will benefit from a deeper understanding of the ontological truths. This view on education became very influential in Europe. This is evident, for instance, in the writings of the Danish professor Jens Kraft (1720--1756) who, in organizing a school for young people expected to become national leaders, insisted on making ontology an essential part of the curriculum. Kraft clearly believed that a deeper understanding of the ontological truths would help the students not only in obtaining a better understanding of the world but also in becoming better people ethically speaking (see [8]).

Book 8 of Lorhard's Ogdoas is devoted to metaphysics or, in the new word that he introduces, ontology. At the end of the 16th century, the predominant view of metaphysics was that found in Suárez's *Disputationes metaphysicae* (which was published in Mainz in 1605 [4]). On Suárez's view, the primary subject matter of metaphysics is being: "the concept of the real being which is the subject of metaphysics..." [p.614]. Suárez's text "fixed the method of instruction in metaphysics for centuries, not only in Catholic schools, but also in Protestant academies and universities" [p.615] and "by the end of the [16th] century, Fonseca, Pereira, and Suárez were standard references in the newly founded Protestant universities" [p.621]. His view was, however, rejected by Clemens Timpler of Heidelberg, whose *Metaphysicae systema methodium* was published in Steinfurt in 1604 and in Hanau in 1606 [6]. (That Timpler knew of Suárez's views is clear from the fact that he references him in bk. 1, cap. 1, q. 7.) Timpler "proposed that the subject-matter of metaphysics is not being, but rather the intelligible, παν νοητον" [p.635]. He says that *metaphysica est ars contemplatiua, quae tractact de omni intelligibili, quatenus ab homine naturali rationis lumine sine ullo materiae conceptu est intelligibile* (metaphysics is a contemplative art which treats of every intelligible, to the extent that it is intelligible by men through the natural light of reason without any concept of matter) [bk. 1, cap. 1].

Timpler's work was enormously influential on Lorhard's ontology. Timpler's *Metaphysicae* is written in a more traditional style than Lorhard's textbook: it is divided in to five books, of which each chapter presents an aspect or a part of his metaphysical views, followed by a number of questions and answers dealing with the philosophical issues arising from the distinctions offered at the beginning of each chapter. (For example, q. 5 of bk. 1, cap. 1 is 'what is the proper and adequate subject matter of Metaphysics?' to which the answer, naturally, is *omne intelligibile* 'every intelligible thing'.) The divisions and distinctions in Timpler's work can be found almost universally without change in Lorhard's ontology, with the exception that Lorhard's text omits all the philosophical commentary (for a comparative taxonomy, see [3]). What is interesting is that in many places where Timpler raises questions about his classification and characterization, Lorhard adopts his distinctions without indicating that they might be questionable.

Following Timpler, Lorhard defined ontology as "the knowledge of the intelligible by which it is intelligible" [p.1]. His ontology is hence a description of the world of intelligibles, i.e., the items, concepts, or objects which are understandable or conceivable from a human perspective. The emphasis on the intelligibility of the world is essential in Timpler's and Lorhard's ontology. When Lorhard followed Timpler's lead and adopted this new proposal about the subject matter of metaphysics, or ontology, he agreed with the idea that we in formulating ontology are concentrating on the knowledge by means of which we can conceive or understand the world. In this way ontology is seen as a description of the very foundation of scientific activity as such.

Lorhard holds that the human rationality must function on basis of what he and Timpler both call 'the natural light of reason' [p.1]. Ontology captures this fundamental understanding of the basic features of the world. Based on this knowledge everything else – to the extent that it is intelligible at all – becomes conceivable. This approach presupposes that there is in fact only one true ontology – the one that reflects to the world as it truly is. The belief was in fact very important for the rise of modern science in the early 17th century. According to J. Needham [5], the confidence that an order or code of nature can in fact be read and understood by human beings was one of the important cornerstones for the rise of modern science in Europe. This strong belief was absent in Eastern civilizations in the early 17th century.

Lorhard, again following Timpler, divides the world of intelligibles into two parts: the universals and the particulars. The set of universals can be further divided into two parts: the set of basic objects, and the set of attributes [p.1]. As mentioned in [9] it should be noted that Lorhard's ontology does not begin with a distinction between physical and abstract, as many modern ontologies do. As we shall see, however, there is a similar distinction integrated in his ontology based on his notion of reality, which he uses several times in the ontology. As a result, a number of important concepts in Lorhard's ontology are mentioned more than once.

The notion of reality is essential when it comes to ontology. The distinction of what is real (*realis*) as opposed what is not real is used no less than 16 times in chapter 8 of *Ogdoas Scholastica*. However, in his book Lorhard uses the term in a rather complex manner, which is apparent from the fact that it contrasted with three different concepts in the text: rational (*rationalis*), imaginary (*imaginaria*), and verbal (*verbalis*). In the following section, we discuss Lorhard's use of the term 'real' as it is introduced in various parts of the text. As we shall see, the term 'real' mainly refers to 'mind-independent' or 'belonging to the external world'.

3 The Reality of Beings and Attributes

According to Lorhard, the essence of a being (*ens*) is that "by which a being is what it is" [p.2]. Some beings are real in the sense that they exist independently of human cognition, whereas other beings depend on human cognition, i.e., they are beings of reason or rationality. The essence of a being of the first kind relates to the external world, whereas the essence of a being of the second kind belongs to the internal (or mental) world of human cognition. According to Lorhard, there is an important duality between the beings themselves and our rational discussion of these being. It appears that he insisted on the necessity of this duality, in the sense that each time we discuss the beings in the world wanting classify them we also have to reflect on the concepts we are using in doing so. Such reflections at the meta-level turn out to form an essential part of Lorhard's work.

3.1 The Reality of Simple Attributes

Lorhard identifies two simple and 'most common' attributes of intelligibles or beings: existence and duration [p.4]. This means that for every intelligible it is correct at least to say that it exists in the world, and in some cases we may be able to say more about the duration of its existence. However, with respect to both existence and time, we make use of the real/imaginary distinction (*realis/imaginaria*). An existent intelligible might be something real in the sense that it exists independently of all human minds (although it is conceivable by the human rationality) or it might be imaginary (i.e., something imagined by the human mind). Also the duration of an intelligible might depend on a single human mind, in which case it is "imaginary", or it might be real, i.e., independent of the human mind as it is a consequence of the properties of the external world. However, even if a duration is real in this sense, it may still have to be determined or measured in relation to human decisions regarding temporal units [p.5].

3.2 The Reality of Beings and Complex Attributes

Real beings, beings which exist "in fact through [their] own essence[s], and further [are] suited to exist apart from cogitation of the mind" [p.3], are distributed across five different classes, each of which are further divided by a positive and a negative characteristic, such as "complex" and "uncomplex", "immaterial" and "material", and so on. This same type of positive/negative division is seen in his classification of conjunctive (non-simple) attributes. He divides conjunctive attributes into eight classes of opposing pairs: Every conjunctive attribute is either a principle/a principiate; a cause/of cause; a subject/an adjunct; a signifier/signified; a whole/a part; the same/diverse; ordered/disordered; prior/posterior [p.17]. In each of these 16 subclasses, Lorhard makes a distinction between real (i.e., mind-independent) attributes and non-real or imaginary attributes which depend on the rationality of the human mind.

We give two specific examples of the reality of complex attributes. As noted above, every conjunctive attribute is either a 'signifier' or a 'signified' [p.17]. If the attribute is in fact a sign, then it is either a natural or an arbitrary sign. If the former, then the reason for the sign relation is something in order of nature [p.28]; if the latter, then the reason is a human decision [p.30]. According to Lorhard, an arbitrary sign is called real if it is manifest in society, e.g., through some institution. Alternatively, an arbitrary sign may just be verbal, i.e., an idea of an individual human being expressed in speech or in writing [p.31].

Another particularly interesting example of the reality of attributes comes up in his discussion of the reality of moral qualities. It seems to have been essential for him to make his students aware of the nature of ethics. This is evident from the fact that not only did he deal in his ontology with the nature of morality, he also has a separate book devoted wholly to ethics in his collection of eight schoolbooks.

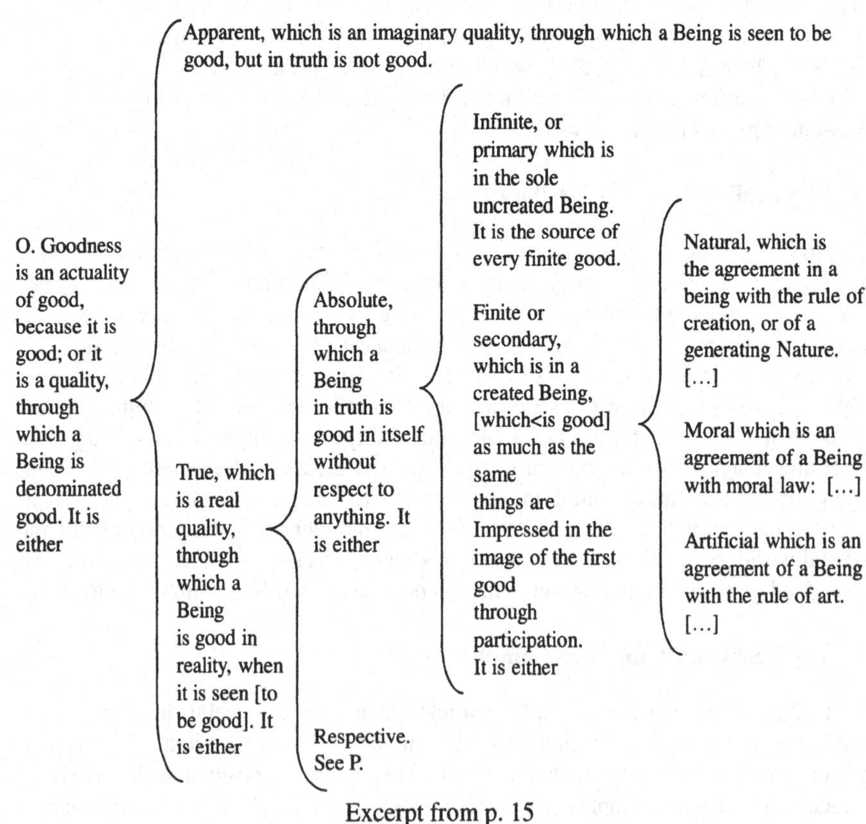

Excerpt from p. 15

In the book on ontology, Lorhard claims that in some cases moral qualities are just apparent (he calls these apparent qualities 'imaginary'). In other cases, however, a being or an intelligible is in fact "good in reality" [p.15]. Here, however, he does not use his standard opposition of *imaginaria* with *realis* but instead contrasts *apparens* with *verus* 'true'. In this way, Lorhard claims that goodness is in fact true in itself, if it is a real (i.e. mind-independent) quality. He treats 'malice' (the contrastive of 'goodness') in a similar way. Malice is in some cases true and in other cases it is merely apparent. If malice is true, then it is truly bad in itself and without respect to anything else. [p.16]

It is worth noting that according to Lorhard there is in fact something which is good in itself and also something which is bad in itself. This means that he accepted the idea of an absolute ethics, which is not a construction of human rationality, but which on the other hand can be understood or realized by humans. The claim is that goodness in the finite world in fact comes from the infinite or eternal good, and the goodness not only occurs in the relations between human beings, but that goodness also may be integrated in the physical world as an important aspect of it.

3.3 The Reality of Structures in the World

Dealing with the more complex attributes of intelligibles, Lorhard used the notion of reality in relation to determinations like 'identity' [p.35]. He points out that there are two different kinds of identity. In some cases the idea of something being identical with something else is just a rational construction, but in other cases a claim of identity between two beings is based on external, mind-independent properties of the beings in question. In these cases the identity is there objectively speaking, and Lorhard refers to it as a 'third thing'.

Similarly, the order of beings in the world may according to Lorhard be real or rational (i.e. just a product of human rationality) [p.39]. If it is real it does not depend on human cognition, but is there independently of human observation and cognition. A purely rational order, on the other hand, is mind-dependent, i.e., it is there as a result of human reason. Lorhard's ontology contains rather elaborate description of the various kinds of order of beings.

Lorhard believed in an essential 'order of nature' related to the origin or creation of the world. This order is structural. It is, however, important to emphasize that the order should not be understood as something static or inescapable. As we shall see, the order should be understood in the perspective of time.

This basic belief in an ordered or structured world was very important for the rise of modern science in the same period. It is, however, just as important with respect to the scientific project that the order of the world can in fact be studied, investigated, and learned by the human mind. This is probably why Lorhard emphasized the importance of the duality between on the one hand the real or external order of the beings in the world and on the other hand the rational order of beings as it can be captured by the human mind. Indeed, he says we obtain a theoretical, rational order of beings, i.e., a theoretical order corresponding to order of the external world, through our "cognition of things" [p.40].

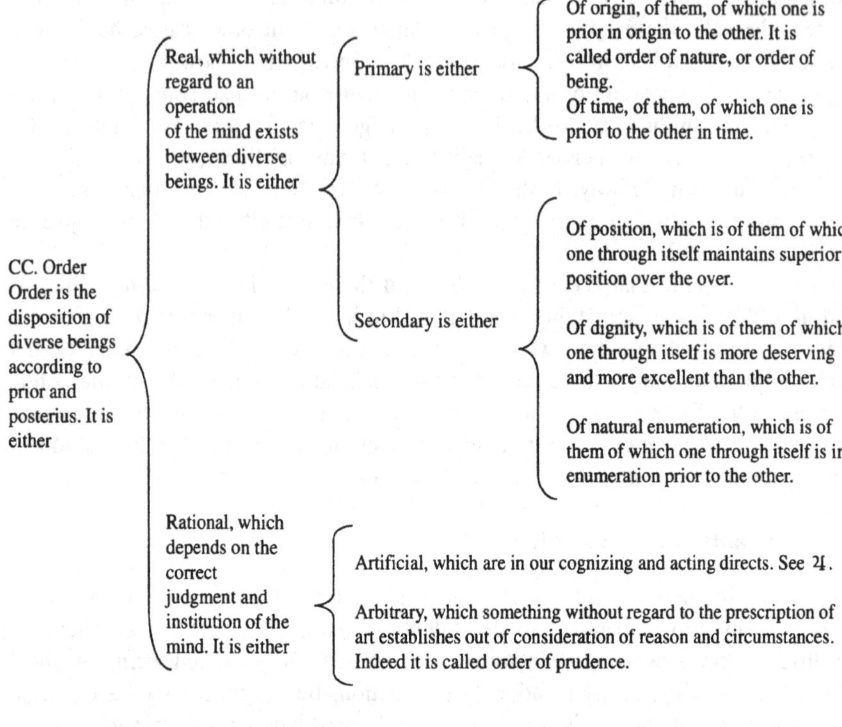

Extract from p. 39

3.4 The Reality of Privation

One of the most philosophically interesting distinctions that occur in Lorhard's ontology is that between being nothing and being negative. In his very first division of intelligibles, he says that an intelligible is either "*Nothing*: This is simply not something" or "*Something*: Whatever is simply not nothing" [p.1]. The remaining 57 pages of Lorhard's book on ontology are devoted to the intelligibles which are something; of the intelligibles which are nothing, nothing further is said. This distinction between being nothing and being something is copied directly from Timpler [bk.1, cap. 2]. Timpler says that an intelligible is anything which is able to be perceived and comprehended in the intellect, and that an intelligible is either something or nothing. In contrast to Lorhard, before continuing Timpler raises a number of questions whose answering seems to be required and which Lorhard doesn't mention at all. The first is "whether 'nothing' can be an intelligible" [q.1]. Timpler gives a positive answer to this question, but it is interesting that he feels the need to argue for it (and indeed gives a number of different arguments), whereas Lorhard simply takes it for granted that 'nothing' is something intelligible. Timpler also asks "whether 'something' and 'nothing' are equivalent to 'being' and 'non being'" [q.3]. He gives a negative answer to this question ("'being' is not always contradictorily opposed to 'non being'" whereas the answer to q.2 is that between 'something' and 'nothing' no middle ground can be attributed), and as an example he says that "privation is called 'non being', and nevertheless is still not nothing".

It is in the discussion of privation that Lorhard diverges from Timpler's metaphysical presentation. Both divide intelligibles which are something into those which are positive and those which are negative. Lorhard describes "something negative" as a privation "which is a negative habit in a being, of which then it is either able or required to be in" [p.58] In contrast, Timpler's discussion of privation is separate from his comments on intelligibles which are something negative. He classifies privation as an accident, namely, something which can be positively attributed to a being. While privation is a removal of something from a being, it can still be affirmed that the removal is present in the being. When Lorhard says that a privation is something negative, he is taking a symmetric view: We do not affirm that a privation is present in a being, but instead we deny that some habit or other is present in a being. For both Timpler and Lorhard, privations, like other types of intelligibles, are divided into those which are mind-independent and those which are mind-dependent. It is interesting here that the choice of words that both Lorhard and Timpler use is not the usual pair of *realis/rationalis* or *realis/imaginaria* but *verus/ficta* 'true/fictional' (though Lorhard does add that fictional privation is also called rational or imaginary, because it "is attributed solely through a fashioning on the mind" [p.58]). The only other time that Lorhard uses 'true' to describe the mind-independent intelligibles instead of 'real' is when he discusses goodness and malice, which we discussed above. There is thus a clear connection between the reality of something negative/privation and the reality of the moral qualities, a status which is not shared with any of the other intelligibles.

4 The Temporal Aspects of Beings

Lorhard's conceptual framework is basically Aristotelian. Following this tradition Lorhard believes that understanding causality is essential for understanding the world.

This point was formulated by Aristotle himself in the following manner:

> Knowledge is the object of our inquiry, and men do not think they know a thing till they have grasped the 'why' of (which is to grasp its primary cause). [1]

In order to establish a theoretical framework corresponding to the structure of the world including its temporal relation he classifies effects based on the four Aristotelian causes [p.23]. An effect will always be one of the following:

Caused by reason of efficiency
Caused by reason of matter
Caused by reason of form
Caused by reason of finality

For some reason Lorhard has chosen to list these four causes in a different order from the one used by Aristotle himself in Physics Book II, Part 3 [1]. In fact, the four causes are listed in the same order later in Lorhard's ontology when de discusses causes of real identity [p.35]. In his book on logic in the *Ogdoas*, Lorhard has also used the Aristotelian order of the four causes. In his logic it is even indicated that the four causes should be conceived as two pairs: efficient and material causes on the one hand, and formal and final causes on the other hand.

From a modern perspective the most controversial part of Lorhard's treatment of causality is probably his reference to final causes. A final cause (*telos*) implies an assumption of a purpose. However, according to Lorhard 'purpose' does not have to refer to human intent. Teleological causes may also be found in nature. There can be little doubt that this approach to purpose (*telos*) in nature should be interpreted in light of the religious assumptions incoporated in Lorhard's ontology.

The temporal aspects of Lorhard's ontology can be found not only in his emphasis on the importance of causality, they are also evident from the fact that the essence of a being introduced in terms of positive characteristics is, according to Lorhard [p.2], an actuality or performance (*actus*). If a being is real, the actuality or performance constituting its essence will reflect features in the external world taking place independently of human cognition. If the being is not real, the actuality or performance in question will depend on human cognition. In both cases the reference to '*actus*' must involve some kind of process. This clearly means that there is an aspect of temporality involved in the essence of beings.

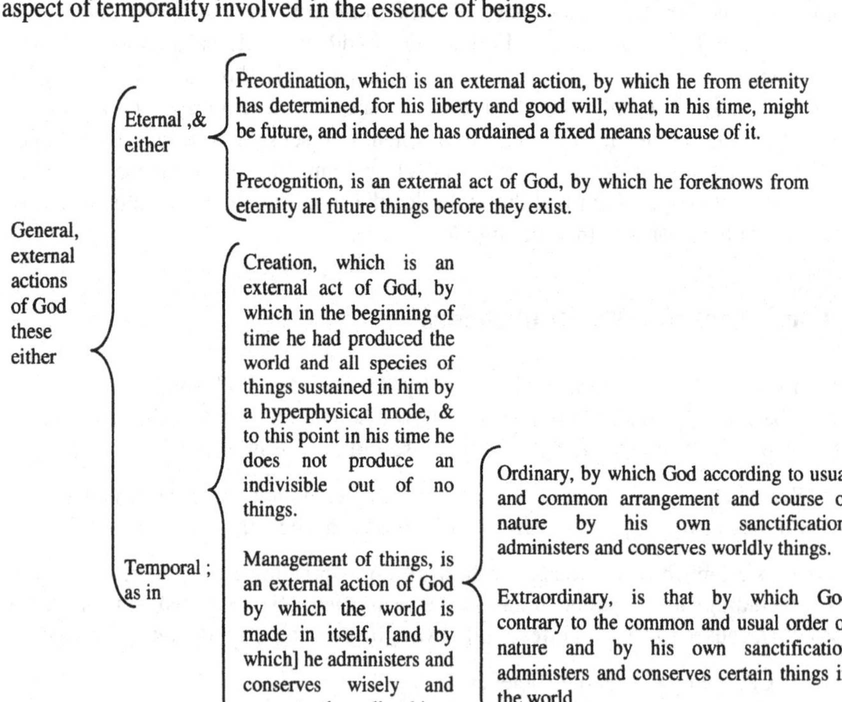

Excerpt from p. 44

It should be noted that according to Lorhard the world as such can also be seen as a result of a number of divine actions, some of which are eternal and some of which are temporal. This is due to Lorhard's religious approach to ontology according which the world as a whole from the very beginning has been under divine supervision and according to which worldly things have been permanently under divine management. The view is that there is "common and usual order of nature" [p.44], which represents

God's management (or actions) most of the time. This order of nature will naturally be the object for scientific investigation. Sometimes, however, God's actions are extraordinarily different (i.e., miraculous).

Lorhard divides God's eternal actions into preordination (i.e., the determination of the general plan for the temporal world) and precognition (i.e., the act by which God knows anything that is going to happen in the future whether it is necessary or contingent) [p.44]. These actions are carried out from a standpoint outside the temporal world. The idea seems to be that there is a non-temporal dimension of existence logically prior to time from which God can relate to the temporal world. From this non-temporal (or eternal) dimension God can either act in eternity by planning the course of events (preordination) or by seeing it (precognition).

The distinction between eternal and temporal is essential in Lorhard's ontology. This distinction may also be referred to as 'uncreated/created' or 'infinite/finite'. Together with 'real/rational', the 'eternal/temporal'-distinction is the most fundamental concept in Lorhard's ontology. In short: it may be held that key concepts in Lorhard's approach to the world of intelligibles are real (as opposed to rational) and temporal (as opposed to eternal).

5 Lorhard's Ontology as a Hypertext on Reality and Temporality of the World

It turns out that Lorhard's ontology can be represented as a hypertext on important aspects of reality and time which are essential for the understanding of the world. As mentioned in the introduction, it is easy to see how Lorhard's textbook can be represented as a modern hypertext because it relies so heavily on cross references. A large number of the pages (especially in the beginning of the text) list sections of the ontology that will be dealt with later. By choosing a graphical style of representation, Lorhard was able to not just present the concepts of his system, but also the structure that connects the concepts. This gives the obvious advantage that a single glance at a page reveals for instance how many subdivisions a given concept has. But the graphical representation also comes at a cost, namely that some of the strands of relations introduced in the early parts must 'wait' for many pages before being charted out. In this respect, it is remarkable that Lorhard throughout the ontology maintains such a close correspondence with Timpler's work, as shown by Lamanna [3]. It would not have been surprising if the diagrammatical style resulted in a different ordering of the material, but this is not the case. This observation adds to the important question of how the diagrams should in fact be read. An analysis of this question may pave the road for designing systems equipped to deal with such complex representations.

5.1 The Didactic Nature of the Representation

As mentioned above, the *Ogdoas* is intended to function as a textbook, and it seems reasonable to assume that it was used as lecture notes by Lorhard and his students. As such, it would be incorrect to view the text exclusively as a hierarchy of concepts or types. Rather, the diagrams take on form as didactic aids suited to address the questions at hand. In support of this view is should be noted that not only the ontology, but all the eight books are written in this style.

It is still an open question how much of the actual layout of the ontology was due to printer's constraints, and how much influence Lorhard had on the questions of layout. For example, the entire section on "Goodness" has been forced onto one page, whereas "Malice", which is structurally simpler, is divided up into multiple lettered sections over more than one page. In the absence of the original manuscript, we cannot say for certain whether such formatting differences have any underlying signification.

5.2 The Nature of Repetitions

An important feature in Lorhard's work is, as we have demonstrated, the use of repetitions. In [9] we mention the extensive use meta comments (or notes) within the ontology, which from a modern point of view cannot be seen as part of the actual hierarchy. Even more pertinent is the repeated divisions into dichotomies such as: created / uncreated (which occur 10 times), generic / specific (9 times), complex / uncomplex (6 times); and as pointed out above, the crucial distinctions of the real, which occurs no less than 15 times, contrasted with rational (7 times), imaginary (6 times), verbal (once), and 'of reason' (once). Quite obviously these terms function not as types in a hierarchy, but closer to the modern notion of metaproperties often discussed in contemporary research, see for example [2]. The use of repeated distinctions adds to the number of steps one has to go through in order to grasp a given concept. In our opinion, the repetitions also add to the difficulty of mentally navigating the ontology. But this solution does address another fundamental problem in ontology engineering, namely the critical problem of the top distinctions. It turns out that the entire structure is affected by these repetitions, and it seems therefore at least reasonable to suggest that the top distinctions de facto chosen by Lorhard (the subject / predicate structure distributed over the universal and the particular) should be seen as balanced by other important top distinctions. Thus, if the entire structure is rearranged according to whether elements are dependent on human cognition or not, the entire ontology could be 'turned upside down', whereby the subject / predicate distinction would be needed as metaproperties. Large portions of the ontology could be treated in the same way if Lorhard's use of created / uncreated was employed as a top distinction, etc.

5.3 Hypertext Arrangements

A contemporary version of Lorhard's text could be a simple hypertext as suggested in [7]. This implementation remains true to the original and preserves the structure in a very direct manner. Possible non-invasive additions could include more navigational aids such as a bi-directional link structure to help maintain the awareness of the big picture. It would also be desirable to have the actual book pages shown alongside the translation. See the sample page here below.

More advanced solutions could take on the challenge of dynamically rearranging the ontology according to the metaproperties discussed above. Such a system should be able to lift specific distinctions from the structure and rearrange the hierarchy

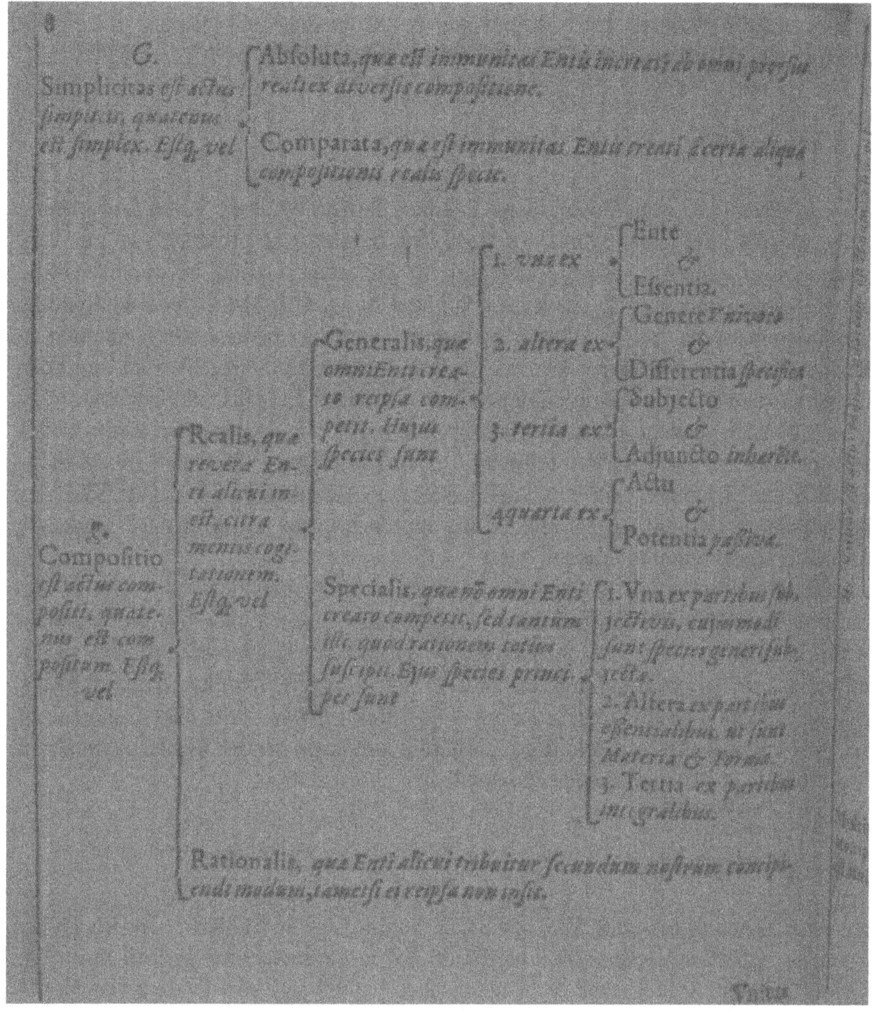

G.

Simplicitas est actus simplicis, quatenus est simplex. Estq; vel

Absoluta, quae est immunitas Entis increati ab omni prorsus reali ex diversis compositione.

Comparata, quae est immunitas Entis creati a certa aliqua compositionis realis specie.

&

Compositio est actus compositi, quatenus est compositum. Estq; vel

Realis, quae revera Enti alicui inest, citra mentis cogitationem. Estq; vel

Generalis, quae omni Enti creato revipsa competit. Hujus species sunt

1. una ex — Ente & Essentia.

2. altera ex — Genere & Differentia

3. tertia ex — Subjecto & Adjuncto

4. quarta ex — Actu & Potentia

Specialis, quae nō omni Enti creato competit, sed tantum illi, quod rationem totius suscipit. Ejus species princi-pes sunt

1. Una ex partibus subjectivis, cujusmodi sunt species generi subjectae.

2. Altera ex partibus essentialibus, ut sunt Materia & Forma.

3. Tertia ex partibus integralibus.

Rationalis, quae Enti alicui tribuitur secundum nostrum concipiendi modum, tametsi ei revipsa non insit.

Page 8 containing one of the *realis* iterations. (Vadianische Sammlung, St. Gallen, Switzerland)

based on all these occurrences. The central idea is to designate a given division as the privileged distinction by means of which the remaining ontology is organized. Important criteria for selecting the privileged distinction is how often it is used, and to what effect, e.g., how big a portion of the original structure is affected by the distinction. Any such selection (except the actual top distinction) would carve up the ontology in two parts: One part consisting of concepts directly affected by the selection, and one part consisting of concepts outside the scope of the current selection. The former of these would consist of fragments to be rearranged and if possible, also merged. The second part should also be listed thereby making the scope of the selection visible. The result would be a dynamic arrangement of original structure that takes the faceted style of representation seriously.

6 Conclusion

We have seen that Lorhard, following Timpler, defined ontology as the fundamental study of the intelligible world. In this way ontology is presented as the very foundation of scientific activity– including the important relations between scientific and religious concepts. We have also pointed out that Lorhard does not use the distinction between 'abstract' and 'concrete' which has become very common in modern ontology as an upper-level distinction. Instead, Lorhard makes an important distinction between what is mind-dependent (what Lorhard calls 'rational') and with is not mind-dependent (what Lorhard calls 'real'). In addition he distinguishes between time, where natural intelligibles exist, and eternity, by which the relations between the scientific and religious matters are understood. In short: The key distinctions in Lorhard's ontology and his approach to the world of intelligibles are rational/real and temporal/eternal.

Lorhard's ontology is a schoolbook using a diagrammatical approach in the tradition after Peter of Ramus. Lorhard accepted the view that ontology was essential as the foundation of science and knowledge in general. In this way ontology in not only about being, but it is about the broader world of intelligibility. Lorhard's work may be seen as a typical representation of the philosophical ideas behind the rise of modern science. And in addition, Lorhard's ontology serves as an example of inspiration to contemporary faceted knowledge representation.

Acknowledgements

We wish to thank the staff at Vadianische Sammlung, St. Gallen, Switzerland, for giving us access to copies of Lorhard's books. In particular, we wish to thank Dr. Rudolf Gamper, Bibliothekar der Vadianischen Sammlung for providing very useful information on the life and works of Jacob Lorhard.

References

1. Aristotle: Physics. Translated by Hardie, R.P., Gaye, R.K., http://classics.mit.edu/Aristotle/physics.2.ii.html
2. Guarino, N., Welty, C.: Ontological Analysis of Taxonomic Relationships. In: Laender, A.H.F., Liddle, S.W., Storey, V.C. (eds.) ER 2000. LNCS, vol. 1920, pp. 210–224. Springer, Heidelberg (2000)
3. Lamanna, M.: Correspondences between the works of Lorhard and Timpler, Bari University (2006), http://www.formalontology.it/essays/correspondences_timpler-lorhard.pdf
4. Lohr, C.H.: Metaphysics. In: Skinner, Q., Kessler, E. (eds.) The Cambridge History of Renaissance Philosophy. Cambridge University Press, Cambridge (1988)
5. Needham, J.: The Grand Titration. Science and Society in East and West (1970)

6. Timpler, C.: Metaphysicae Systema Methodicum. 1604, Steinfurt
7. Uckelman, S.L.: Diagraph of Metaphysics or Ontology - Jacob Lorhard, 1606 (2008), http://staff.science.uva.nl/~suckelma/lorhard/lorhard-english.html
8. Øhrstrøm, P., Andersen, J., Schärfe, H.: What has Happened to Ontology. In: Dau, F., Mugnier, M.-L., Stumme, G. (eds.) ICCS 2005. LNCS (LNAI), vol. 3596, pp. 425–438. Springer, Heidelberg (2005)
9. Øhrstrøm, P., Uckelman, S.L., Schärfe, H.: Historical and Conceptual Foundation of Diagrammatical Ontology. In: Priss, U., Polovina, S., Hill, R. (eds.) ICCS 2007. LNCS (LNAI), vol. 4604, pp. 374–386. Springer, Heidelberg (2007)

Revelator's Complex Adaptive Reasoning Methodology for Resource Infrastructure Evolution

Mary Keeler[1] and Arun Majumdar[2]

[1] CyberCORE
mkeeler@myuw.net
[2] VivoMind Intelligence
arun@vivomind.com

Abstract. Countless archives around the world, including C.S. Peirce's manuscripts at Harvard, wait for an infrastructure that makes possible the collaborative work to re-create them as "e-resources." Meanwhile, the U.S. National Science Foundation's new Cyberinfrastructure Initiative, calls for deeper understanding of *infrastructure*, enabling the engagement of researchers as participants in sustained e-resource development. Revelator is a game conceived as a methodology for that collaborative participation, in an evolving infrastructure of Conceptual Structures technology. Our conception of Revelator's game context derives from C. S. Peirce's theory of knowledge evolution in semeosis and his pragmatic methodology for inquiry, and incorporates J.H. Holland's complex adaptive systems modeling. Revelator's role is to engage participants in "the game of inquiry," which reveals significant patterns and paths structured by complex logical relations among the conditional propositions that represent players' conjectures as plays in the game. Revelator offers a general methodology for building trustworthy collaborative research, in an evolving infrastructure of knowledge technology and e-resources.

1 Introduction

Peirce's late manuscripts (written during his last decade, 1903-1914) hold remarkable evidence that his ideas anticipated major 20th century developments, including game and model theory [Keeler 2007; Pietarinen], and the more recent advancements of evolutionary epistemology and self-organizing complex adaptive systems. Unfortunately, these manuscripts are largely inaccessible (for complicated reasons briefly explained here) archived in the Houghton Library at Harvard University. In spite of its severely limited access, this mature theoretical work in "semeotic" (his preferred spelling [*CP* 8.377]), metaphysics, and pragmatic methodology has suggested fruitful directions to researchers in cognitive and neuroscience, machine intelligence, and complexity theory [Deacon 2004; Sowa 2006]. Revelator is conceived to apply Peirce's comprehensive ideas in a methodology for collaborative inquiry that incorporates advancing technology, not only for access to such archives in the form of "e-resources," but for building a new infrastructure to sustain their continued effective use and evolution. Key motivations and elements in the design of Revelator are explained in this paper, leaving implementation design features for the next effort.

P. Eklund and O. Haemmerlé (Eds.): ICCS 2008, LNAI 5113, pp. 88–103, 2008.

Encouragingly, the new Cyberinfrastructure Initiative at the U.S. National Science Foundation extends previous knowledge engineering initiatives, but now recognizes the need for a deeper understanding of *infrastructure* that would enable its developers to "think strategically and act opportunistically" in creating information that "builds synergy among learning, research, and societal issues." The Initiative further stresses that resources in the new cyberinfrastructure must be more than merely PDFs; they must involve ongoing participation that calls for "more sophisticated mutual-stakeholder consortia and more metrics for e-resource development testbeds to assess impacts and guide iterative design in e-research communities" [Atkins]. Infrastructure developers, in partnership with users, as human actors must become architectural elements within the system, rather than entities outside its boundary, as in the traditional UML use-case approach.

Harvard historian Mario Biagioli compares 20th century computer-network infrastructure and emerging e-resources to the period before published books and institutionalized libraries in the 15th century. Many network-based resources are now being created, across all disciplines, without the functional infrastructure to support collaboration and continuing innovation based upon them. Biagioli observes that the values of *scientific community* historically were formed and progressed with the evolution of an infrastructure that created "interdependence by peer-based allocation of resources." He concludes: "distrust is maximum when infrastructure is minimum" [in Kahin].

Resource archives around the world are waiting for an infrastructure that makes possible: the collaborative work to re-create them as e-resources, to keep them perpetually viable as technology evolves, and to ensure their use in the responsible intellectual evolution of research communities. Revelator is a general methodology for building trustworthy collaborative research, in an evolving infrastructure of e-resources and advancing technology. Revelator's design combines C. S. Peirce's theory of knowledge evolution with his work on improving the economy of inquiry in a game format that somewhat resembles familiar intellectual games, such as bridge, chess, and crossword puzzles [Keeler 2007]. Revelator's purpose is to reveal complex relations among conditional propositions (in the form of "if-then" rules) by which players represent their conjectures as plays in the game, and to enforce their dynamic validation and the verification of evidence they entail. Its game conception provides a framework for conceptual structures technology (including Semantic Web, Conceptual Graphs, Formal Concept Analysis, ISO Common Logic, and other approaches) to evolve in iterative cycles of innovation, testing, and advancement.

The strategic challenge of the game is to construct hypotheses collaboratively, aided by logical processing and diligent evidence checking. Evidence checking cannot be founded on belief based methods, such as Bayesian techniques, but must derive from methods such as Dempster-Shafer based evidential assessment. Structured "plays" in Revelator become players' "logical agents," which must adapt to promote the emergence of model-driven multi-agents that represent collaboratively formulated robust hypotheses. J.H. Holland's models of complex adaptive systems and the emergent behavior of agent-based mechanisms are translated into Peirce's logical mechanisms, to model the logical behavior of *complex adaptive reasoning* among conditional propositions.

This paper explains the conception of Revelator as a context that can demonstrate the value of Peirce's mature *dynamic* logic in a methodology that will enable collaborators to build the *self-organizing complex adaptive infrastructure* required for e-resource creation and continuity. Revelator's methodology can then serve in the evolution of conceptual structuring tools, toward the Cyberinfrastructure vision of "more sophisticated mutual-stakeholder consortia" with the "metrics for e-resource development testbeds to assess impacts and guide iterative design," for mutual reliance among "players in the game."

2 A Resource for an Evolving Infrastructure

Perhaps no more that than 20 percent of Peirce's estimated 100,000-page archive at Harvard has been published anywhere, in spite of three multi-volume editions appearing in the last century. Especially Peirce's writings from his final years (1903-1914, some 40,000 pages, or nearly half the collection archived at Harvard) have never been at all well represented in any print edition. Since their high-acid paper is in delicate condition, the curator of this collection at Harvard's Houghton Library estimates that these manuscripts may not survive beyond the next two decades. Scholars, who already have severely limited access to this material, may never be able to study his most intensive theoretical work in semeosis, metaphysics, and pragmatic methodology, including his advanced systems of graphical notation for the study of logic, less than 10 percent of which has ever been published [Clark, Roberts, and Zellweger have described some of these in detail and indicated the significance of the "gold mine" that remains in his later work, 1997].

These manuscript pages clearly exhibit the difficulties of representing Peirce's work in traditional print media, most obviously in their progressively more graphical and colorful features. They are full of invented symbols and complicated graphics with crucially meaningful color, in both text and diagrams [see a more extensive account in Keeler & Kloesel: 1997]. These pages also become progressively more difficult to decipher, as his handwriting deteriorated with age. Peirce's prolific editorial marginalia indicate his own recognition that traditional publishing would severely limit effective availability of his work. Its representational complexity includes text enclosed in graphical figures, graphics embedded in text, text contoured around graphics, whole pages of graphics with no text at all, and graphical figures with as many as four colors, "since four tinctures are necessary to break the continuity between any two parts of any ordinary surface" [CSP-MS 295: 44].

Those who have experienced the effort of scholarly research of his manuscript collection have become aware of other difficulties due to Peirce's compositional style and the fate of his corpus after his death. His writings cannot conveniently be arranged usefully in topical order, as the first editors presented a small selection of them in the *Collected Papers of Charles Sanders Peirce*. This edition represents fewer than 150 of his unpublished works (and only one-fifth of these are complete), but is still the most complete portion of this material in print. The misleadingly named collection includes fragments of some manuscripts that are separated among at least three of the eight volumes, and some series of papers are scattered through seven, to accord with the editors' topics rather than with Peirce's originally intended sequence.

Because of his broadly polymathic training and abilities, Peirce's work in mathematics, logic, and experimental science intricately influenced his work in philosophy and the humanities, which in turn shaped his views on scientific inquiry. Scholars can never be sure where, in the entire collection, Peirce will mention a particular topic and weave it into an unexpected context, such as in the imaginary dialogues he often invents, to play the part of his own critics as a strategy to explore the implications of an idea thoroughly. During his final years, he forecasts in a letter to one of his many correspondents what he imagines will be the fate of his own work, and why print media are inappropriate.

> Much of my work will never be published. If I can before I die, get so much made accessible as others may have a difficulty in discovering, I shall feel that I can be excused from more. My aversion to publishing anything has not been due to want of interest in others but to the thought that after all a philosophy can only be passed from mouth to mouth, where there is opportunity to object & cross-question & that printing is not publishing unless the matter be pretty frivolous. (*SS* 44: 1904)

Perhaps the most significant challenge in creating any useful resource, of his work is that the Houghton archive contains several thousand "lost pages," pages separated from their original manuscript context. When the collection was moved to Harvard, shortly after Peirce's death, they were not properly stored for several years, during which they fell into disarray (apparently, some even distributed as scratch paper in war-time paper shortages). The available (but incomplete) monochrome microfilm copies offer little aid in their proper replacement which, along with careful consideration of their content, depends on careful matching of any discriminating features that might be clues as to where they belong (features such as color and width of ruled lines, shade and type of paper, watermark, pattern of torn edges, weathering effects, and so on).

M. Keeler's previous work [2005], presents a hypothetical scenario of Peirce scholars in collaboration, trying to find the proper placement for a "lost page," and suggests a game as an effective context for this process. Keeler [2006] proposed Revelator as the game context for conceptual structures tool evolution; and [2007] introduced Holland's modeling of complex adaptive systems (*cas*) in terms of multi-agent mechanisms as a framework for the operation of Revelator as game of pragmatic inquiry. The following sections of this paper further relate Revelator's conception to Peirce's ideas, which have emerged in current methodological trends, to bring its game methodology closer to implementation.

Clearly, the full value of Peirce's surviving writings cannot be revealed by the traditional methods in print media, and his writings challenge us to construct a new sort of e-resource that can continue to evolve through collaborative inquiry:

> ... All that you can find in print of my work on logic are simply scattered outcroppings here and there of a rich vein which remains unpublished. Most of it I suppose has been written down; but no human being could ever put together the fragments. I could not myself do so. [MS 302 (1903)]

3 Peirce's Ideas Toward an Evolutionary Infrastructure

We are, according to Peirce, primally creatures who can form hypotheses, which he considered to be "spontaneous conjectures of instinctive reason" [*CP* 6.475 (1908)].

His theory of inquiry (or logic in semeosis) attempts to explain how instinct evolves into intellect and to examine our reasoning capability from its most vague to its most precise. He hoped to build the logical instruments by which to examine how this conceptual growth occurs. His theory of continuity in reasoning and communication (as semeosis) implies that without pragmatic inquiry, which progressively resolves the diversity of views among inquirers, "knowledge" will be established by dogmatic authority or popular opinion. His pragmatic methodology was to clarify how our necessarily fallible knowledge can progress resourcefully (with increasing validity and verifiability), in the "bootstrapping operation of inquiry," and to justify its application: "as we remain disposed to self-criticism and to further inquiry, we have in this disposition an assurance that if the truth of any question can ever be got at, we shall eventually get at it" [CSP-MS 83: 4-5 (1900); also Hookway 2002; and see Note 1: Peirce's fallibilism is *not* Popper's falsificationism].

D. Anderson adeptly explicates Peirce's analysis of his methodology:

> A hypothesis must explain the phenomena in question. An analysis of its logical purport, of its would-bes, allows an inquirer to determine this. Deduction then develops the implications of the would-bes, and induction tests for the reality of the generality that is the hypothesis or, more accurately, the object of the hypothesis, and thus "gives us the only approach to certainty concerning the real that we can have. [56; *CP* 8.209 (1905)]

By 1893, after two decades of intensive development of his logic, Peirce could respond to the question: "How do concepts evolve?"

> We can answer for ourselves after having worked a while in the logic of relatives. It is not by a simple mental stare, or strain of mental vision. It is by manipulating on paper, or in the fancy, formulæ or other diagrams -- experimenting on them, experiencing the thing. Such experience alone evolves the reason hidden within us and as utterly hidden as gold ten feet below ground -- and this experience only differs from what usually carries that name in that it brings out the reason hidden within and not the reason of Nature, as do the chemist's or the physicist's experiments. [*CP* 4.86]

In his 1902 manuscript, (which the *CP* editors named) the "Minute Logic," he maintained that the Greek notion of *episteme,* or "comprehension," as "the ability to define a thing in such a manner that all its properties shall be corollaries from its definition," which is the notion that "all deduction is corollarial deduction," had misguided scientific inquiry [*CP* 1.232]. The *living process* of inquiry, he insisted, "is busied mainly with conjectures, which are either getting framed or getting tested." In referring to his earlier intensive biological studies, he remarked on the amazing chemical complexity of protoplasm and, thereafter, began to use the metaphor of protoplasm for inquiry, which "has to be in a liquid state in order that the operations of metabolism may go on," concluding: "in all its growth and reproduction, it preserves its specific characters" [*CP* 2.198]. Peirce eventually came to regard the most marked characteristic of genuine inquiry as "an incessant state of metabolism and growth," in contrast to systematized classifications of knowledge as "nothing but the exudation of living science." In 1903, he made the following addition to his early article, "How to Make Our Ideas Clear," published in 1878.

> Pragmatism makes thinking to consist in the living inferential metaboly of symbols whose purport lies in conditional resolutions to act. As for the ultimate purpose of thought, which must be the purpose of everything, it is beyond human comprehension; but according to the stage of approach which my thought has made to it ... [with the aid of many others listed] it is by the indefinite replication of self-control upon self-control ... [*CP* 5.402]

Peirce's mature theory of logic-in-semeosis accommodates conceptual relativity, and recognizes that our ultimate concern is not to establish consensus that would simply resolve diverse opinions, but to reach consensus about interpretations that could then continue to be tested and modified in further, concerted experience of the (always hypothetical) existential conditions. "There would not be any such thing as truth unless there were something which is as it is independently of how we may think it to be" [*CP* 7.659 (1903)]. The effect of his pragmatic methodology is, he stressed, "simply to open our minds to receiving any evidence" [*CP* 8.259 (1904)]. A significant implication for software engineering is to abandon the traditional Church-Turing computation models and the associated closed-world proof theoretic concepts in favor of interactionism and iterations of approximate or speculative computations in an open-world paradigm [see Note 2].

Of the many works Peirce never completed but returned to in later years is "Lessons from the History of Science" (c. 1896), in which he identifies three main theories of evolution: "Darwin's ('purely fortuitous and insensible variations in reproduction'); Lamarck's: (successive minute variations, or force of habit-taking, accomplished by striving individuals), and cataclysmal evolution: sudden changes in the environment lead to disadvantages in established ways, and sporting to find new ways to adapt" [*CP* 1.104]. He concludes that all three modes have acted in organic evolution, the last being the most efficient, and the way that scientific inquiry mainly evolves.

> It advances by leaps; and the impulse for each leap is either some new observational resource, or some novel way of reasoning about the observations. Such novel way of reasoning might, perhaps, be considered as a new observational means, since it draws attention to relations between facts which would previously have been passed by unperceived. [*CP* 1.109]

Nevertheless, he finds a better account of the evolution of consciousness in Lamarckian theory, which can explain a mediation between the other two theoretical modes.

> Thus, the first step in the Lamarckian evolution of mind is the putting of sundry thoughts into situations in which they are free to play. As to growth by exercise, ... it consists of the flying asunder of molecules, and the reparation of the parts by new matter. It is, thus, a sort of reproduction. It takes place only during exercise, because the activity of protoplasm consists in the molecular disturbance which is its necessary condition. Growth by exercise takes place also in the mind. Indeed, that is what it is to learn. But the most perfect illustration is the development of a philosophical idea by being put into practice. The conception which appeared, at first, as unitary splits up into special cases; and into each of these new thought must enter to make a practicable idea. This new thought, however, follows pretty closely the model of the parent conception; and thus a homogeneous development takes place. The parallel

> between this and the course of molecular occurrences is apparent. Patient attention will be able to trace all these elements in the transaction called learning.

Though not explicitly, Peirce's work has influenced Holland's, whose publication of his research on induction [1986] was partially dedicated to Peirce, and whose research group at the University of Michigan includes Arthur Burks, one of the editors of the *Collected Papers* edition. Peirce's pragmatic view of habit-taking as learning in the evolution of mind [Keeler 2000, 2003] foreshadows Holland's models of adaptation and emergence.

Holland's models of emergent behavior are based on a conjecture that: "the mechanisms of selection in the creative process are akin to those of evolutionary selection, simply running on a faster time-scale." Recent studies in "evolutionary epistemology" have further developed this conjecture into theoretical accounts of scientific inquiry as "the cutting edge of the self-organizing adaptive processes that have been developing (no doubt irregularly and riskily) over evolutionary time" [Hooker: 42]. Holland's work joins what these theorists call a "fundamental revolution in the conceptual foundations of all the sciences, one with important consequences also for the professions: the shift from linear, reversible, and compositionally reducible mathematical models of dynamics to nonlinear, irreversible, and functionally irreducible complex dynamic systems models, especially for complex adaptive systems." Unfortunately, as Hooker assesses, the value and conceptual importance of the ideas deriving from this revolution have as yet scarcely touched those who "still model rational agents (explicitly or by tacit presumption) in terms of simple logical structure," as in most of current science [Hooker: 3].

As Holland and many evolutionary epistemologists point out, the explicit pattern of incrementally revising hypotheses (of hypothesize, test, and revise), which ideally abstracts inquiry's process, fails to account for the need to identify a target to start with, which in actual inquiry is identified as an unexplained phenomenon, controversial question or complex issue. In Revelator, this is a wholly human component, engineered into the system, outside the Church-Turing computation model but within the agent interaction model [Wegner, Goldin]. A game begins when human players are registered into the Revelator space, and meet the conditions of interacting through structured dialogs within the system language (a Common Logic Controlled English for rules based on the ISO 24707) for engaging in inquiry in the first place.

4 Revelator's Context for Complex Adaptive Reasoning

Revelator is to be played by any group of inquirers who collaborate to construct robust hypotheses from their individual conjectures, which might solve some puzzling question. Like the game of bridge, Revelator resembles a laboratory experiment in which experts carry out a dialogic, goal-directed, and limited but intellectually complex activity, using lean vocabulary and rigid bidding conventions. Formal, collaborative inquiry is conducted to improve ordinary solitary inquiry, constrained by the individual's sensory and cognitive restrictions and often limited commitment to investigation [Haack 2003].

Hypotheses become robust as they improve our anticipation: when consequences we expect appear to follow from certain conditions, as we guessed they would. In the game of Revelator, players express their guesses as explicit conjectures, by formulating each as a conditional proposition whose antecedent specifies a course of action to be performed and whose consequent describes certain consequences to be expected. This "IF ... THEN" form specifies what conditions a player thinks should justify each conjecture, as a rule. In using the conditional form for conjectures, players consciously distinguish the possibility that something is in fact true from how they think they know it is true. Inquiry does not select which conjectures are true, but constructs knowledge of how *justified* a conjecture is: represented "facts" are conditionally dependent on how we perceive and conceive them [Haack 1993]. In expressing this conditional dependency, players become responsible for their conjectures, the significance of which together with that of other conjectures that are found justified, must wait to be *revealed* in the evolution of further inquiry.

Players (and their software agents) keep track of the interactions among conjectures as the game evolves, especially those that are revealed to be inferentially or interpretationally incompatible, indicating that more investigation is needed. Unlike classic game players, Revelator players dynamically develop new strategies, in the form of *more general conjectures* calculated to incorporate other players' conjectures. Determining progress in a game of inquiry is like determining the reasonableness of entries in a crossword puzzle, established by their pervasive, nonlinear, interconnected mutual support without vicious circularity. An especially successful Revelator play resembles a long central crossword entry that makes other entries significantly easier to fill-in; but it must also score well with *experiential anchoring* (verified conditions of evidence) and be supported by other conjectures also anchored in evidence as integrated components. Such a "breakthrough" may even make further breakthroughs feasible, generalizing over many dependent conjectures. Conversely, discovering a wrong conjecture that supports many dependent conjectures may lead to a "breakdown" in the game of inquiry: when a key conjecture is confirmed unverifiable or unreliable by all players. A conjecture is more justified, the more jointly supported it is by other conjectures and their evidence, and the more independently secure is its evidence, but also the more comprehensively relevant evidence it takes into account. Devastating evidence (unnoticed because of failures to look closely enough, to check from different angle, etc.) can "wipe-out" an entire construct of conjectures (without creating a breakdown) [further explained in Keeler 2007].

Relations among conjectures in collaborative inquiry are far more complex than relations among entries in a crossword puzzle, and justification for conjectures must be ascertained in stages, by degrees, not categorically. Revelator's reasoning context distinguishes error- from ignorance-related aspects of fallibility, as inference- and evidence-related interactions, to reveal their pervasive interdependence. As do game players, researchers in collaborative inquiry often jointly uncover possibilities unsuspected by any one participant, and even begin to recognize certain kinds of conceptual patterns that become "building blocks" for longer-term, subtle strategies (something like "forks," "pins," and "discovered attacks" in chess), which are reliable enough to serve as stable strategies in the evolution of further inquiry. Within Revelator's game context, conjectures behave as players' agent-strategies in what Holland identifies as complex adaptive systems (*cas*) [1995]. The building blocks for evolving the stable strategies in *cas* are interacting agents, and any agent must adapt to other adaptive agents, just as a player's contributed conjectures must adapt to other conjectures in a game of complex adaptive reasoning (*car*).

In *cas* agents adapt their behavior by changing their rules as experience accumulates; in the same way, complex conjectures must change as inferences and evidence accumulate. Similarly in Revelator, agents have persistent behavioral identities, with internal state spaces whose future reactions and interactions influence the external state-space of Revelator, which itself will correspondingly feedback to the agents, influencing their internal state-space through iterative cycles of interactions. This highly dynamical model suggests physical metaphors (such as, damping the feedback) to guide agents. Fallibility, then, serves as gravity does in physical systems of building blocks, driving the strategic "dynamics" of inferential and evidential constraints [see Note 1]. Players can explore possible future trends and continually bring the state of their multi-agent conceptual model up to date as new conjectures are contributed, to improve its *faithfulness*. Because Revelator is explicitly a *game* of inquiry, players remain aware that: "uncertainty lies in the model's *interpretation*, the mapping between the model and the world" [Holland 1998: 44]. They continually test that mapping, as valid inferences and verified evidence accumulate in agent-rules, within their private state spaces, that must adapt to their changing conceptual environment of other agent-rules, in harmony with the external system state-space.

The conception of Revelator institutes Peirce's *dialogic* structure of semeosis in the *dynamic* context of a game, where players model their collaborative interpretation process as *car*, with the goal of revealing the "emergent phenomena of knowledge." According to Holland, emergent phenomena are recognizable and recurring, or *regular,* in a flux of interactions, and the point of modeling complex systems is to understand the origin of these regularities and relate them to one another. He cautions that even when the underpinning laws of dynamics are known, the crucial step of extracting the regularities from incidental and irrelevant details may be difficult (recognizing the patterns of play in chess took centuries of study); but he encourages that computers now make possible more complex and dynamic models, mathematical descriptions help to discern patterns in the process of modeling, and computer-generated games and maps can reveal patterns and regularities once inaccessible for exploration. Holland observes: "In both evolutionary and creative exploration we encounter patterns and lines of development (strategies) that emerge under selection, in a flux of change. And in both cases emergent building blocks, which he calls constrained generating procedures (*cgp's*), propagate their effect in cumulative ways, through recombination and interaction" [1998: 218]. He even speculates that there could be a "game" with the rigor of a *cgp* that permits insightful combinations of the powerful symbols: "a vision that has held me since the days when I first read Hesse's masterpiece" [220]. Peirce's logic has this role in the semeosis of inquiry.

5 Knowledge Emergence in *CAR*

Holland's methodology entails building models of *cas*, to study the nature of the mechanisms and interactions required for emergence. Revelator appropriates the same framework as a methodology for building models of *car*, to study the emergence of robust hypotheses (or possible knowledge). Holland summarizes his understanding of emergence.

> Emergence is above all a product of coupled, context-dependent interactions. Technically these interactions, and the resulting system are *nonlinear*. The behavior of the overall system *cannot* be obtained by *summing* the behaviors of its constituent parts. We can no more truly understand strategies in a board game by compiling statistics of the movements of its pieces than we can understand the behavior of an ant colony in terms of averages. Under these conditions, the whole is indeed more than the sum of its parts. However, we *can* reduce the behavior of the whole to the lawful behavior of its parts, *if* we take the nonlinear interactions into account. [1998: 122]

Holland's framework covers two major steps: "(a) discovery of relevant building blocks, and (b) construction of coherent, relevant combinations of those building blocks" [1998: 217]. He compares these to the "standard building blocks of language," where the creative challenge is to find the "salient patterns in the tree of combinations" [1998: 218]. In Revelator, players' sets of conjecture-agents (as mechanisms) are the building blocks and their combinations (as generators) are the potentially selected candidates for robust hypotheses. Such pools of agents correspond to Peirces notion of an organic and fluid metabolic system, rather than to rigid workflow of processes, which treat the "human" like a "hardwired" component, using business process languages, such BPML, and BPEL workflows that never "evolve," since the human is considered static (with fixed inputs, outputs and capabilities).

In the study of emergent phenomena, Holland identifies a minimum of three levels (mechanisms, agents, and aggregates) to be "more revealing and more productive as a first step," but suggests that when only two levels are formally considered the higher levels can be treated as recursions of these basic relations [1998: 239]. In Revelator the corresponding levels are: conjecture components (antecedent and consequent), conjectures (rules as agents), and hypotheses (aggregates of tested rules as multi-agents). Notice that these are generatively related: conjectures must have the mechanism of components, and hypotheses must have the agency of conjectures.

Holland's *cas* are models of complex adaptive systems, with *cgp's* as mechanisms for adaptation and *cgp-v's* (variable constrained generating procedures, with embedded genetic algorithms) as strategic mechanisms for anticipatory (or improved) adaptation. In Revelator, collaborative reasoning (considered as a complex adaptive system) is modeled in *car*, with language and logic supplying the mechanisms: the *cpg's* for strategic structuring in adaptation, and the *v-cpg's*, motivated by support, security, and relevance relations among agent conjecture mechanisms, for improving the adaptive reasoning process.

When players *create* rules in a game of Revelator, with each responsible and legal play in the form of a conditional proposition, these *rule-mechanisms* may be logically structured to become the agents from which players must select and construct *generators* (*cgp's)* as "winning combinations," and as these multiagent-rules become more secure with validation and verification, they take on dynamic (pragmatic) "trajectories" (operating as *cgp-v's*). In playing Revelator (as in conducting any inquiry) players *create* their game environment by the conjectures (rules) they contribute. Winning involves strategically selecting and combining those conjectures-mechanisms that reveal adaptive, higher-order behavior hidden in the complexity of their conceptual environment. These emerge as robust hypotheses explaining the successful inferences and evidence.

The selective exploration of different possible combinations is quite like finding the strategies in playing any other game. Like good plays in chess, sophisticated actions in complex adaptive reasoning depend on crediting *foresight* (or *pragmatic* anticipation). Players (with naturally limited capacity for negotiating logical complexities) must manage to identify possible generators of higher-level organization, because these are the "levers" that make "breakthroughs" possible.

Under Holland's framework, the *car* process would start with a complex pattern of related conjectures (in rule form) from which players may have no idea what might emerge. In the "mechanism selection process," induction must "mediate," as Holland says, in players' choice of patterns of interest from those that emerge from deductive processing. Knowing what details to ignore is *not* a matter of derivation or deduction; it *is* a matter of the experience and discipline of any artistic or creative endeavor. When deduction and induction work effectively together, they reveal repeated elements and symmetries that suggest higher-level mechanisms (which are higher level rules) [230].

Conveniently, Holland's methodology conforms to the evolutionary epistemological view of rules *as hypotheses* that must undergo testing and confirmation. Instead of viewing rules as a set of facts and implications which must be kept consistent with one another by consistency checking, as he says, "the object is to provide contradictions rather than to avoid them ... [and] rules amount to alternative, competing hypotheses. When one hypothesis fails, competing rules are waiting in the wings to be tried" [53]. Revelator takes this view one step back, in viewing the reasoning process itself, with its conjecture-rules serving as the competing predecessors to any surviving hypotheses. Its game context, then, is a generative context for the abductive, deductive, and inductive stages of reasoning.

Holland's technique for resolving the competition among rules is experience-based (closely related to the concept of building confirmation statistically): a rule's winning ability depends on its usefulness in the past. Each rule is assigned credit strength that over time comes to reflect the rule's usefulness to the system, which changes the system's performance as it gains experience (for adaptation, by credit assignment). Revelator's technique is the same, except that credit strength is represented in player scoring, which measures the survival value of contributions, not by simply summing each player's conjectures but, ultimately, by counting only those that survive breakthroughs, breakdowns, and wipe-outs.

Holland stresses that an agent-rule's value is then based on its *interactions* rather than on some predetermined fitness function [1995: 97]. The goal is the improvement of relations among rules, not some pre-determined optimality [1998: 216]. Ultimately, "Only persistent patterns will have directly traceable influence on future configurations in generated systems. The rules of the system, of course, assure causal relations among all configurations that occur, but the persistent patterns are the only ones that lend themselves to a consistent observable ontogeny" [1998: 225]. The goal in the game of Revelator is also improvement of relations among rules (as conjecture-agents), but because these are inferences that entail evidential references, they remain "tethered" to causal relations beyond the game itself, as signs in semeosis.

6 Conclusions: Complex Adaptive Infrastructure?

Revelator's complex adaptive reasoning context for inquiry resembles in several ways the "skills-building" features of familiar intellectual games, and each feature is a candidate for automated evidential reasoning support. First, the game would formalize the strategic process of inquiry, explicitly and engagingly. With conceptual structures tool support for both input of conjecture-rules and output of graphical representation of results at each stage, the user's cognitive burden in complex reasoning could be greatly reduced. Second, it would encourage collaborators to engage in the conceptual discipline of formulating model hypotheses. Rarely, in traditional research, do inquirers express their ideas in the clear, pragmatic form that is inferentially relatable to other ideas, and that explicitly refers to evidence for justification. Again, the conceptual-structures tool mode of interaction could enforce this convention, and by means of ontological search could increase the efficiency of finding, representing, and checking evidence on the Web. Third, Revelator's context would induce responsible conduct among players in the orderly succession of plays, which could also be enforced by tool control that would not allow one play to follow another until the first has completed an attempted adaptive process, and its tentative implications are observed by all players (promoting global common knowledge sharing).

Finally, the game context would encourage competition within a stable pattern of cooperation (as described by game theory [Axelrod]). Scores representing all contributions would constitute an objective evaluation of each player's provenance in the collaborative construction process. Revelator would delegate to automated conceptual processing the burden of inferring intricate logical relations, which would create an automatic credit path (perhaps represented in an ontology) that promotes fair competition among inquiring players. Furthermore, an extensive attribution system would result, so that players interested in similar aspects of the target question could easily find and keep track of one another's ideas.

We have identified some basic evaluation measures for Revelator's automated operations, as well as new metrics for technology evaluation [Majumdar 2007]. The key software requirements are:

1) To accept highly controlled and constrained English propositions, as plays;
2) To provide an evidential measurement system that is homogenous for humans and agents in order to enable interoperable active knowledge capture based on quantitative evidential assessments; and
3) To provide an environment that co-evolves a space of emerging hypotheses.

Requirements (1) and (2) form the core of the Revelator environment, while requirement (3) represents the need for *car*, combining Holland's model with a biochemical model for evolutionary computation to satisfy the Peircean metabolic view of inquiry. Many functions, for example, persistence of various hypotheses and data can be handled by wrapping and encapsulating other tools, such as knowledge base systems (e.g., WebKB [Martin]), within the Revelator architecture [see Note 3].

Evolutionary epistemologists typically agree that classical logic has been the epistemic paradigm for 20th century science, which still generally prevents philosophers and scientists, alike, from conceiving the role of intelligence in the complex process of adaptation (as *adaptive adaptivity*, or strategic methods of adapting). Some observe that Holland's methodology replaces "the established static, narrowly logical/AI models with a more dynamic process." In their view, the essence of science (exemplifying its open-endedness) has been the co-evolution of method, theory, and technology. The dynamics of this co-evolution form a set of "generic positive feedback/feedforward loops," in which methods are conjectural, risk-taking, resource-distributing strategies [Hooker 30-33]. In this sense, Revelator's architecture recognizes the need to integrate traditional e-resource tools, such as common RDBM systems, while extending new technologies, such as agents and more recent computational metaphors (leading to methods) such as biochemical calculi to implement Revelator in successive, evolutions by spiral iterations of development (starting at the beginning of the spiral, and not its end).

Peirce's pragmatic methodology, also in contrast to the classical view, conceives logic as "the *art of devising methods of research — the method of methods,"* which will not tell you what experiments you ought to make ... but it tells you how to proceed to form a plan of experimentation," to reduce risk and cost [*CP* 7.59 (1882)]. How does this strategic procedure evolve?

> We can answer for ourselves after having worked a while in the logic of relatives. It is not by a simple mental stare, or strain of mental vision. It is by manipulating on paper, or in the fancy, formulæ or other diagrams -- experimenting on them, experiencing the thing. Such experience alone evolves the reason hidden within us and as utterly hidden as gold ten feet below ground -- and this experience only differs from what usually carries that name in that it brings out the reason hidden within and not the reason of Nature, as do the chemist's or the physicist's experiments. [*CP* 4.86 (1882)]

We intend Revelator to be an evolving implementation of Peirce's methodology, which demonstrates logic's proper role in dynamic complex adaptive reasoning. The creation of e-resources from resources such as Peirce's manuscripts requires complex collaborative operations, including optical scanning, digitization, annotating and encoding the minutae of language and other forms of expression, which must conform to current evolving technology, digital preservation, intellectual property, and many other standardizing management constraints. The infrastructure required for this *continuous strategic planning process* would be most usefully modeled, itself, as a dynamic self-organizing complex adaptive reasoning system with an embedded technological complex adaptive system. Revelator brings Peirce's ideas to that task, with the hope of eventually building the complex adaptive e-resource to pursue study of his later work, which will continue to advance its methodology in support of improving knowledge technology research, as has been urged [Hovy 2005; Keeler 2006]. In that effort, A.-V. Pietarinen predicts that we will realize the value of graphical systems of logic when we see their "dynamic and dialogical character revealed in the apparatus of extensive games" [171].

In anticipating an evolving new infrastructure, we might recall what Peirce observed as he looked back to the 18th century: "But Kant had not the slightest suspicion of the inexhaustible intricacy of the fabric of conceptions, which is such that I do not flatter myself that I have ever analyzed a single idea into its constituent elements" [6.523 (1908)]. Knowledge must be conceptually constructed by intricate collaborative inquiry, in an evolving fabric of e-resources — not merely found on the Web.

Notes

General Note: "MS" references are to Peirce's manuscripts archived at the Houghton Library, Harvard; for *CP* references, *Collected Papers of Charles Sanders Peirce,* 8 vols., ed. Arthur W. Burks, Charles Hartshorne, and Paul Weiss (Cambridge: Harvard University Press, 1931-58).

Note 1: In Peirce's theory of inquiry, inductive fallibility is a metaphysical condition, not to be confused with Popper's falsification, which is strictly a deductive procedure (see Haack, *Evidence and Inquiry*, p. 131).

Note 2: A non-Turing model renders statistical methods, which rely on analyzing known data sets, useless; the goal of Revelator is to discover and reveal the unknown by incremental development of stable structures, formed as hypotheses during the runtime of Revelator. For a recent more detailed explanation of interactionism and a refutation of the traditional computing paradigms see Goldin and Wegner, and also Wegner and Goldin.

Note 3: For a formal theory to support the requirements of a *cas* environment, see the recent works of Regev and Shapiro [e.g., "Cellular Abstractions: Cells as Computation," *Nature* (2002)], which develop biochemical computational models based on the concurrent pi-calculus that are particularly well suited as a formal basis for Revelator; and also Parrow's, "An Introduction to p-Calculus" in *Handbook of Process Algebra* (2001), and also Regev, Panina, Silverman, Cardelli, and Shapiro's "BioAmbients: An Abstraction for Biological Compartments," in *Theoretical Computer Science* (Special Issue on Computational Methods in Systems Biology, 2004).

References

Anderson, D.: Strands of System. Purdue University Press (1995)

Atkins, D.: http://mitworld.mit.edu/video/434 (2007)

Axelrod, R.: The Evolution of Cooperation. Basic Books (1984)

Boehm, B.W.: A Spiral Model of Software Development and Enhancement. IEEE Computer Soc. 21(5), 61–72 (1988)

Clark, G.: New Light on Peirce's Iconic Notation for the Sixteen Binary Connectives. In: Houser, N., Roberts, D., Van Evra, J. (eds.) Studies in the Logic of Charles S. Peirce, pp. 304–333. Indiana University Press (1997)

Deacon, T.: Memes as Signs in the Dynamic Logic of Semiosis: Beyond Modular Science and Computation Theory. In: Wolff, K.E., Pfeiffer, H.D., Delugach, H.S. (eds.) ICCS 2004. LNCS (LNAI), vol. 3127, pp. 17–30. Springer, Heidelberg (2004)

Feibleman, J.K.: On the Future of Some of Peirce's Ideas. In: Wiener, P.P., Young, F.H. (eds.) Studies in the Philosophy of Charles S. Peirce. Harvard University Press, Cambridge (1952)

Haack, S.: Evidence and Inquiry: Towards Reconstruction in Epistemology. Blackwell, Malden (1993)

Haack, S.: Defending Science: within Reason. Prometheus Books (2003)

Holland, J.H., Holyoak, K.J., Nisbett, R.E., Thagard, P.R.: Induction: Processes of Inference, Learning, and Discovery. The MIT Press, Cambridge (1986)

Holland, J.H.: Adaptation in Natural and Artificial Systems. MIT Press, Cambridge (1992)

Holland, J.H.: Hidden Order: How Adaptation Builds Complexity. Basic Books (1995)

Holland, J.H.: Emergence: from Chaos to Order. Basic Books (1998)

Hookway, C.: Truth, Rationality, and Pragmatism: Themes from Peirce. Clarendon Press, Oxford (2002)

Hooker, C.A.: Reason, Regulation, and Realism: Toward a Regulatory Theory of Reason and Evolutionary Epistemology. SUNY Press (1995)

Hovy, E.: Methodology for the Reliable Construction of Ontological Knowledge. In: Dau, F., Mugnier, M.-L., Stumme, G. (eds.) ICCS 2005. LNCS (LNAI), vol. 3596, pp. 91–106. Springer, Heidelberg (2005)

Kahin, B., Jackson, S.: Cyberinfrastructure for Collaboration and Innovation, Selected Papers from the Conference, January 29-30 (2007)

Keeler, M., Kloesel, C.: Communication, Semiotic Continuity, and the Margins of the Peircean Text. In: Greetham, D. (ed.) The Margins of the Text, pp. 269–322. University of Michigan Press (1997)

Keeler, M.: Pragmatically Yours. In: Ganter, B., Mineau, G. (eds.). LNCS (LNAI), vol. 1876, pp. 82–99. Springer, Heidelberg (2000)

Keeler, M.: Hegel in a Strange Costume: Reconsidering Normative Science in Conceptual Structures Research. In: de Moor, A., Lex, W., Ganter, B. (eds.). LNCS (LNAI), vol. 2746, pp. 37–53. Springer, Heidelberg (2003)

Keeler, M.: Games of Inquiry for Collaborative Concept Structuring. In: Dau, F., Mugnier, M.-L., Stumme, G. (eds.) ICCS 2005. LNCS (LNAI), vol. 3596, pp. 396–410. Springer, Heidelberg (2005)

Keeler, M.: Collaboratory Testbed Partnerships as a Knowledge Capture Challenge. In: Proceedings of the Third International Conference on Knowledge Capture, pp. 203–205. ACM Press, New York (2005)

Keeler, M.: Building a Pragmatic Methodology for KR Tool Research and Development. In: Schärfe, H., Hitzler, P., Øhrstrøm, P. (eds.) ICCS 2006. LNCS (LNAI), vol. 4068, pp. 314–330. Springer, Heidelberg (2006)

Keeler, M.: Revelator Game of Inquiry: A Peircean Challenge for Conceptual Structures in Application and Evolution. In: Priss, U., Polovina, S., Hill, R. (eds.) ICCS 2007. LNCS (LNAI), vol. 4604, pp. 443–459. Springer, Heidelberg (2007)

Keeler, M.: Revelator's Challenge: Games of Inquiry. In: Proceedings of the Fourth International Conference on Knowledge Capture, pp. 197–198. ACM Press, New York (2007)

Majumdar, A., Keeler, M., Sowa, J., Tarau, P.: Semantic Distances as Knowledge Capture Constraints. In: Proceedings of The First International Workshop on Knowledge Capture and Constraint Programming. Fourth International Knowledge Capture Conference, October 2-5 (2007)

Martin, P.: WebKB (2008), http://www.webkb.org/webkb.html

Pietarinen, A.-V.: Signs of Logic: Peircean Themes on the Philosophy of Language, Games, and Communication. Springer, Heidelberg (2006)

Roberts, D.: A Decision Method for Existential Graphs. In: Houser, N., Roberts, D., Van Evra, J. (eds.) Studies in the Logic of Charles S. Peirce, pp. 334–386. Indiana University Press (1997)

Sowa, J.: Common Logic Controlled English (2008), http://www.jfsowa.com/clce/specs.htm

Sowa, J.: Peirce's Contributions to the 21st Century. In: Schärfe, H., Hitzler, P., Øhrstrøm, P. (eds.) ICCS 2006. LNCS (LNAI), vol. 4068, pp. 54–69. Springer, Heidelberg (2006)

Wegner, P., Goldin, D.: Coinductive Models of Finite Computing Agents. Electronic Notes in Theoretical Computer Science 19 (1999)

Zellweger, S.: Untapped Potential in Peirce's Iconic Notation for the Sixteen Binary Connectives. In: Houser, N., Roberts, D., Van Evra, J. (eds.) Studies in the Logic of Charles S. Peirce, pp. 334–386. Indiana University Press (1997)

Conceptual Spider Diagrams

Frithjof Dau[1] and Andrew Fish[2,*]

[1] University of Wollongong
dau@uow.edu.au
[2] University of Brighton, UK
Andrew.Fish@brighton.ac.uk

Abstract. Conceptual Graphs are a common knowledge representation system which are used in conjunction with an explicit type hierarchy of the domain. However, this means the interpretation of information expressed in conceptual graphs requires the combined use of information from different sources, which is not always an easy cognitive task. Though it is possible to explicitly represent the type hierarchy with Conceptual Graphs with Cuts, this less natural expression of the type hierarchy information is not as easy to interpret and soon takes up a lot of space. Now, one of the main advantages of Euler diagram-based notations like Spider diagrams is the natural diagrammatic representation of hierarchies. However, Spider diagrams lack facilities such as the ability to represent general relationships between objects which is necessary for knowledge representation tasks. We bring together the most pertinent features of both of these notations, creating a new hybrid notation called Conceptual Spider Diagrams. We provide formal syntax and semantics of this new notation, together with examples demonstrating its capabilities.

1 Introduction

Contemporary knowledge processing systems that include inferential abilities are typically based on some variant of formal logic where the information is internally stored in a particular format according to the sentences of the logic used. Such formal logics and their reasoning mechanisms have been thoroughly investigated and form a solid background for knowledge processing systems. However, the representation of knowledge as formulae has drawbacks if they are to be used for communication: in particular, they can be hard to comprehend by readers who are untrained in mathematics.

In contrast with the usual formal logics, human reasoning is often multimodal, involving information obtained from sentences, diagrams, sound, nuance or moving pictures for instance. The research field of *diagrammatic reasoning* investigates all forms of human reasoning and argumentation wherever diagrams are involved. Diagrams are often deemed to be easier to comprehend than symbolic notations [20, 23, 25], especially when they make good use of spatial relationships which are not utilized in symbolic notations; in particular it has been argued that they are useful for knowledge representation systems [8, 19].

* Funded by UK EPSRC grant EP/E011160: Visualisation with Euler Diagrams.

P. Eklund and O. Haemmerlé (Eds.): ICCS 2008, LNAI 5113, pp. 104–118, 2008.

There are two major families of mathematically well-elaborated diagrammatic reasoning systems, called conceptual graphs (CGs) and spider/constraint diagrams (SDs/CDs), which both have their roots in the works of Charles Sanders Peirce. Firstly, Sowa's CGs are based on Peirce's Existential Graphs. Various fragments of Sowa's CGs have been developed in a precise manner based on graph theory (see [2] for an overview), and in these it is possible to represent and carry out reasoning with relations of arbitrary arity. Secondly, SDs and CDs are based on Euler diagrams which are closely related to Euler circles and Venn-Peirce diagrams, and provide explicit means to easily represent certain relationships between sets (such as subset and disjointness). Various systems of SDs and CDs have been developed and formalized using algebraic means.

Now, although SDs and CDs are effective at expressing and reasoning with the relationships between sets, since the spatial relationships which encode them are well-matched [10], their usefulness in representing and reasoning with arbitrary relations is not so clear. On the other hand, CGs use a convenient representation of relations, which were in fact designed for this purpose, but their only means to express relationships between sets is to employ an underlying type-hierarchy. This has the drawbacks that: it only allows the expression of subset-superset relationships; it is a different representation of information, formally separated from the conceptual graphs and is often not even displayed with the CGs. This paper provides a step towards unifying these two diagrammatic systems, drawing on the specific advantages of each system and overcoming some of their disadvantages. We will use the underlying notation of SDs in order to make the type-hierarchy[1] of CGs both more explicit and more expressive, and we will augment this notation with relations of arbitrary arity as is done in CGs. The resulting notation will be called *conceptual spider diagrams*.

Due to limited space, we will assume that the reader is familiar with CGs; see [2, 26] for details. In section 2 we provide an exposition of the Euler diagram variants with a particular emphasis on SDs. We compare features of the SD and CG notations in section 3, identifying good and bad properties, as they relate to knowledge representation. An introduction to our hybrid notation via a collection of simple examples is provided in section 4, where the usefulness of the combined features becomes clear. A formalisation of the syntax and semantics of Conceptual Spider Diagrams is given in section 5. This opens up many interesting avenues of future research, and some of these are mentioned, together with our concluding remarks, in section 6.

2 Euler Diagram Based Systems

In seminal work [24], Shin produced a sound and complete formal diagrammatic reasoning system based on an extension of Venn diagrams, and logical reasoning systems based on Euler diagrams are now commonplace [4, 11, 15, 29, 31].

[1] To be more precise: this paper tackles the type-hierarchy of the *concepts*; relations are not addressed yet since a convenient way for diagrammatically depicting subset-superset relationships between relations goes beyond the abilities of SDs.

Spider diagrams (SDs) and constraint diagrams (CDs) are diagrammatic reasoning systems based on Euler diagrams which are applied in a multitude of areas including: file-information systems, library systems, statistical data representation and for logical software specification and reasoning systems. In this section we give a brief review of this family of diagrams, starting from Euler diagrams, passing through Venn-Peirce diagrams and arriving at SDs and CDs.

Euler first introduced Euler circles [3] in which sets are depicted by circles, and the spatial relationships between the circles mimic the set-theoretic relationships. The modern variant of these which we describe are called Euler diagrams: they use simple closed curves in the plane to represents sets. Two curves do not overlap (or more precisely, the interiors of the discs bounded by the curves are disjoint) if and only if the corresponding sets are disjoint, and if one circle is contained in another circle (or, more precisely, the interior of one is contained in the interior of the other) then we have a subset relationship between the corresponding sets. An example of an Euler diagram is shown on the left of Fig. 1. The interpretation is that there are three sets A, B and C such that B is a subset of A and that B and C are disjoint; note that no information about the relationship between A and C is provided.

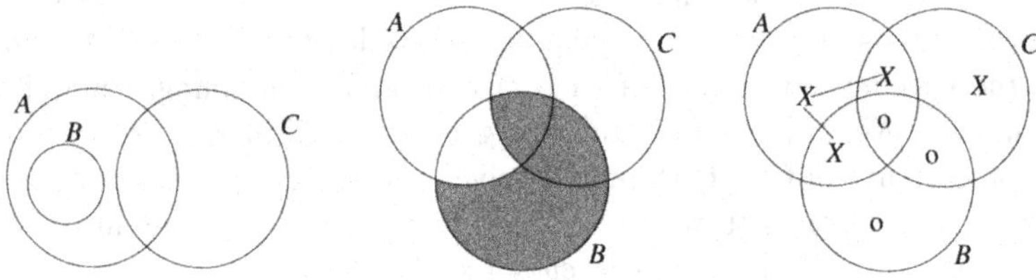

Fig. 1. An Euler diagram, a Venn diagram, and a Venn-Peirce diagram

There are collections of set intersections that cannot be represented using simple closed curves due to natural topological constraints [21], and so shading in a region is used in order to express emptiness. Then, even restricting to Venn diagrams [32], which are a subclass of Euler diagrams in which all possible set intersections are represented, one can depict any collection of set intersections. The middle diagram in Fig. 1 is a Venn diagram with exactly the same meaning as the Euler diagram. Using shading within the Euler diagram system allows a greater flexibility of representation than just using Venn diagrams with shading, but it is still not possible to express that sets are non-empty.

Peirce augmented Venn diagrams by X-sequences, which denote an element of the set corresponding to the region in which the sequence is placed. In his system, the shading of Venn diagrams is replaced by placing an 'o' in the region. The righthand diagram of Fig. 1 is a Venn-Peirce diagram which again expresses that $B \subseteq A$ and $B \cap C = \emptyset$ holds by the use of the o's; it also expresses that there is an element in $A \backslash (A \cap B \cap C)$, depicted by the X-sequence with three Xs, and that there is an element in $C \backslash (A \cup B)$.

Since the first paper on SDs [9] appeared, several elaborations and variations of the ideas have been published [33]. Two examples of SDs are shown below.

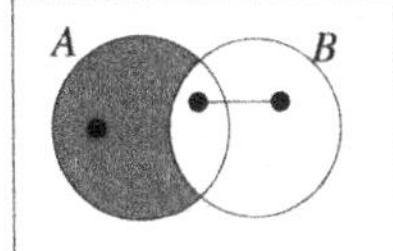

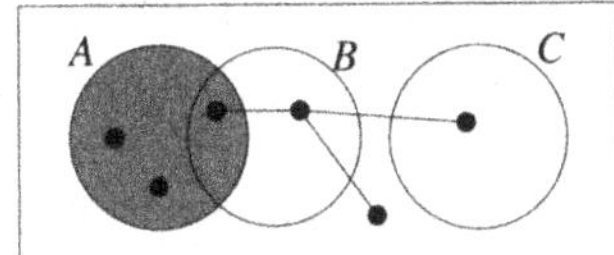

By convention a bounding rectangle is used to depict the UNIVERSE OF DISCOURSE, which is the ground set of the respective models in which the diagrams are evaluated; the letter U is used to denote this universe. The lefthand diagram above has two CONTOURS – the curves labeled A and B – representing the sets A and B, and four ZONES representing all possible set intersections involving A and B. The diagram contains two spiders, and the shaded zone representing $A - B$ is the HABITAT of the first spider, whilst the habitat of the second spider is the REGION representing B which is composed of two zones representing $A \cap B$ and $B - A$. The interpretation is that there are two sets A and B, the set $A - B$ contains exactly one element, and the set B contains at least one element.

So, there are two important differences between the semantics of Venn-Peirce diagrams and those of SDs: for SDs different spiders necessarily denote different objects, whilst the objects represented by different X-sequences in Venn-Peirce diagrams are not necessarily distinct; the shading in SDs means that a region does not contain *more* elements than the elements represented by the spiders touching that region, whilst for Venn-Peirce diagrams the 'o' indicates emptiness independently of any impinging X's. Now, in the righthand diagram there is another contour representing a set C and it does not overlap with the contours representing A and B and so C and $A \cup B$ are disjoint. Moreover, there are three distinct elements u, v and w (represented by the spiders) such that $u, v \in A$ and $w \in U - (A - B)$. Due to the shading the set $A - B$ contains exactly u and v, and $A \cap B$ either contains no elements or exactly w.

The above diagrams are called UNARY SDs, but spider diagrams can also be propositionally combined using the logical operators $\sqcap$ ('and') and $\sqcup$ ('or'): an example which uses these conjunctors is shown below.

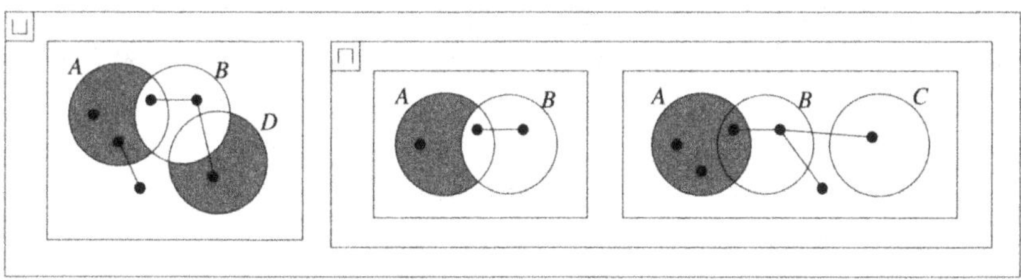

Similar to CGs, an important facet of SDs is that they are not only diagrammatic *representation* languages, but that they also facilitate diagrammatic *reasoning.* The system of SDs [16] is equipped with a sound and complete calculus. Automatic theorem provers for Euler diagrams and SDs have been developed and implemented [7, 28], with the goal of aiding the user's understanding of proofs. A tableaux system for SDs has also been implemented by Patrascoiu [22].

Finally, CDs were first proposed by Kent [18] as a notation for visualizing object-oriented invariants in software modelling and they can be used to express operation pre-conditions and post-conditions for modelling purposes [14]. They extend SDs, allowing the explicit representation of universal quantification via an extra type of spiders, called *universal* spiders which are indicated by asterisks, and arrows which provide information about binary relations. An example of a CD is shown below, where the core information expressed is that "every person can only borrow books that are in the collections of libraries they have joined".

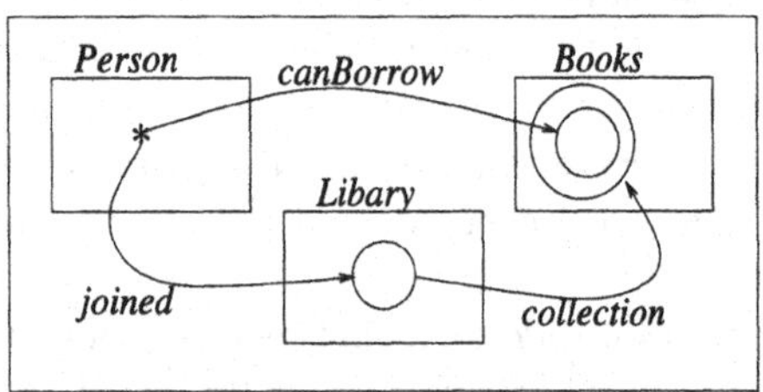

Since Kent's informal idea was proposed, several papers have provided various levels of mathematical formalism for Kent's vision. Problems such as the ambiguities in ordering of quantifiers have been solved by augmenting with an explicit reading tree in [5] where formal syntax and semantics for constraint diagrams are provided. Although some rules have been developed [4], a sound and complete calculus has not yet been developed for this system. However, for certain restricted fragments of the system, sound and complete calculi have been developed (e.g. in [27] no universal quantifiers were allowed as well as severe restrictions on the allowable relations and in [30] the restriction to order all of the universal quantifiers after all of the existential ones was imposed).

We elaborate briefly on the definition of relations between spiders in these systems. Arrows place restrictions on the relations determined by their labels. In particular if an arrow labelled R has source a spider s and target a spider t then the semantic phrase "$S.R = T$" is associated with this arrow (where S is the object represented by s and similarly for T, and $S.R := \{x \mid (S,x) \in R\}$).

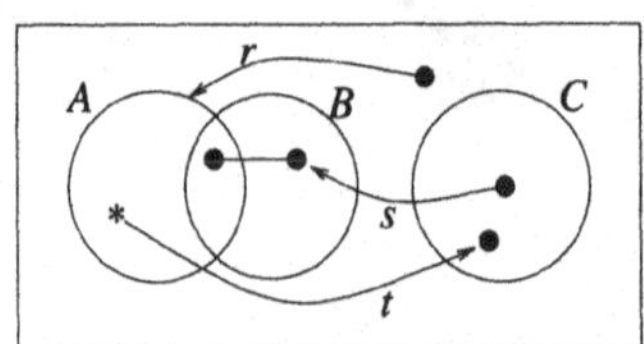

Using the semantics in [5], but adopting one of the conventions of [30] of reading the universal spiders after the existential ones (removing the need for an explicit depiction of a reading tree), a reading of the diagram is given by:

$$\overbrace{C \cap (A \cup B) = \emptyset}^{\text{Euler notation}} \wedge \overbrace{\exists u \in U - (A \cup B \cup C). \exists b \in B. \exists c_1, c_2 \in C:}^{\text{objects denoted by (existential) spiders}}$$

$$(c_1 \neq c_2 \wedge \underbrace{u.R = A \wedge c_1 S = b \wedge \forall x \in A - B : x.T = \{c_2\}}_{\text{reading the arrows}}).$$

3 Desirable Features of a Hybrid Notation

We investigate the advantages and disadvantages of some features of CGs and SDs that are useful for knowledge representation reasoning system with the aim of building a hybrid notation using the most pertinent features.

CGs are based on an underlying type hierarchy, which is usually a partially ordered set that indicates the subtype/supertype relationships between the types. This type hierarchy is handled as some sort of background information, not explicitly appearing in the actual CGs, but it is used for reasoning purposes.[2] This separation of information can make reasoning more difficult for humans. On the other hand, SDs provide an effective method for expressing the relationships between sets, since they are based on Euler diagrams. Thus if one combined the Euler diagrams features with a CG then one could make the type hierarchy of CGs explicit thereby making it visible to the user and aiding them in reasoning tasks; since SDs depict relationships between sets in a very iconic manner, the user will not only read off only the information which was needed to construct the type hierarchy, but other information which can be deduced from the type hierarchy can often easily been read of the SDs as well. This advantage of icons is described by Peirce in [12], 2.279, where he writes that "a great distinguishing property of the icon is that by the direct observation of it other truths concerning its object can be discovered than those which suffice to determine its construction." In the modern research field of diagrammatic reasoning, Shimojima coined the term *free ride* for this property of iconic representation [23].

Another advantage of bringing the type hierarchy to the foreground, using SDs is that we extend the expressiveness of the type hierarchies. Whilst the existing type hierarchies used only express subtype/supertype relationships between the types, taxonomies often contain disjointness information as well, and since disjointness constraints as well as subset relations can be naturally expressed with SDs, we can overcome this limitation.

When building a hybrid system, it pays to consider the semantic options carefully. The choice of shading over o's increases the expressiveness by allowing

[2] If projections are chosen for reasoning, we have appropriate conditions for the projections which reflect the type hierarchy. If transformation rules are used instead, we usually have type-generalization rules among these transformation rules.

the expression of upper bounds on the cardinality of sets. The choice of using spiders over X-sequences is an interesting choice, with the spider semantics adhering closely to diagrammatic design principles (distinct spiders meaning distinct objects) whilst the use of X-sequences closely matches the usual symbolic logic rules. We will actually investigate both options for the semantics of spiders (hereafter we will refer to the SD-semantics or the VP-semantics of spiders instead of spiders versus X-sequences).

When considering relations between objects, we notice that in CDs, the semantics for arrows is significantly different to the semantics for binary relations in FOL or CGs. For example, if we have two spiders standing for objects u and v, an arrow between these two spiders does not mean that the corresponding objects stand in relation R (i.e. uRv). Instead, we have the stronger condition $u.R = \{v\}$ (i.e. uRv and there is no object $w \neq v$ with uRw). Moreover, constraint diagrams do not allow relations with an arity > 2 (although this could be addressed by using multi-sourced or targetted arrows). Thus we choose to augment SDs with a different means to express relations between objects, adapted from the handling of relations in CGs.

4 Hybrid System Examples

Before we come to the formalization, we will exemplify our approach with the following well known toy example from the CG-community:

Cat : * —1— on —2— Mat : *

The meaning of this CG is 'there is a cat and there is a mat, and the cat is on the mat'; according to the usual semantics of CGs, it is not guaranteed that the cat and the mat are different objects. We assume that we have an underlying type hierarchy, with types $\top$, Animal, Cat, Thing, Mat, and Rug. The type-hierarchy is a partially ordered set, indicating only the supertype/subtype relations, and is shown on the left of Fig. 2. On the right of the figure an extension of this type-hierarchy is depicted, where disjointness-constraints are added.

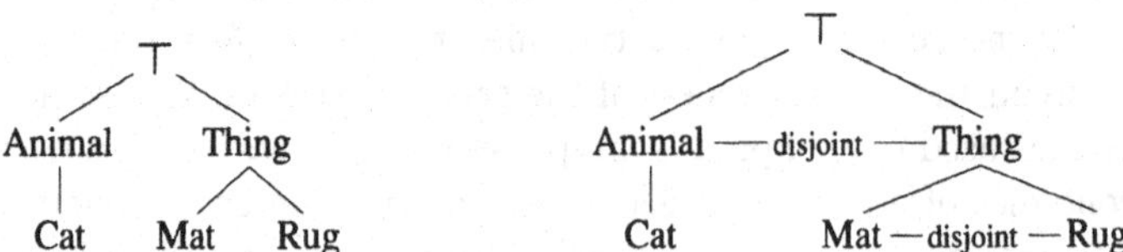

Fig. 2. The underlying type hierarchy: without, and then with disjointness constraint

Now, if one wished to *explicitly* display this type hierarchy in CGs, we need the facility to express supertype-subtype relations and disjointness constraints. This can be done with CGs with cuts, as shown in Figure 3, but obviously this CG

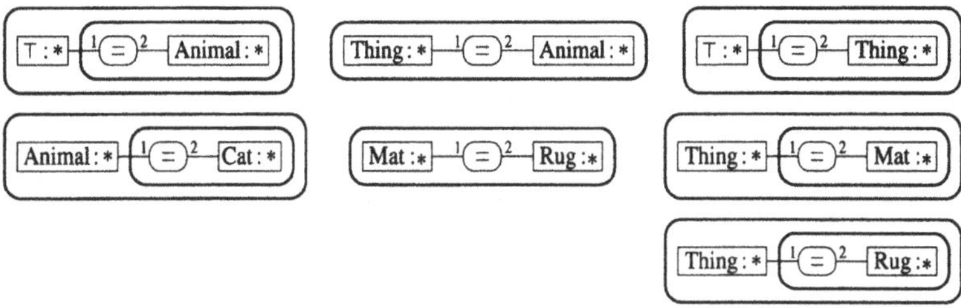

Fig. 3. The underlying type hierarchy with disjointness constraints as a CG with Cuts

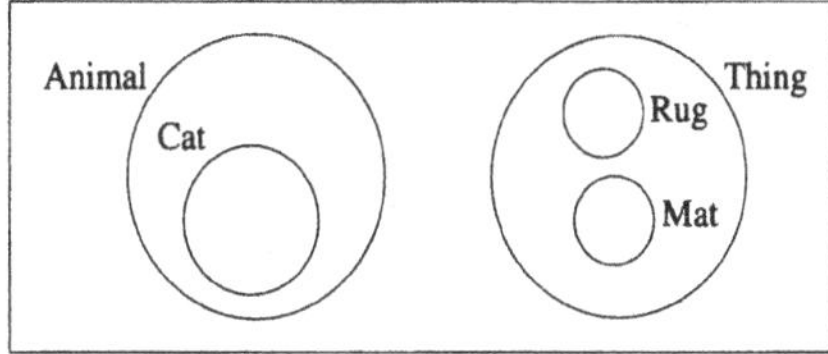

Fig. 4. The type hierarchy with disjointness constraints as an Euler Diagram

suffers from readability problems. Alternatively, we can depict the type hierarchy with an Euler Diagram, as shown in Fig. 4. The ⊤-concept corresponds to the universe of discourse, so there is no need to have a contour for it (although one could think of the rectangular bounding box for the diagram as signifying this concept). This diagram is significantly more readable than the CG of Fig. 3, and this may be partly due to free rides: some information which is only implicitly given in the CG in Fig. 3 or from the diagram in Fig. 2 (e.g. that the types Cat and Mat are disjoint) can immediately be read off from the Euler diagram.

We now demonstrate our proposed hybrid notation for the cat-on-mat example. In the lefthand diagram of Fig. 5, the Euler diagram of Fig. 4 is augmented with two spiders, denoting a cat and a mat, and the relationship 'on' between them. This diagram contains the complete type hierarchy plus the information that there is a cat on a mat. In the righthand diagram, we show only the fragment of the type hierarchy that is necessary to express the 'there is a cat on a mat' statement (in real-life scenarios, it will be a choice of the knowledge engineer which part of the type-hierarchy is actually displayed). This diagram is still strictly more expressive than the initial CG of this section: since the contours for cat and for mat do not overlap, this diagram also contains the information that *the cat and the mat are different.* We note that in contrast to the spiders, the location of the relation-symbol 'on' is not semantically important.

The next diagram shown in Fig. 6 is more sophisticated, and since it involves distinct spiders touching the same region, we consider both the SD-semantics and the VP-semantics. As well as the hierarchy information, it expresses that Yoyo is a cat, there is an unnamed cat (which is different from Yoyo in the SD-semantics, but possibly the same as Yoyo in the VP-semantics), and there exist no other cat (this holds for both semantics given the interpretation of shading),

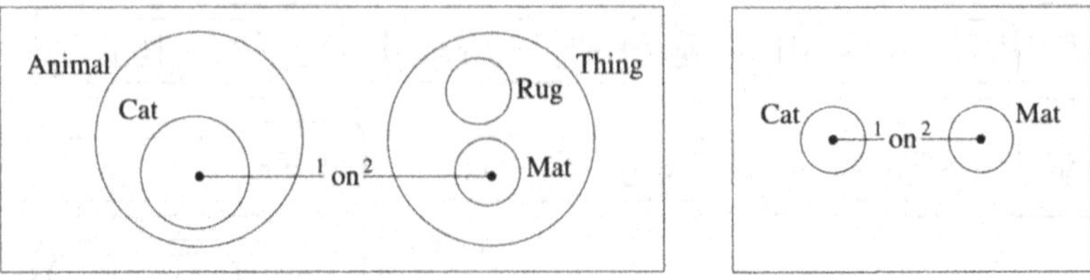

Fig. 5. Two hybrid diagrams for 'a cat is on a mat'

Yoyo is on a mat, and the unnamed cat is either on a rug or a mat (which is different from Yoyo's mat in the SD-semantics, but possibly the same in the VP-semantics).

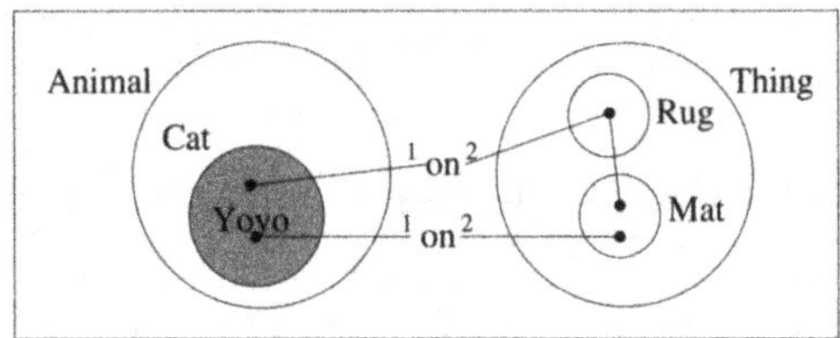

Fig. 6. A more sophisticated hybrid diagram

If we express this information by means of a CG with cuts, even without expressing the type-hierarchy information, then the outcome is significantly less readable. The diagram below shows this, using the SD-semantics; the corresponding CG for the VP-semantics can be obtained by removing the two vertical, negated identity edges in the middle of the CG making the diagram only slightly more comprehensible.

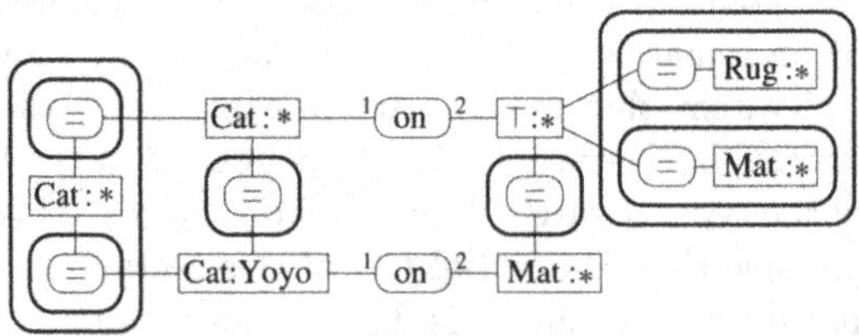

5 Formalisation

When considering the formalization of diagrammatic logic systems, it is essential to realize that there are two levels of notation. Certain graphical features of the diagram may be varied without changing the meaning of the diagram. For example, the shape of the contours in Euler diagrams or the shape of cuts in CGs with cuts is not semantically important; nor does the positioning of the boxes of a CG matter. Thus we have to deal with an *abstract syntax* which prescinds the meaningful information from such graphical properties, and a *concrete syntax*, which corresponds to the actual drawings of diagrams. This vital distinction has been discussed in the CG-community [1] as well as in the SD- and CD-community [13]; both papers argue that it is essential to elaborate the abstract syntax in a formal manner, but differ in their opinions concerning the concrete syntax. In [1] diagrammatic systems like existential or conceptual graphs are considered, where every diagram at the abstract level has a graphical representation (i.e. a corresponding concrete diagram), and argues that a mathematical formalization of the concrete syntax is *not* needed. However, for Euler diagrams the situation is different, since not all collections of sets intersections can be graphically represented without introducing some extra set intersections (or zones). That is, there are abstract Euler diagrams which are not *drawable* using simple closed curves, without introducing shaded zones, and although this introduction is not particularly problematic in logical reasoning systems it might be very undesirable in information display systems. So unsurprisingly, [13] argues that the concrete syntax as well as the relationship between the abstract and concrete syntax has to be *mathematically* formalized as well.

The methodology usually employed is to build Euler diagram based reasoning systems at the abstract level and then generate concrete diagrams (under various sets of wellformedness conditions) when it is possible to do so. On the positive side, from a human interaction point of view, the automatic generation of concrete Euler diagrams from abstract Euler diagrams for a strict set of wellformedness conditions is possible [6]. When extending to spider and constraint diagrams no extra substantial representational difficulties occur. The conceptual spider diagrams that we introduce are also extensions of spider diagrams and so they have to cope with these same difficulties, but the augmentation to express relations does not add any complexity to this issue either. For this reason, we introduce only the abstract syntax of conceptual spider diagrams (although there is no real problem in also defining the concrete syntax formalization).

5.1 Conceptual Spider Diagrams

Definition 1 (Alphabet). *An* ALPHABET *is a triple* $\mathcal{A} := (\mathcal{O}, \mathcal{C}, \mathcal{P})$ *of disjoint sets* $\mathcal{O}$, $\mathcal{C}$, $\mathcal{P}$ *of* OBJECT NAMES, CONCEPT NAMES *and* PREDICATE NAMES, *respectively. To each predicate name* P, *we assign its* ARITY $ar(P)$. *Let* $*$ *be another sign, the* GENERIC MARKER. *We set* $\mathcal{O}^* := \mathcal{O} \dot\cup \{*\}$.

In the examples provided earlier, object names or the generic marker $*$ were used as the labels of spiders, concept names were the labels of contours, and predicate names were the labels of predicate edges. An abstract zone represents a particular set intersection, and as such it will be defined to be a pair of disjoint, finite sets (a, b), where the set of contour labels a are those of the containing sets and the set of contour labels b are the excluding sets (i.e. the rest of the contour labels). In a unary diagram specifying the containing set a is sufficient (since a and b form a partition of the set of concept names $\mathcal{C}$), but when considering non-unary diagrams, or when reasoning with diagrams, the contour label sets may vary and so the pair of sets is required. Let $\mathcal{Z}$ and $\mathcal{R} := \mathfrak{P}(\mathcal{Z})$ denote the sets of all zones and regions respectively, where $\mathfrak{P}$ denotes the power set.

Definition 2 (Conceptual Spider Diagram). *An* UNITARY CONCEPTUAL SPIDER DIAGRAM *(abbreviated* UNITARY CSD*), d, over an alphabet $\mathcal{A}$ is a tuple $(L, Z, Z^*, S, \eta, E, \nu, \kappa)$ whose components are as follows:*

1. *$L := L(d) \subseteq \mathcal{C}$ is a finite set of concept names.*
2. *$Z := Z(d) := \{(a, L - a) : a \subseteq L\}$ (with $Z(d) \subseteq \mathcal{Z}$) is a set of zones s.t.*
$$(i) \quad \forall l \in L\, \exists (a,b) \in Z : l \in a \qquad \text{and} \qquad (ii) \quad (\emptyset, L(d)) \in Z\,.$$
We define $R(d) = \mathfrak{P}(Z) - \{\emptyset\}$ to be the set of regions in d.
3. *$Z^* := Z^*(d) \subseteq Z$ is a set of* SHADED ZONES. *We define $R^*(d) = \mathfrak{P}(Z^*) - \{\emptyset\}$ to be the set of* SHADED REGIONS *in d.*
4. *$S := S(d)$ is a finite set of* SPIDERS *with $S(d) \cap (\mathcal{C} \cup \mathcal{Z} \cup \mathcal{R}) = \emptyset$.*
5. *A function, $\eta := \eta_d : S(d) \to R(d)$ which returns the habitat of each spider.*
6. *$E := E(d)$ is a finite set of* PREDICATE EDGES *with $E(d) \cap (\mathcal{C} \cup \mathcal{Z} \cup \mathcal{R}) = \emptyset$.*
7. *A function, $\nu := \nu_d : E(d) \to \bigcup_{n \in \mathbb{N}} (S(d))^n$ which returns for each edge an n-tuple ($n \in \mathbb{N}$) of spiders. Let $E^n := E^n(d) := \{e \in E(d) \mid \nu(e) \in (S(d))^n\}$.*
8. *A function, $\kappa := \kappa_d : S(d) \cup E(d) \to \mathcal{O}^* \cup \mathcal{P}$ which returns for each spider $s \in S(d)$ its label $\kappa(s) \in \mathcal{O}^*$, and which returns for each predicate edge $e \in E(d)$ its label $\kappa(e) \in \mathcal{P}$, where we have $ar(\kappa(e)) = n$ for $e \in E^n(d)$ (i.e., $\kappa(e)$ has the appropriate arity).*

Define CONCEPTUAL SPIDER DIAGRAMS (CSD) *as follows:*

- *Every unitary conceptual spider diagrams is a conceptual spider diagram.*
- *if $\mathcal{D}_1$ and $\mathcal{D}_2$ are finite bags (multisets) of conceptual spider diagrams, then $\vee(\mathcal{D}_1 \uplus \mathcal{D}_2)$ and $\wedge(\mathcal{D}_1 \uplus \mathcal{D}_2)$ are conceptual spider diagrams, where $\uplus$ is the union for bags.*

Note that there is an empty CSD, $\bot := (\emptyset, \emptyset, \emptyset, \emptyset, \emptyset, \emptyset, \emptyset, \emptyset)$. To provide an example for Def. 2, we return to the diagram of Fig. 6. For the types, we use to following abbreviations: A(nimal), C(at), T(hing), M(at), R(ug). The abstract syntax of the concrete diagram of Fig. 6 is given by:

$$\begin{aligned}
L(d) &:= \{A, C, T, M, R\} \\
Z(d) &:= \{(\{A\}, \{C, T, M, R\}), (\{A, C\}, \{T, M, R\}), (\{T\}, \{A, C, M, R\}), \\
&\qquad (\{T, M\}, \{A, C, R\}), (\{T, R\}, \{A, C, M\}), (\{\}, \{A, C, T; M; R\})\} \\
Z^*(d) &:= \{(\{A, C\}, \{T, M, R\})\} \\
S(d) &:= \{s_1, s_2, s_3, s_4\} \\
\eta_d(s_1) &:= \eta_d(s_2) := \{(\{A, C\}, \{T, M, R\})\} \\
\eta_d(s_3) &:= \{(\{T, M\}, \{A, C, R\}), (\{T, R\}, \{A, C, M\})\} \\
\eta_d(s_4) &:= \{(\{T, M\}, \{A, C, R\})\} \\
E(d) &:= \{e_1, e_2\} \\
\nu_d(e_1) &:= (s_1, s_3) \text{ and } \nu_d(e_2) = (s_2, s_4) \\
\kappa_d(s_1) &:= \kappa_d(s_3) := \kappa_d(s_4) := *, \ \kappa_d(s_2) := \text{Yoyo}, \ \kappa_d(e_1) := \kappa_d(e_2) := \text{on}.
\end{aligned}$$

The semantics is based on classical Tarski-style interpretations.

Definition 3 (Interpretation). *An* INTERPRETATION *is a pair* $\mathcal{I} := (U, I)$ *consisting of an* UNIVERSE *and an* INTERPRETATION MAPPING I *which maps object names to elements of* U*, concept names to subsets of* U*, and predicate names of arity* n *to subsets of* U^n.

A point to note is that we are conforming to the general semantics of SDs, but contrasting to the usual semantics of FOL and CGs, by allowing *empty* universes. We are now prepared to provide the semantics for CSDs, and in fact we provide two slightly different readings of CSDs: the SD-semantics, where different spiders denote different objects; and the VP-semantics, where different spiders might denote the same object.

Definition 4 (Semantics). *Let a unary CSD* $d := (L, Z, Z^*, S, E, \nu, \kappa)$ *over an alphabet* $\mathcal{A} := (\mathcal{O}, \mathcal{C}, \mathcal{P})$ *and an interpretation* $\mathcal{I} := (U, I)$ *be given.*

- *First, we canonically extend* I *to zones* (a, b) *and regions* r *by setting*

$$I(a,b) := \bigcap_{C \in a} I(C) \cap \bigcap_{C \in b} \overline{I(C)} \qquad \text{and} \qquad I(r) := \bigcup_{z \in r} I(z)$$

- *We say that the* PLANE TILING CONDITION *holds iff we have* $\bigcup_{z \in Z} I(z) = U$ *(this condition reflects the Euler diagram features of CSDs).*
- *Any mapping* $val : S \to U$ *with* $val(s) = I(\kappa(s))$ *for* $\kappa(s) \neq *$ *is called* VALUATION *(of the spiders). If we have* $val(s) \in I(\eta(s))$ *for each spider* $s \in S$*, we say that val satisfies the* SPIDERS CONDITION*. If val is injective, we say that val satisfies the* STRANGERS CONDITION*. If for each edge* $e \in E$ *with* $\nu(e) = (s_1, \ldots, s_n)$*, we have* $(val(s_1), \ldots, val(s_n)) \in I(\kappa(e))$ *we say that val satisfies the* PREDICATES CONDITION.
- *We say that* $\mathcal{I}$ SATISFIES d IN THE VP-SENSE *iff the plane tiling condition holds and if there exists a valuation which satisfies the spiders condition and the predicates condition. We write* $\mathcal{I} \models_{VP} d$*. If the valuation additionally satisfies the strangers condition then* $\mathcal{I}$ SATISFIES d IN THE SD-SENSE*, and we write* $\mathcal{I} \models_{SD} d$.

6 Conclusion

We have provided a novel formal diagrammatic system called conceptual spider diagrams which utilizes useful features from CGs and SDs. A demonstration of its potential has also been provided via examples. This is a step towards a unification of CGs and SDs, and this "best of breed" approach looks promising. It also raises many interesting avenues of future research.

Firstly, the usability of the system of CSDs has to be thoroughly compared to SDs and to CGs. Although the examples in section 4 give an indication that that CSDs are easier to read than CGs, this benefit might get lost if we have type hierarchies with many overlapping concepts, since this could lead to underlying

Euler diagrams where the contours are mutually intersecting to a large extent (thereby rendering the diagrams more cluttered [17] and hence more difficult to read). One would imagine that the more disjointness-constraints a type-hierarchy has the easier the corresponding Euler diagram is to read, but the associated tradeoff needs to be scrutinized carefully.

Secondly, it should be investigated if the CSD representation can be extended so that not only the type-hierarchy of concepts but also the type-hierarchy of relations can be diagrammatically depicted in a convenient way.

Thirdly, sound and complete calculi for the system have to be developed. Since we have two different semantics, we will require two slightly different calculi: these calculi should have many rules in common, differentiating only in rules which reflect the difference between the two semantics. Since CSDs are a hybrid of SDs and CGs, it is desirable that the calculi do not differ too much from the existing calculi for SDs and CGs.[3] Furthermore, if CSDs are to be used in practice it is likely that only limited subsets of all the concepts in a given type-hierarchy are going to used at any one time (as we have shown in the righthand diagram of Fig. 5). As an example of the types of rules for CSDs, recall that in the calculi for SDs, there are rules which allow the addition or removal of contours; it would be reasonable to have rules in the CSD calculi which allow the addition or removal of contours *with respect to a given type-hierarchy*. Another thing to keep in mind is that a set of rules designed with proving completeness of the system in mind might differ dramatically from a set of rules designed for human interaction. Trying to capture all of these features means that the rules for the calculi will have to designed very carefully.

In the long term the intention is to develop CSDs as a fully fledged, formal, diagrammatic reasoning system, unifying the existing systems of SDs, CDs and CGs, utilising the most appropriate features of each notation.

References

[1] Dau, F.: Types and tokens for logic with diagrams: A mathematical approach. In: Wolff, K.E., Pfeiffer, H.D., Delugach, H.S. (eds.) ICCS 2004. LNCS (LNAI), vol. 3127, pp. 62–93. Springer, Heidelberg (2004)

[2] Dau, F.: Formal, diagrammatic logic with conceptual graphs. In: Hitzler, P., Scharfe, H. (eds.) Conceptual tructures in Practice. CRC Press (Chapman and Hall/Taylor & Francis Group (2008)

[3] Euler, L.: Lettres a une princesse dallemagne sur divers sujets de physique et de philosophie. Letters 2, 102–108 (1775) (Berne, Socit Typographique)

[4] Fish, A., Flower, J.: Investigating reasoning with constraint diagrams. In: Visual Language and Formal Methods 2004, Rome, Italy. ENTCS, vol. 127, pp. 53–69. Elsevier, Amsterdam (2005)

[3] For the semantics in the SD-sense, a promising approach is to take one of the existing adequate calculi for SDs, and augment it with rules which allow us to generalize or even remove relations (such rules usually appear in calculi for simple CGs, as they have been developed by Prediger, Dau, or Chein and Mugnier).

[5] Fish, A., Flower, J., Howse, J.: The semantics of augmented constraint diagrams. Journal of Visual Languages and Computing 16, 541–573 (2005)

[6] Flower, J., Fish, A., Howse, J.: Euler diagram generation. Journal of Visual Languages and Computing (accepted, 2007)

[7] Flower, J., Masthoff, J., Stapleton, G.: Generating readable proofs: A heuristic approach to theorem proving with spider diagrams. In: Blackwell, A.F., Marriott, K., Shimojima, A. (eds.) Diagrams 2004. LNCS (LNAI), vol. 2980, pp. 166–181. Springer, Heidelberg (2004)

[8] Gaines, B.R.: An interactive visual language for term subsumption languages. IJCAI, 817–823 (1991)

[9] Gil, J., Howse, J., Kent, S.: Formalizing spider diagrams. In: IEEE Symposium on Visual Languages, pp. 130–137 (1999)

[10] Gurr, C.: Effective diagrammatic communication: Syntactic, semantic and pragmatic issues. Journal of Visual Languages and Computing 10(4), 317–342 (1999)

[11] Hammer, E., Shin, S.J.: Euler's visual logic. History and Philosophy of Logic, 1–29 (1998)

[12] Hartshorne, W., Burks(eds.): Collected Papers of Charles Sanders Peirce, Cambridge, Massachusetts, pp. 1931–1935. Harvard University Press

[13] Howse, J., Molina, F., Shin, S.-J., Taylor, J.: On diagram tokens and types. In: Proceedings of 2nd International Conference on the Theory and Application of Diagrams, Georgia, USA, April 2002, pp. 146–160. Springer, Heidelberg (2002)

[14] Howse, J., Schuman, S.: Precise visual modelling. Journal of Software and Systems Modeling 4, 310–325 (2005)

[15] Howse, J., Stapleton, G., Taylor, J.: Spider diagrams. LMS Journal of Computation and Mathematics 8, 145–194 (2005)

[16] Howse, J., Stapleton, G., Taylor, J.: Spider diagrams. LMS Journal of Computation and Mathematics 8, 145–194 (2005)

[17] John, C., Fish, A., Howse, J., Taylor, J.: Exploring the notion of clutter in Euler diagrams. In: 4th International Conference on the Theory and Application of Diagrams, Stanford, USA, pp. 267–282. Springer, Heidelberg (2006)

[18] Kent, S.: Constraint diagrams: Visualizing assertions in object-oriented models. In: OOPSLA, pp. 327–341. ACM Press, New York (1997)

[19] Kremer, R.: Visual languages for konwledge representation. In: Proc. of 11th Workshop on Knowledge Acquisition, Modeling and Management (KAW 1998), Banff, Alberta, Canada, Morgan Kaufmann, San Francisco (1998)

[20] Larkin, J.H., Simon, H.A.: Why a diagram is (sometimes) worth ten thousand words. Cognitive Science 11(1), 65–100 (1987)

[21] Lemon, O., Pratt, I.: Spatial logic and the complexity of diagrammatic reasoning. Machine GRAPHICS and VISION 6(1), 89–108 (1997)

[22] Patrascoiu, O., Thompson, S., Rodgers, P.: Tableaux for diagrammatic reasoning. In: Cox, P., Smedley, T. (eds.) Proceedings of the 2005 International Workshop on Visual Languages and Computing, September 2005, pp. 279–286 (2005)

[23] Shimojima, A.: On the Efficacy of Representation. PhD thesis, The Department of Philosophy, Indiana University (1996)

[24] Shin, S.-J.: The Logical Status of Diagrams. Cambridge University Press, Cambridge (1994)

[25] Shin, S.-J.: The Iconic Logic of Peirce's Graphs. Bradford Book, Massachusetts (2002)

[26] Sowa, J.F.: Conceptual structures: information processing in mind and machine. Addison-Wesley, Reading, Mass (1984)

[27] Stapleton, G., Howse, J., Taylor, J.: A decidable constraint diagram reasoning system. Journal of Logic and Computation 15(6), 975–1008 (2005)
[28] Stapleton, G., Masthoff, J., Flower, J., Fish, A., Southern, J.: Automated theorem proving in Euler diagrams systems. Journal of Automated Reasoning (2007)
[29] Stapleton, G., Thompson, S., Howse, J., Taylor, J.: The expressiveness of spider diagrams. Journal of Logic and Computation 14(6), 857–880 (2004)
[30] Stapleton, G.: Reasoning with Constraint Diagrams. PhD thesis, Visual Modelling Group, Department of Mathematical Sciences, University of Brighton (2004)
[31] Swoboda, N., Allwein, G.: Using DAG transformations to verify Euler/Venn homogeneous and Euler/Venn FOL heterogeneous rules of inference. Journal on Software and System Modeling 3(2), 136–149 (2004)
[32] Venn, J.: On the diagrammatic and mechanical representation of propositions and reasonings. Phil. Mag (1880)
[33] VMG. The visual modeling group homepage, university of brighton, `http://www.cmis.brighton.ac.uk/Research/vmg/`

An Algorithmic Study of Deduction in Simple Conceptual Graphs with Classical Negation

Michel Leclère and Marie-Laure Mugnier

LIRMM, CNRS - Université Montpellier 2,
161, rue Ada, F-34392 Montpellier cedex, France
{leclere,mugnier}@lirmm.fr

Abstract. Polarized conceptual graphs (PGs) are simple conceptual graphs added with a restricted form of negation, namely negation on relations. Classical deduction with PGs (in short PG-Deduction) is highly intractable; it is indeed Π_P^2 complete. In [LM06] a brute-force algorithm for solving PG-Deduction was outlined. In the present paper, we extend previous work with two kinds of results. First, we exhibit particular cases of PGs for which the complexity of PG-Deduction decreases and becomes not more difficult than in simple conceptual graphs. Secondly, we improve the brute-force algorithm with several kinds of techniques based on properties concerning the graph structure and the labels.

1 Introduction

Simple conceptual graphs (SGs) [CM92] constitute the kernel of conceptual graphs (CGs) [Sow84]. They can be used as such, to represent facts or queries. They are also basic bricks for more complex constructs, corresponding to more expressive conceptual graphs, for instance rules or constraints [BM02]. Full conceptual graphs are obtained when negation is added to SGs without restriction. Several works inspired from Peirce's existential graphs, a diagrammatical system for logics, have studied full conceptual graphs, in particular [Sow84, WL94, Dau03]. Full conceptual graphs have the expressive power of FOL. We think that they are too complicated at the end-user level, for modeling applications, building knowledge-based systems and understanding how they work; they are also too complex from a computational viewpoint since deduction becomes non decidable. We thus prefer to add a limited form of negation to SGs, namely atomic negation (i.e. negation whose scope is an atom). Atomic negation allows us to express knowledge of form "this kind of relation does not hold between these entities", "this entity does not have that property" or "this entity is not of that type".

Polarized Graphs. SGs plus atomic negation yield polarized graphs[1] (PGs), which are equivalent to the FOL fragment of existentially closed conjunctions

[1] This name is borrowed to [Ker01].

P. Eklund and O. Haemmerlé (Eds.): ICCS 2008, LNAI 5113, pp. 119–132, 2008.

of positive and negative literals. Several works have pointed out difficulties introduced by atomic negation (and inequality) in classical FOL [Mug00, Ker01, Kli05]. In [LM06, ML07], we discussed several semantics of negation in relation with deduction checking but also with querying PGs: negation with closed-world assumption, negation in classical FOL, and negation in intuitionistic logic. In the first and in the third case, negation can be processed without complexity overhead. In the classical case, deduction checking in PGs (PG-Deduction) becomes highly intractable: indeed, it is Π^2_P-complete (Π^2_p is co-NP^{NP}), whereas deduction in SGs is NP-complete (see f.i. [Mug07] for a proof of Π^2_P-completeness).

Contribution. In [LM06, ML07], we proposed a brute-force algorithm for (classical) PG-Deduction. In the present paper, we extend this previous work with two kinds of results. First, we exhibit particular cases of PGs for which the complexity of PG-Deduction decreases and becomes not more difficult than in SGs. These particular cases rely on the notion of pair of *exchangeable* relation nodes (that appear in the graph to be deduced). Secondly, we improve the brute-force algorithm with several kinds of techniques based on properties concerning the graph structure and the labels. Finally, let us mention that this paper extends another work of ours on the containment problem of conjunctive queries with negation, in the context of databases [LM07]. Indeed, this problem can be seen as a particular case of PG-Deduction, where relation types are not partially ordered. Furthermore, the notion of exchangeable relation nodes defined here generalizes that of opposite literals in [LM07].

The sequel of this paper is organized as follows. Section 2 is devoted to preliminary definitions and results. Exchangeable pairs of relation nodes and related special cases are studied in section 3 and algorithmic improvments in section 4. These improvements are based first on a limitation of the "completion vocabulary", and secondly on a specific exploration of the search space.

2 Polarized Graphs

In this section, we define notations and recall some definitions and results of [LM06] about polarized conceptual graphs. We assume that the reader is familiar with the basics of conceptual graphs (cf. for instance [ML07] for definitions consistent with the present paper).

Basic notations and results. A conceptual graph *vocabulary* contains at least a poset (partially ordered set) of concept types, a poset of relation types and a set of individual markers. We denote by $\mathcal{V}$ a vocabulary, and by T_R its set of relation types. Φ is the classical translation from conceptual graphs (and vocabularies) to first-order logic (FOL). $\Phi(G)$ denotes the logical formula assigned by Φ to a graph G, and $\Phi(\mathcal{V})$ denotes the set of formulas translating the concept and relation type posets. We use the symbol $\models$ to denote both the logical entailment and the deduction (as both notions are equivalent in FOL). *Projection* is the fundamental mechanism to reason with simple conceptual graphs (SGs). Since it is indeed a graph homomorphism, and the term "projection" can be misleading because of the operation with the same name in databases, we prefer to call it *SG homomorphism*, or simply *homomorphism* if there is no ambiguity. If there is a homomorphism from G to H, we say that G can be *mapped* to H. A SG is *normal* if it does not possess two concept nodes with the same individual marker. Under natural assumptions, every SG has a unique normal form. SG homomorphism is *sound* and *complete* with respect to Φ, i.e.: for all SGs G and H on a vocabulary $\mathcal{V}$, if there is a homomorphism from G to H then $\Phi(\mathcal{V}), \Phi(H) \models \Phi(G)$ (soundness, [Sow84]) and if $\Phi(\mathcal{V}), \Phi(H) \models \Phi(G)$ then there is a homomorphism from G to the normal form of H (completeness, [CM92]).

Polarized conceptual Graphs (PGs). PGs are built from SGs by "polarizing" their relation nodes. Beside positive relation nodes, there are now negative relation nodes.

Definition 1 (Polarized Graph (PG)). *A* polarized graph *(PG) is defined similarly to a SG except that relation nodes are labeled not only by a type but also by a polarity (denoted + or −). A* positive *(resp.* negative*) relation node is labeled by $+r$ (resp. $-r$), where r is a relation type. $+r$ can also be noted r.*

A negative relation node with label $-r$ and arguments $(c_1, ..., c_k)$ expresses that "there is no relation of type r between $c_1, ..., c_k$" (or if $k = 1$, "c_1 does not possess the property r"); it is logically translated by Φ into the literal $\neg r(e_1, ..., e_k)$, where e_i is the term assigned to c_i. PGs are equivalent to the FOL fragment composed of existentially closed conjunctions of (positive and negative) literals (without functions). In the following, we note $+r(c_1, ..., c_k)$ (resp. $-r(c_1, ..., c_k)$), a relation node with label $+r$ (resp. $-r$) and argument list $c_1, ..., c_k$, where the $c_1, ..., c_k$ are not necessarily distinct nodes.

Definition 2 (inconsistent PG). *A PG is said to be* inconsistent *if its normal form contains two relation nodes $+r(c_1, ..., c_k)$ and $-s(c_1, ..., c_k)$ with contradictory labels, i.e. with $r \leq s$. Otherwise it is said to be* consistent.

It can be immediately checked that any PG G on a vocabulary $\mathcal{V}$ is inconsistent iff[2] $\Phi(\mathcal{V}) \cup \{\Phi(G)\}$ is inconsistent. The order on relation labels is extended as follows: we set $-r_1 \leq -r_2$ if $r_2 \leq r_1$.

Definition 3 (Extended order on relation labels). *Given two relation labels l_1 and l_2, $l_1 \leq l_2$ if, either l_1 and l_2 are both positive labels, say $l_1 = (r_1)$ and $l_2 = (r_2)$, and $r_1 \leq r_2$, or l_1 and l_2 are both negative labels, say $l_1 = (-r_1)$ and $l_2 = (-r_2)$, and $r_2 \leq r_1$.*

Given this extended order on relation labels, homomorphism can be used without changing its definition. Recall that homomorphism is logically sound and complete for SGs. For PGs, one part of the property still holds:

[2] If and only if.

Property 1. Given two PGs G and H on a vocabulary $\mathcal{V}$, if there is a homomorphism from G to H then $\Phi(\mathcal{V}), \Phi(H) \models \Phi(G)$.

Thus, homomorphism remains sound. But it is no longer complete. Indeed, we might have $\Phi(\mathcal{V}), \Phi(H) \models \Phi(G)$ and no homomorphism from G to H, as illustrated by Fig. 1. The formulas assigned to G and H by Φ (here we ignore the atoms associated with concept nodes) are respectively $\Phi(G) = \exists x \exists y (p(x) \wedge \neg p(y) \wedge r(x, y))$ and $\Phi(H) = p(a) \wedge r(a,b) \wedge r(b,c) \wedge \neg p(c)$. $\Phi(G)$ can be deduced from $\Phi(H)$ using the tautology $p(b) \vee \neg p(b)$ (indeed, every model of $\Phi(H)$ satisfies either $p(b)$ or $\neg p(b)$; if it satisfies $p(b)$, then x and y are interpreted as b and c; in the opposite case, x and y are interpreted as a and b; thus every model of $\Phi(H)$ is a model of $\Phi(G)$).

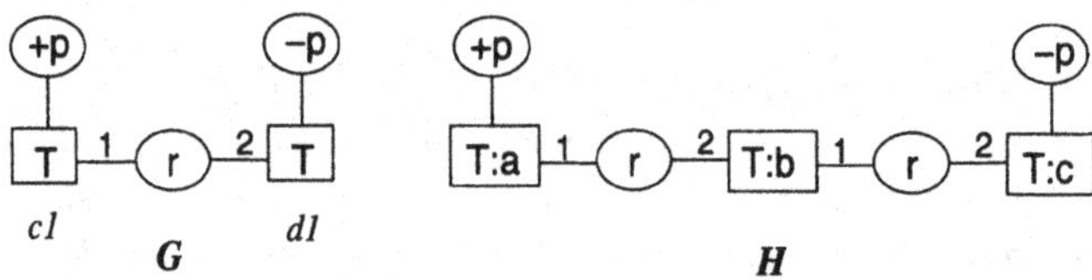

Fig. 1. Atomic negation and homomorphism

More generally, negation introduces disguised disjunctive information that cannot be taken into account by homomorphism. This disjunctive information is related to the law of the excluded-middle which holds in classical logic: given a proposition P, either P is true, or $\neg P$ is true. This leads to reasoning by cases: if a relation is not asserted in a fact, either it is true or its negation is true. We thus have to consider all ways of *completing* the knowledge asserted by a PG. Let us look again at the example in Fig. 1. H does not say whether the unary relation p holds for b. We thus have to consider two cases : either a relation node with label $+p$ or a relation node with label $-p$ can be attached to b. Let H_1 and H_2 be the graphs respectively obtained from H (Fig. 2). There is a homomorphism from G to H_1 and there is a homomorphism from G to H_2. From the homomorphism soundness, we conclude that G can be logically deduced from H.

The next definition specifies the notion of *completion* of a PG.

Definition 4 (Complete PG). *A* complete PG *on a vocabulary $\mathcal{V}$ with relation type set T_R is a consistent (normal) PG satisfying the following completion condition: for each relation type r of arity k in T_R, for each k-tuple of concept nodes $(c_1, \ldots, c_k)$, where $c_1, \ldots, c_k$ are not necessarily distinct nodes, there is a relation $+s(c_1, \ldots, c_k)$ with $s \leq r$ or (exclusive) there is a relation $-s(c_1, \ldots, c_k)$ with $r \leq s$. A PG is* complete w.r.t. a subset of relation types $T \subseteq T_R$ *if the completion condition considers only elements of T. If a PG G^c that is complete w.r.t. T is obtained by adding relation nodes to a graph G, it is called a T-*completion *of G (or simply a* completion *of G if T is implicit).*

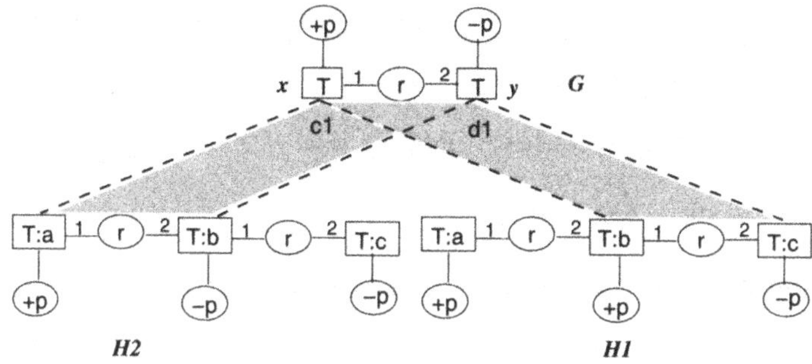

Fig. 2. When the law of the excluded-middle intervenes

Property 2. If a relation node is added to a complete PG, either this relation node is redundant (there is already a relation node with the same argument list and a label less or equal to it) or it makes the PG inconsistent.

A complete PG is obtained from a consistent PG G by repeatedly adding positive and negative relation nodes as long as adding a relation brings new information and does not yield an inconsistency. Since a PG is a finite graph defined over a finite vocabulary, the number of different complete PGs that can be obtained from it is finite. Let us now define deduction on PGs.

Definition 5 (PG-Deduction problem). *The* PG-Deduction *problem takes two PGs G and H as input, with H being consistent, and asks whether G can be* PG-deduced *from H, i.e. whether for each complete PG H^c obtained from H, there is a homomorphism from G to H^c.*

Theorem 1. *Let G and H be two PGs on a vocabulary $\mathcal{V}$, with H being consistent and normal. Then G can be PG-deduced from H iff $\Phi(\mathcal{V}), \Phi(H) \models \Phi(G)$.*

Proof. See [ML07] (Appendix B). □

From now on, we can thus identify PG-deduction and logical deduction on the associated formulas, and we will sometimes simply say "deduction" (in this case, it is assumed that H is normal). Note that each completion of H is normal if (and only if) H is normal. Algorithm 1 is a brute-force algorithmic schema for checking deduction (see Sect. 4 for algorithmic improvements). An immediate observation for generating the completions H^c is that we do not need to consider all relation types but only those appearing in G. The algorithm generates all complete PGs relative to this set of types and for each of them checks whether G can be mapped to it. A complete graph to which G cannot be mapped can be seen as a counter-example to the assertion that G is deducible from H.

3 Special Cases with Lower Complexity for PG-Deduction

The existence of a PG homomorphism from G to H is a sufficient condition for G to be deducible from H. However it is not a necessary condition, as we have seen

Algorithm 1. PG-Deduction

```
Data: PGs G and H, s.t. H is consistent (and normal)
Result: true if G can be PG-deduced from H, false otherwise
begin
    Compute ℋ the set of complete PG obtained from H w.r.t. relation types in
    G;
    forall H^c ∈ ℋ do
        if there is no homomorphism from G to H^c then
            return false ;   // H^c is a counter-example
    return true;
end
```

before. In this section, we study the question "when is a homomorphism from G to H a necessary condition for G to be deducible from H?". Answers to this question yield particular cases where the theoretical complexity of PG-Deduction decreases. We shall also identify special subgraphs of G for which there must be a homomorphism to H when G is deducible from H. These subgraphs can be used as filters or guides during the completion algorithm.

Let us first identify relation nodes in G which might play a role in the problem complexity, in the sense that they may lead to use the law of the excluded-middle.

Definition 6 (Opposite relation labels and nodes). *Two relation labels are said to be* opposite *if they have opposite polarities and the type s of the negative label is less than the type r of the positive label (i.e. $s \leq r$). By extension, two relation nodes are said to be opposite if they have opposite labels $+r$ and $-s$.*

Opposite and contradictory relation nodes should not be confused. Let us consider for instance the binary relation types $relativeOf$ ("x is a relative of y") and $motherOf$ ("x is the mother of y"). One has $motherOf \leq relativeOf$. Relation nodes labeled $-relativeOf$ and $+motherOf$ are *contradictory,* since for all x and y, if x is the mother of y then x is a relative of y. Relation nodes labeled $+relativeOf$ and $-motherOf$ are *opposite.* The intuitive idea behind this notion is that, if one considers one of the types $relativeOf$ or $motherOf$ (let us call it t), $+relativeOf \geq +t$ and $-motherOf \geq -t$. The notion of exchangeable relation nodes generalizes this idea: two opposite relation nodes in G are "exchangeable" if their arguments can have the same images by homomorphisms to (necessarily distinct) completions of H. More precisely:

Definition 7 (Exchangeable relation nodes). *Two relation nodes $+r(c_1, ..., c_k)$ and $-s(d_1, ..., d_k)$ in G are* exchangeable *with respect to H if (1) they are opposite, (2) there are two completions of H, say H_1 and H_2, and two homomorphisms h_1 and h_2, respectively from G to H_1 and from G to H_2, such that for all $i : 1, ..., k$, $h_1(c_i) = h_2(d_i)$.*

See, for instance, the PG G in Fig. 1. Let us consider the opposite relation nodes $r_1 = p(c_1)$ and $r_2 = -p(d_1)$. These nodes are exchangeable, as can be seen in Fig. 2: there is a homomorphism h_1 from G to a completion H_1 of H and there is a homomorphism h_2 from G to another completion H_2 of H, such that $h_1(c_1) = h_2(d_1)$ (and is the concept node in H with marker b). It can be checked that the definition of exchangeable relation nodes is strictly more restrictive than the definition of opposite relation nodes.

The following property will be used to prove other properties:

Property 3. Let G and H be two PGs, where H is consistent and G is PG-deducible from H. Let G' be a subgraph of G without pair of exchangeable relation nodes. Then there are a completion H^c of H and a homomorphism from G to H^c that maps G' entirely to H.

Proof. Consider any H^c maximal completion of H (for each n-ary type p and each n-tuple u of concept nodes in H, either one has $+p(u)$ or one has $-p(u)$). Assume that there is no homomorphism from G to H^c that maps G' to H. For each homomorphism from G to H^c, there is at least one added relation $\sim p(u)$ in H^c which is the image of a relation in G' and such that H does not contain a node $\sim q(u)$ with $\sim p \geq \sim q$. Let R be the set of *all* such relation nodes in $H^c \setminus H$ for all homomorphisms from G to H^c. Let us inverse the polarity of the nodes in R. The graph obtained cannot be inconsistent[3] and it is of max size: thus it is again a maximal completion of H. Let $H^{c'}$ be this maximal completion. As G' does not possess exchangeable relation nodes, there is no homomorphism from G to $H^{c'}$ that maps a relation node in G' to a node in R. If there is no homomorphism from G to $H^{c'}$ that maps G' entirely to H, let R' be the set of all relation nodes $\sim p(u)$ in $H^{c'} \setminus H$ which are images of a node in G' and such that H does not contain a node $\sim q(u)$ with $\sim p \geq \sim q$. As previously, reverse the polarity of all nodes in R', which yields the graph $H^{c''}$. Add the nodes of R' to R. We thus build a sequence of maximal completions of H and a set R of relation nodes of these completions not belonging to H (nor redundant with nodes of H). As R grows strictly from one completion to another, this sequence is finite. The last graph of this sequence is a completion satisfying the property. □

The next property can be seen as a corollary of Prop. 3 (however, its direct proof is simpler; in particular, instead of any maximal completion of H, one can consider a maximal completion obtained by adding only positive relation nodes):

Property 4. Let G and H be two PGs, where H is consistent and G has no pair of exchangeable relation nodes w.r.t. H. If G is PG-deducible from H, then there is a homomorphism from G to H.

We thus obtain a case for which PG-Deduction has the same complexity as homomorphism checking (and is NP-complete):

[3] Indeed, assume we obtain two contradictory relation nodes $-q(u)$ and $p(u)$, with $q \geq p$. One of these nodes does not belong to R, otherwise G would have exchangeable nodes. Let x be this node and y be the node that belongs to R. The label of x in H^c is necessarily more general than the label of y in H^c (note that both nodes were comparable and had the same polarity in H^c). Thus, by inverting the label of y, it is impossible to obtain an inconsistency.

Property 5. Let G and H be two PGs, where H is consistent and G has no pair of exchangeable relation nodes w.r.t. H. G is PG-deducible from H iff there is a homomorphism from G to H.

Note also that G is deducible from H iff each connected component of G is deducible from H. Thus in previous property, the condition on G can be replaced by "each connected component of G has no pair of exchangeable relation nodes". If G is acyclic (and more generally has bounded treewidth, or bounded hypertreewidth when seen as a hypergraph) then homomorphism checking is polynomial ([MC92] for acyclicity, and f.i. [GLS01] for more general notions), hence PG-Deduction.

A desirable property is that recognizing exchangeable relation nodes is not difficult compared to checking PG-deduction. It is indeed the case: checking whether G has exchangeable relation nodes, or checking whether a pair of relation nodes in G is exchangeable, is NP-complete [Tho07]. More precisely, it has the same complexity as homomorphism checking (from G to H), and is polynomial when G is acyclic.

If G is PG-deducible from H, for each subgraph of G without exchangeable relation nodes, there must be a homomorphism from this subgraph to H. Moreover, there must be such a homomorphism that is potentially extensible to a homomorphism from the entire G to a completion of H. We call it a *compatible* homomorphism. See Fig. 3: all concept nodes are assumed to have the same label $(\top, *)$ and relation types are incomparable. There are three homomorphisms from G^- to H: $h_1 = \{x \to t, y \to u, z \to w\}$, $h_2 = \{x \to t, y \to w, z \to v\}$, $h_3 = \{x \to u, y \to w, z \to v\}$. To check the compatibility, we have to consider $s(y, x)$ and $r(x, z)$. h_1 is not compatible because it cannot be extended to $r(x, z)$ due to the presence of $-r(t, w)$ in H.

Definition 8 (compatible homomorphism). *Given two PGs G and H, and G' any subgraph of G, a homomorphism h from G' to H is said to be* compatible *(w.r.t. G) if for each relation node x of G that does not belong to G' but has all its arguments in G', say $c_1, ..., c_k$, there is no relation node y with argument list $h(c_1), ..., h(c_k)$ in H and with a label contradictory to that of x (i.e. with label $-r$ if the label of x is $+s$, or with label $+s$ if the label of x is $-r$, s.t. $s \leq r$).*

Property 6. If G is PG-deducible from H, then there is a compatible homomorphism from every subgraph of G without exchangeable relation nodes to H.

Proof. Let G' be any subgraph of G without exchangeable relation nodes. From property 3, there is a homomorphism from G to a completion of H which maps G' entirely to H. By restricting the domain of this homomorphism to G', we have a homomorphism from G' to H which is compatible w.r.t. G. □

One can remark some easily identifiable subgraphs without exchangeable relation nodes: the *positive subgraph* of G, denoted G^+, is the subgraph obtained from G by selecting all concept nodes and only the positive relation nodes. The *negative subgraph* G^- of G is the dual notion, i.e. the subgraph obtained from G by selecting all concept nodes and only the negative relation nodes. Negative and

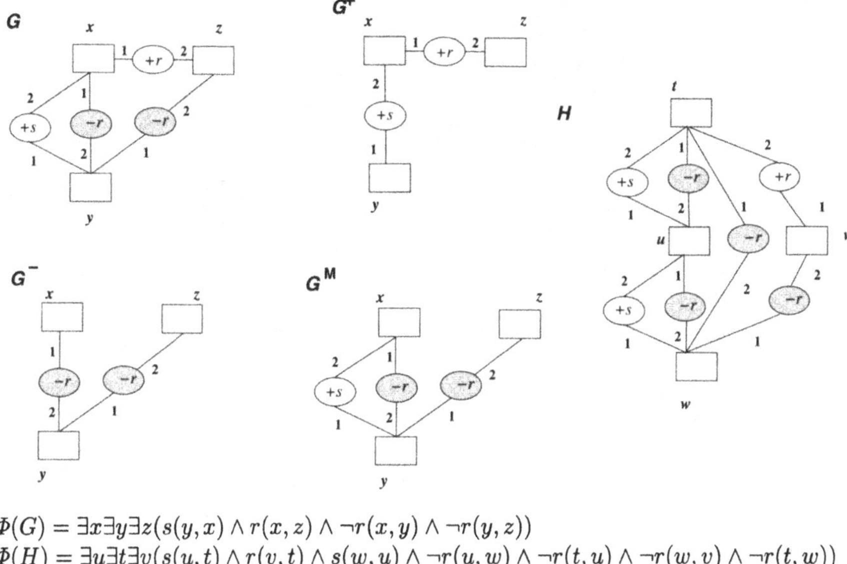

$$\Phi(G) = \exists x \exists y \exists z (s(y,x) \wedge r(x,z) \wedge \neg r(x,y) \wedge \neg r(y,z))$$
$$\Phi(H) = \exists u \exists t \exists v (s(u,t) \wedge r(v,t) \wedge s(w,u) \wedge \neg r(u,w) \wedge \neg r(t,u) \wedge \neg r(w,v) \wedge \neg r(t,w))$$

Fig. 3. Special subgraphs of G (G is deducible from H)

positive subgraphs are particular cases of subgraphs without opposite relation labels. A subgraph of G without opposite relation nodes and maximal for the inclusion is easily built by selecting, for each relation type r appearing in G, either all its positive occurrences or all its negative occurrences, while satisfying the following constraint: if one selects the positive (resp. negative) occurrences of r, then the same choice must be done for all subtypes (resp. supertypes) of r. F.i. in Fig. 3, several subgraphs without opposite relation nodes of G are pictured. G^+ and G^- are respectively the positive and negative subgraphs of G. G has two subgraphs without opposite nodes maximal for inclusion: G^+ and G^M.

4 Algorithmic Improvements

Let us say that a concept or relation label l_x occurring in G has a support in H if there is a label l_y in H with $l_y \leq l_x$ (and l_y is said to support l_x). By extension, we say that a node x in G has a support in H if there is a node y in H s.t. the label of x is supported by the label of y (and y is said to support x). A first observation is that if a node in G has no support in H then G is not deducible from H. This is trivial for concept nodes. For relation nodes, if this node is negative (resp. positive), consider the positive (resp. negative) completion of H. There is no homomorphism from G to this completion.

4.1 Limitation of the Completion Vocabulary

Let us call "completion vocabulary" the set of relation types used to build completions of G. The size of the completion vocabulary determines the number of completions of G. The number of completions of G is itself a key element in the complexity of deduction checking. It is thus essential to decrease as much as possible the number of relation types involved in completion. One can observe that the completion vocabulary can be restricted to the relation types occurring in G, and furthermore to the relation types occurring in opposite relation nodes:

Property 7. G is PG-deducible from H iff G can be mapped to each completion of H w.r.t. relation types occurring in opposite relation nodes of G (i.e. r and s such that there are nodes in G with labels $+r$ and $-s$ and $s \leq r$).

Proof. Let T_R be the set of relation types in the vocabulary, and let T be the set of relation types occurring in opposite relation nodes in G. ($\Leftarrow$) We prove that if G can be mapped to each T-completion of H then it can be mapped to each T_R-completion of H. Indeed, let H^c be any T_R-completion of H. Let $H^{c'}$ be the graph obtained from H^c by replacing all relation nodes with types outside T with a set of relation nodes built as follows: let r be a node labeled by $+t$ (resp. $-t$) such that $t \notin T$. Let $\{t_1, ..., t_n\}$ be the types in T greater than t (resp. less than t). If r is positive, consider the minimal elements of this set, otherwise consider the maximal elements of this set. Let S be the obtained set. Replace r with $|S|$ relation nodes, each labeled by a type in S, with the same polarity and the same arguments as r. $H^{c'}$ is a T-completion of H. By construction, there is a homomorphism, say h_1, from $H^{c'}$ to H^c (which is the identity on concept nodes). By hypothesis, there is a homomorphism, say h, from G to $H^{c'}$. The composition of these homomorphisms $h_1 \circ h$ is a homomorphism from G to H^c.

($\Rightarrow$) Let G be deducible from H and assume that H^c is a T-completion of H such that there is no homomorphism from G to H^c. We show that this assumption leads to a contradiction. From H^c, we build the following T_R-completion of H, say $H^{c'}$. For all types t in $T_R \setminus T$, let us add only $(+t)$ nodes if $(+t)$ does not support any node in G; otherwise, add only $(-t)$ nodes if $(-t)$ does not support any node in G (if neither $(+t)$ nor $(-t)$ support nodes in G, relation nodes typed t can be added with any polarity); if both $(+t)$ and $(-t)$ support nodes in G, there are opposite nodes in G with label $(+r)$ and $(-s)$ and $r \geq t \geq s$, thus r and s belong to T, and nodes labeled T would be redundant in $H^{c'}$ thus are not needed to obtain a completion. In all cases, nodes are added only if they do not lead to an inconsistency. Since G is deducible from H, there is a homomorphism from G to $H^{c'}$. By construction, no node in G can be mapped to an added node. Thus this homomorphism is a homomorphism from G to H^c, which contradicts the hypothesis on H^c. □

We can even restrict the completion vocabulary to the types of *exchangeable* relation nodes in G.

Theorem 2. *G is PG-deducible from H iff G can be mapped to each completion of H w.r.t. relation types occurring in exchangeable relation nodes of G w.r.t. H (i.e. relation types r such that there is a pair of exchangeable relation nodes in G with one of the two labeled $+r$ or $-r$).*

Proof. Let *exchangeable*(G) denote the types occurring in exchangeable relation nodes in G. ($\Leftarrow$) Same as in the proof of Prop. 7, where T is replaced by *exchangeable*(G). ($\Rightarrow$) Let H^c be a completion of H w.r.t. exchangeable(G) such that there is no homomorphism from G to it. As in the proof of Prop. 7, we build a completion of H, say $H^{c'}$, as follows: for any type t occurring in G but not in exchangeable(G), let us add only $(-t)$ nodes if t supports only positive nodes, $(+t)$ nodes if t supports only negative nodes. Let us add it positively if it supports both forms. A homomorphism from G to $H^{c'}$ is a homomorphism to H^c plus the nodes added positively for types supporting both forms. Let us now inverse the polarity of these latter nodes if they are images of nodes in G. No node in G can be mapped to these nodes, otherwise their type would be in exchangeable(G). Thus, we have a homomorphism from G to H^c, which contradicts the hypothesis on H^c. □

4.2 Space Algorithm

Consider the space of consistent graphs obtained from H by adding relation nodes (with types in the completion vocabulary). This space is ordered[4] as follows: given two graphs H_1 and H_2 in this space, $H_2 \leq H_1$ if for each relation node x in H_1, there is a relation node with the same list of arguments in H_2 and a label less than or equal to the label of x. H is the greatest element of this space, and the smallest elements are (necessarily) completions of H. The question "is there a homomorphism from G to each completion H^c" can be reformulated as "is there a *covering set* of all H^c, i.e. a subset of incomparable graphs of this space $\{H_1, ..., H_k\}$ such that (1) there is a homomorphism from G to each H_i ; (2) for each H^c, there is a H_i with $H^c \leq H_i$". The brute-force algorithm (Algorithm 1) takes the set of all completions of H as covering set. The next algorithm (Algorithm 2) searches the space in a top-down way starting from H and tries to build a covering set with partial completions of H. Reasoning by cases is applied at each step: for a given relation type r with arity k and a tuple $(t_1, ..., t_k)$ of concept nodes such that neither $+r$ nor $-r$ is supported by the label of a relation node on $(t_1, ..., t_k)$ in the current partial completion, two graphs are generated according to each case. Note that if $+r$ or $-r$ is supported by a $\sim s$ in the current completion, then adding $+r(t_1, ..., t_k)$ or $-r(t_1, ..., t_k)$ to it would lead to a redundancy or an inconsistency.

The algorithm is justified by the following property:

Theorem 3. *G is PG-deducible from H iff:*
1. There is a homomorphism h from G to H **or**
2. G is PG-deducible from H' and H'' where H' (resp. H'') is obtained from H by adding the positive relation node $+r(t_1, ..., t_k)$ (resp. the negative relation node $-r(t_1, ..., t_k)$) where r is a relation type of arity k occurring in G (and more specifically r belongs to the completion vocabulary) and $t_1, ..., t_k$ are concept nodes of H such that neither $+r$ nor $-r$ is supported by the label of a relation node on $(t_1, ..., t_k)$ in H.

[4] Or preordered, if redundant relations can be added.

Proof. (sketch) (⇒) Any completion of H' or H'' is a completion of H. (⇐) Condition 1 corresponds to property 1. For condition 2, check that $\{H', H''\}$ is a covering set (of completions of H). □

Subalgorithm 3 is supposed to have direct access to data available in the main algorithm 2. The choice of r and $t_1, ..., t_k$, in Algorithm 3, can be guided by a compatible homomorphism from a special subgraph of G.

Algorithm 2. Check by space exploration

```
Data: PGs H and G, with H being consistent
Result: true if G is PG-deducible from H, false otherwise
begin
  | Result ← Filtering();
  | if (Result ≠ undetermined) then
  |   L return Result
  | Let R be the completion vocabulary;
  | return RecCheck(H); // See Algorithm 3
end
```

The Filtering subalgorithm performs "simple" tests corresponding to necessary or sufficient conditions of deduction that would allow us to conclude without entering the completion steps:

1. If a concept or relation node of G has no support in H, then return false.
2. If there is a homomorphism from G to H, then return true.
3. Compute some subgraphs of G without exchangeable relation nodes (for instance a subgraph without opposite relation nodes maximal for the inclusion). If one of these subgraphs cannot be mapped to H by a compatible homomorphism then return false.

The following property ensures that Algorithm 3 does not generate the same graph several times, which is a crucial point for complexity. Otherwise the algorithm could be worse than the brute-force algorithm in the worst-case.

Property 8. The subspace explored by Algorithm 3 is a (binary) tree.

Indeed, at each recursive call, $\{H', H''\}$ is a covering set inducing a bipartition of the covered space: each PG in this space is below exactly one of these two PGs.

Property 9. The time complexity of Algorithm 2 is in $\mathcal{O}(2^{(n_G)^k \times |\mathcal{R}|} \times hom(G, H^c))$, where n_G is the number of concept nodes in G, k is the maximum arity of a relation, $\mathcal{R}$ is the completion vocabulary and $hom(G, H^c)$ is the complexity of checking the existence of a homomorphism from G to H^c. Its space complexity is in $\mathcal{O}(max(size(G), size(H), (n_G)^k \times |\mathcal{R}|))$.

Proof. The size of a completion of H is bounded by $2^{(n_G)^k \times |\mathcal{R}|}$. Property 8 ensures that the number of graphs generated is at most twice the number of completions

Algorithm 3. RecCheck(H)

```
Data: Consistent PG H        Access: G, R
Result: true if G is PG-deducible from H, false otherwise
begin
    if there is a homomorphism from G to H then
        return true
    if H is complete w.r.t. R then
        return false
    (r, t1, ..., tk) ← ChooseRelationTypeToAdd;
    /* r is a relation type of R, t1, ..., tk are concept nodes in H and
       neither +r nor -r is supported by the label of a relation node
       on (t1, ..., tk) in H                                              */
    Let H' be obtained from H by adding the relation node r(t1, ..., tk);
    Let H'' be obtained from H by adding the relation node -r(t1, ..., tk);
    return (RecCheck(H') AND RecCheck(H''))
end
```

of H (in the worst case, all leaves of the generated tree of graphs correspond to complete graphs). If the relation types are not ordered, all completions have the same size, which is $\sum_{r \in \mathcal{R}} (n_G)^{arity(r)}$; checking whether a graph is complete can then be done in constant time if the number of relation nodes in the graph is incrementally maintained. When relation types are ordered, the size of completions varies according to the order in which relation types are considered. One solution is to count the addition of a relation node $\sim r(t_1, ..., t_k)$ not for one, but for n, where n is the number of types s in $\mathcal{R}$, such that $\sim s$ is supported by the new node and was not before. Computing n at each node addition can be roughly bound by $|\mathcal{R}|^2$, which can be reasonably considered as less than $hom(G, H^c)$. For space complexity, see that the tree is explored in depth-first way. □

5 Further Work

The proposed algorithm for checking deduction on PGs is simple to describe and to implement. Its theoretical worst-case complexity is not better than that of the brute-force algorithm but, not surprisingly, first experiments show that its running time is much better. Further work will involve an experimental comparison of several heuristics. These heuristics concern in particular the choice of special subgraphs without exchangeable relation nodes in the filtering phase and the choice of the next relation to add in the completion phase (cf. the ChooseRelationTypeToAdd subalgorithm).

References

[BM02] Baget, J.-F., Mugnier, M.-L.: The Complexity of Rules and Constraints. JAIR 16, 425–465 (2002)

[CM92] Chein, M., Mugnier, M.-L.: Conceptual Graphs: Fundamental Notions. Revue d'Intelligence Artificielle 6(4), 365–406 (1992)

[Dau03] Dau, F. (ed.): The Logic System of Concept Graphs with Negation And Its Relationship to Predicate Logic. LNCS (LNAI), vol. 2892. Springer, Heidelberg (2003)

[GLS01] Gottlob, G., Leone, N., Scarcello, F.: Hypertree decompositions: A survey. In: Sgall, J., Pultr, A., Kolman, P. (eds.) MFCS 2001. LNCS, vol. 2136, pp. 37–57. Springer, Heidelberg (2001)

[Ker01] Kerdiles, G.: Saying it with Pictures: a logical landscape of conceptual graphs. PhD thesis, Univ. Montpellier II / Amsterdam (November 2001)

[Kli05] Klinger, J.: Local negation in concept graphs. In: ICCS, pp. 209–222 (2005)

[LM06] Leclère, M., Mugnier, M.-L.: Simple conceptual graphs with atomic negation and difference. In: ICCS, pp. 331–345 (2006)

[LM07] Leclère, M., Mugnier, M.-L.: Some algorithmic improvements for the containment problem of conjunctive queries with negation. In: Schwentick, T., Suciu, D. (eds.) ICDT 2007. LNCS, vol. 4353, pp. 404–418. Springer, Heidelberg (2006)

[MC92] Mugnier, M.L., Chein, M.: Polynomial algorithms for projection and matching. In: Pfeiffer, H.D., Nagle, T.E. (eds.) Conceptual Structures: Theory and Implementation. LNCS, vol. 754, pp. 49–58. Springer, Heidelberg (1993)

[ML07] Mugnier, M.-L., Leclère, M.: On querying simple conceptual graphs with negation. Data Knowl. Eng. 60(3), 468–493 (2007)

[Mug00] Mugnier, M.-L.: Knowledge Representation and Reasoning based on Graph Homomorphism. In: Ganter, B., Mineau, G.W. (eds.) ICCS 2000. LNCS, vol. 1867, pp. 172–192. Springer, Heidelberg (2000)

[Mug07] Mugnier, M.-L.: On the π_p^2-completeness of the containment problem of conjunctive queries with negation and other problems. Research Report 07004, LIRMM (2007)

[Sow84] Sowa, J.F.: Conceptual Structures: Information Processing in Mind and Machine. Addison-Wesley, Reading (1984)

[Tho07] Thomazo, M.: Complexité et propriétés algorithmiques de la déduction dans les graphes polarisés. First year research report, ENS Cachan (2007)

[WL94] Wermelinger, M., Lopes, J.G.: Basic conceptual structures theory. In: Mugnier, M.-L., Chein, M. (eds.) ICCS 1994. LNCS (LNAI), vol. 835, pp. 144–159. Springer, Heidelberg (1994)

Flexible Querying of Fuzzy RDF Annotations Using Fuzzy Conceptual Graphs

Patrice Buche[1], Juliette Dibie-Barthélemy[1,2], and Gaëlle Hignette[1,2]

[1] INRA Department of Applied Mathematics and Computer Science, Mét@risk, 16 rue Claude Bernard, F-75231 Paris Cedex 05
[2] UFR Informatique, AgroParisTech, 16 rue Claude Bernard, F-75231 Paris Cedex 05
{Patrice.Buche,Juliette.Dibie,Gaelle.Hignette}@agroparistech.fr

Abstract. This paper presents a flexible querying system of fuzzy RDF annotations which consists in translating fuzzy RDF annotations into fuzzy conceptual graphs and using an "approximate"-projection operation in order to compare fuzzy query graphs with fuzzy annotation graphs. The fuzzy sets in the query graphs having a semantic of preferences are compared with the fuzzy sets in the annotation graphs having a semantic of similarity or imprecision. These comparisons deliver several scores which are used by our flexible querying system to sort the answers according to a total order even if these scores are not commensurable.

1 Introduction

The aim of the "Semantic Web" is to obtain a more pertinent querying of the documents available on the Web using semantic annotations associated with them. Previous works (see for example [1]) have proposed to translate XML/RDF annotations, RDF being the standard language to express annotations recommended by the W3C, into conceptual graphs (CGs). Thanks to this translation, the querying and inferencing capabilities enabled by the CG formalism can be used to query an XML/RDF database. Our paper proposes to extend this kind of work to support flexible querying of fuzzy annotations. The context of our proposal is the design of a data warehouse opened on the Web in which local databases, including structured and semi-structured sources of information represented respectively as relational and CG databases, are completed by data sources retrieved from the Web and stored in an XML database. A dedicated flexible querying system has been proposed in [2] to query simultaneously those sources using a given domain ontology. This ontology is a central element of the querying system because, in order to permit a unified querying of the sources, it is previously used: (i) to index manually the data stored in the local databases and (ii) to compute automatically fuzzy annotations of Web data tables (see [3]) which are stored in an XML database. The aim of this paper is to use the CG formalism in order to query the XML database that contains fuzzy annotations. The main originality of our proposal is the support of (i) fuzzy annotation graphs representing similarity and imprecision and (ii) fuzzy query graphs representing user's preferences. Moreover, our querying system is able to maintain a total order of the answers using not commensurable scores. In section

P. Eklund and O. Haemmerlé (Eds.): ICCS 2008, LNAI 5113, pp. 133–146, 2008.

2, we present briefly the ontology and the fuzzy semantic annotation process of Web data tables. In section 3, we present the fuzzy CG base corresponding to fuzzy RDF annotations. In section 4, we propose a flexible querying of this fuzzy CG base.

2 Semantic Annotation of Data Tables

We have proposed in [3] a method to annotate automatically data tables from the Web according to a given domain ontology. The annotation process consists in identifying the relations represented by the Web data tables according to the relations given in the ontology. Examples presented in this paper correspond to our current application domain in food microbiology. In the following, we present successively the structure of the ontology and the fuzzy annotations which are generated using the ontology.

2.1 Structure of the Ontology

The ontology has been structured in a generic way such as being applicable in many application domains. It is composed of datatypes -numeric types and symbolic types- and of relations that allow datatypes to be linked. Numeric types are used to define the numeric data. A numeric type is described by the name of the type, the units in which data of this type is usually expressed, and the interval of possible values for this type. For example, the type *Temperature* can be expressed in the units {°C, °F} and has no restriction on values while the type *pH* has no unit and has a range of $[0, 14]$. Symbolic types are used when the data of interest are represented as a string. A symbolic type is described by the name of the type and the type hierarchy (which is the set of possible values for the type, partially ordered by the subsumption relation). For example, *Food Product* and *Microorganism* are symbolic types. Relations are used to represent semantic links between datatypes. A relation is described by the name of the relation and its signature. For example, the relation *GrowthParameterAw* represents the growth limits of a microorganism concerning water activity[1] of any food product. This relation has for domain the symbolic type *Microorganism* and for range the numeric type *aw*. Figure 1 shows the simplified structure of our domain ontology.

This ontology has been expressed in OWL distinguishing two types of knowledge: (i) generic knowledge, expressed as OWL classes and properties, which define the structure of the ontology: for instance, the class *numericalType* (resp. the class *Relation*) which is the superclass of all numerical types (resp. relations); (ii) domain-dependant knowledge, expressed as classes and constraints : for instance, the class *GrowthParameterAw* is a subclass of the class *relation* and the class *aw* is a subclass of the class *numericalType*.

[1] noted aw and which corresponds to an index of the water, comprised in the interval $[0, 1]$, that is available in the food to be used by microorganisms.

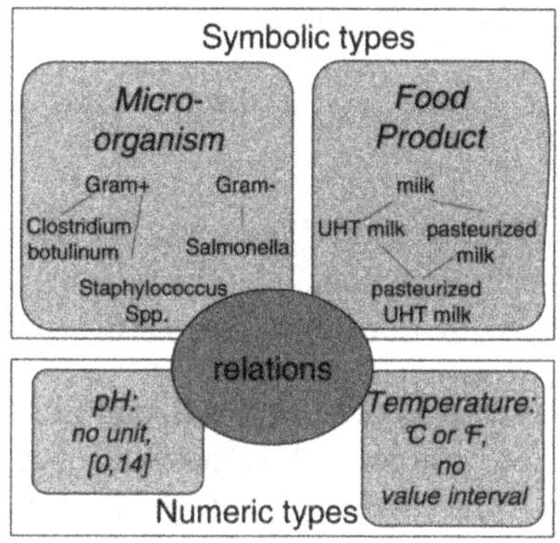

Fig. 1. Structure of the domain ontology

2.2 Fuzzy Semantic Annotations

The semantic annotation process takes as inputs an OWL ontology structured as expressed in section 2.1 and a Web data table expressed in an XML document using XHTML tags. The aim of the annotation process is to annotate the semantic relations represented by the rows of the Web data table with relations of the ontology. Each row of the Web data table is associated with the XHTML tag *tr*. A unique identifier is associated with each row and is represented as the XML attribute *URI* of the tag *tr*. Figure 2 presents an example of a Web data table in which the semantic relation *GrowthParameterAw* has been identified. The first line of the Web data table indicates that *Clostridium* has a growing range between 0.943 and 0.97 which is optimal in the range [0.95, 0.96].

Organism	aw minimum	aw optimum	aw maximum
Clostridium	0.943	0.95-0.96	0.97
Staphylococcus	0.88	0.98	0.99
Salmonella	0.94	0.99	0.991

Table 1: Cardinal values

Fig. 2. Example of a Web data table

The annotation process generates RDF descriptions which represent the semantic relations of the ontology recognized in each row. Some of these RDF descriptions include values expressed as fuzzy sets. We use the representation of fuzzy sets proposed in [4,5].

Definition 1. A **fuzzy set** f on a definition domain $Dom(f)$ is defined by a membership function μ_f from $Dom(f)$ to $[0, 1]$ that associates the degree to which x belongs to f with each element x of $Dom(f)$. We call kernel (resp. support) of the fuzzy set, the set of elements x with $\mu_f(x) = 1$ (resp. $\mu_f(x) \neq 0$).

We distinguish two kinds of fuzzy sets: (i) discrete fuzzy sets and (ii) continuous fuzzy sets.

Definition 2. A **discrete fuzzy set** f is a fuzzy set associated with a symbolic type of the ontology. Its definition domain is the type hierarchy. It is represented in RDF as a resource typed by the OWL class DFS.

Definition 3. A **continuous fuzzy set** f is a trapezoidal fuzzy set associated with a numeric type of the ontology. A trapezoidal fuzzy set is defined by its four characteristic points which correspond to min(support(f)), min(kernel(f)), max(kernel(f)) and max(support(f)). Its definition domain is the interval of possible values of the type. It is represented in RDF as a resource typed by the OWL class CFS.

The fuzzy values used to annotate Web data tables may express two of the three classical semantics of fuzzy sets (see [6]): similarity or imprecision. For each resource typed DFS or CFS, the OWL property *HasForSemantic* defines the semantic of the fuzzy set. In the current version of the annotation process, a fuzzy set having a semantic of similarity is associated with each cell belonging to a symbolic column. It represents the ordered list of the most similar values of the ontology associated with the value present in the cell. A fuzzy set having a semantic of imprecision may be associated with cells belonging to numerical columns. It represents an ordered disjunction of exclusive possible values. Figure 3 presents a part of the RDF descriptions corresponding to the recognition of the relation *GrowthParameterAw* in the first row of the table shown in figure 2. The first description expresses that the first row (having the URI *uriRow1* in the XML document) is an instance of the *GrowthParameterAw* relation recognized with a pertinence score of 1.0. This pertinence score is computed by the annotation process and expresses the degree of certainty associated with the relation recognition. The domain of the relation, which is an instance of the symbolic type *Microorganism*, is annotated by a discrete fuzzy set. This fuzzy set, typed by the OWL class *DFS*, has a semantic of similarity and indicates the list of closest values of the ontology compared to the value *Clostridium*. Two values (*Clostridium Perfringens* and *Clostridium Botulinum*) belong to this fuzzy set with a membership degree of 0.5. The range of the relation, which is an instance of the numeric type *aw*, is annotated by a continuous fuzzy set. This fuzzy set, typed by the OWL class *CFS*, has a trapezoidal form and a semantic of imprecision. It indicates the possible growth limits ([0.943, 0.97]) and the possible optimal growth limits ([0.95, 0.96]) represented respectively as the support and the kernel of the fuzzy set.

We have presented in this section the fuzzy semantic annotation process of Web data tables according to a given ontology. The output of this process are RDF descriptions which represent the semantic relations that have been identified in the Web data tables according to the relations of the ontology. We are

```
<onto:GrowthParameterAw rdf:about="uriRow1">
        <onto:HasForScore>1.0</onto:HasForScore>
        <onto:AssociatedDomain rdf:resource="uriRow1/Micro:1" />
        <onto:AssociatedRange rdf:resource="uriRow1/Aw:1" />
</onto:GrowthParameterAw>
<onto:Microorganism rdf:about="uriRow1/Micro:1">
        <onto:IsAnnotatedBy rdf:resource="uriRow1/Micro:1/DFS:1" />
</onto:Microorganism>
<onto:DFS rdf:about="uriRow1/Micro:1/DFS:1">
        <onto:HasForSemantic>similarity</onto:HasForSemantic>
        <onto:HasForElement rdf:resource="uriRow1/Micro:1/DFS:1/elt:1"/>
        <onto:HasForElement rdf:resource="uriRow1/Micro:1/DFS:1/elt:2"/>
</onto:DFS>
<onto:ClostridiumPerfringens rdf:about="uriRow1/Micro:1/DFS:1/elt:1">
        <onto:HasForMembershipDegree>0.5</onto:HasForMembershipDegree>
</onto:ClostridiumPerfringens>
<onto:ClostridiumBotulinum rdf:about="uriRow1/Micro:1/DFS:1/elt:2">
        <onto:HasForMembershipDegree>0.5</onto:HasForMembershipDegree>
</onto:ClostridiumBotulinum>
<onto:Aw rdf:about="uriRow1/Aw:1">
        <onto:IsAnnotatedBy rdf:resource="uriRow1/Aw:1/CFS:1" />
</onto:Aw>
<onto:CFS rdf:about="uriRow1/Aw:1/CFS:1">
        <onto:HasForUnit>NONE</onto:HasForUnit>
        <onto:HasForSemantic>imprecision</onto:HasForSemantic>
        <onto:HasForMinSupport>0.943</onto:HasForMinSupport>
        <onto:HasForMaxSupport>0.97</onto:HasForMaxSupport>
        <onto:HasForMinKernel>0.95</onto:HasForMinKernel>
        <onto:HasForMaxKernel>0.96</onto:HasForMaxKernel>
</onto:CFS>
```

Fig. 3. Example of RDF annotations generated from the Web data table of figure 2

now interested in querying these RDF annotations that contain fuzzy values. For that, we propose to extend the flexible query processing on fuzzy CGs defined in [2]. We first present the fuzzy CG base obtained from the RDF annotations and then the query processing of such a base.

3 The Fuzzy CG Base Corresponding to RDF Annotations

The model of CG we use [7] relies on (i) a support made of a concept type lattice, a relation type set possibly organized in hierarchy, a set of individual markers enabling the designation of instances and a conformity relation between markers and types, (ii) a base of CGs built on this support.

In a querying perspective, we propose to translate the fuzzy RDF annotations presented in section 2 into fuzzy CGs. A previous work has already proposed a mapping from RDF to CG [1]. It proposes to translate i) the RDF descriptions into a base of CGs, ii) the hierarchy of classes appearing in an RDF schema or an OWL ontology into a concept type hierarchy, and iii) the hierarchy of properties appearing in a RDF schema or an OWL ontology into a relation type set of the CG model. In this section, our aim is to extend the translation rules proposed in [1] in order to be able to translate fuzzy RDF descriptions into fuzzy

CGs. We first recall the main rules defined in [1], then we propose extensions to translate fuzzy values from RDF descriptions into CGs and finally we present the terminological knowledge generated by these rules.

3.1 Translation Rules from RDF Descriptions into CGs

A RDF description is a triple of the form <Resource, Property, Value>. For example, the RDF description which says that the resource having for URI *uriRow1* has a score of 1.0 can be written in XML/RDF syntax as:

```
<rdf:Description about='uriRow1'>
    <onto:HasForScore>1.0</onto:HasForScore>
</rdf:Description>
```

The translation from RDF into CG consists in considering a RDF description as an instance of a *Resource* concept type and the associated properties as relations of this concept. The individual marker of a *Resource* concept is the URI of the resource itself. Literal values are instances of the *Literal* concept type. For example, the previous example can be translated into the following CG:

[Resource:uriRow1] - { -> (HasForScore) -> [Literal: 1.0]}

RDF descriptions can be typed according to a predefined ontology called a RDF schema or an OWL ontology. For example, if we want to type the previous RDF description as an instance of the *GrowthParameterAw* class defined in a RDF schema or an OWL ontology, the XML/RDF syntax becomes:

```
<onto:GrowthParameterAw about='uriRow1'>
    <onto:HasForScore>1.0</onto:HasForScore>
</onto:GrowthParameterAw>
```

Remark 1. *In the previous examples, each markup is prefixed with an XML namespace in order to identify what is RDF-related ("rdf" prefix) and what is domain ontology specific ("onto" prefix). For the sake of readability, we skip these prefixes for CG in the paper, but they are mandatory in the implementation.*

When the RDF description is typed with a class defined in a RDF schema or an OWL ontology, a corresponding concept type must be created in the concept type lattice. For example, a *GrowthParameterAw* subtype of the *Resource* concept type is created and the corresponding CG is:

[GrowthParameterAw:uriRow1] - { -> (HasForScore) -> [Literal: 1.0]}

3.2 Translation Rules from RDF Fuzzy Values into Fuzzy CGs

In this section, we focus on the translation of the RDF descriptions representing fuzzy values. We have seen in section 2.2 that they are typed with the OWL classes DFS and CFS. The class DFS corresponds to fuzzy sets defined on a symbolic definition domain which is a type hierarchy. The class CFS corresponds to fuzzy sets defined on a numerical definition domain. In [8], where comparison to previous works can be found, we have proposed an extension of the CG model to represent and compare fuzzy values. A fuzzy set can appear in two ways in a concept vertex: (i) as a *fuzzy type* when the definition domain of the fuzzy set is hierarchized; a fuzzy type is a fuzzy set defined on a subset of the concept type set; (ii) as a *fuzzy marker* when the definition domain of the fuzzy set is *numerical*; a fuzzy marker is a fuzzy set defined on a subset of the set of individual markers. In order to translate the RDF descriptions representing fuzzy values into fuzzy CGs using the formalism of [8], we introduce the two following translation rules:

Rule 1. A RDF description typed with the class DFS and all its associated descriptions typed with the property *HasForElement* are translated into a generic concept vertex with a fuzzy type.

Rule 2. A RDF description typed with the class CFS and its associated descriptions typed with the properties *HasForMinSupport, HasForMaxSupport, HasForMinKernel, HasForMaxKernel* are translated into an individual concept vertex with a fuzzy marker.

Moreover, we propose, in the translation, to replace the *Literal* concept type presented in section 3.1 by the *Numval* concept type because we are only concerned by crisp or fuzzy numerical values. For example, the fuzzy CG of figure 4 corresponds to the translation of the fuzzy RDF annotations of figure 3.

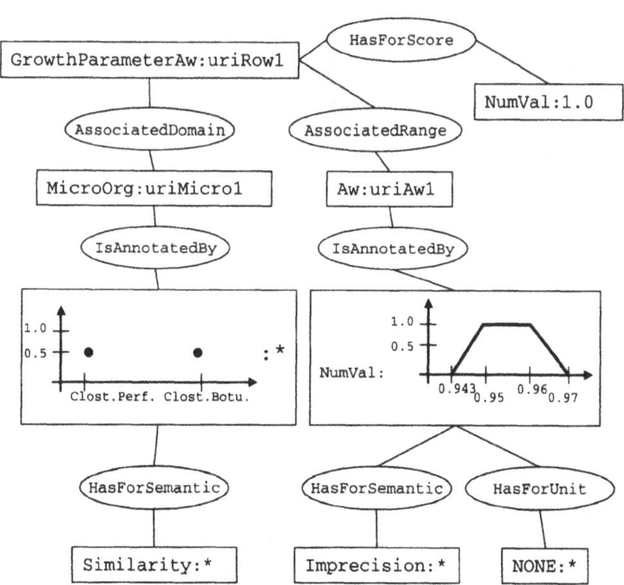

Fig. 4. Translation of the RDF annotations of figure 3 into a fuzzy CG

3.3 The Terminological Knowledge

We now briefly present the support corresponding to the ontology described in section 2. This support has been generated using the translation rules presented above. The **concept type set** is used to represent the main part of the ontology, since it is a partially ordered set designed to contain the concepts of a given application. It is built as follows. A concept type t_a is associated with each type a of the ontology. If a is a symbolic type, then a concept type t_{v_i} is associated with each element v_i of the definition domain of a. The t_a's and t_{v_i}'s are inserted into the concept type set, w.r.t. the partial order of the definition domain. Furthermore, a concept type t_r is associated with each relation r of the ontology. The **set of individual markers** is used to store the definition domain of each type a that is numerical. It is assumed to be $\mathbb{R}$ (see [8]) and contains the numerical values associated with the concept type $Numval$. The **set of relation types** is used to represent the generic relations of the ontology (*HasForScore*, *Associatedkey*, *AssociatedResult*, *IsAnnotatedBy*, *HasForSemantic*, *HasForUnit*). In the following section, we present the querying of the fuzzy CGs corresponding to the fuzzy RDF annotations.

4 Flexible Query Processing of a Fuzzy CG Base

In this section, we propose to extend the query processing presented in [2] in order to be able to manage fuzzy CGs including similarity data and imprecise data. Such an extension allows the end-user to query simultaneously and uniformly each source of information (relational, CG and XML) of a data warehouse opened on the Web (as described in [2]) using a given domain ontology. The query processing proposed in [2] is done through the MIEL query language. It relies on the vocabulary defined in an ontology (as the one defined in 2.1) and proposes a mechanism of query enlargement by means of expression of preferences- represented by fuzzy sets- in the values of the selection attributes. We present the notion of views, queries and answers used in the MIEL query language and then the query processing of the fuzzy CG base corresponding to the fuzzy RDF annotations.

4.1 The Views

In the MIEL query language, a query is asked in a view which corresponds to a given relation of the ontology. A view is a pre-written query allowing the system to hide the complexity of the database schema. A view is characterized by its set of queryable attributes and by its actual definition. Each queryable attribute corresponds to a type of the relation associated with the view. In the CG querying system, a view is defined by means of a *view graph*. A view graph is a pre-defined query graph which has to be instantiated in order to become an actual query graph.

Definition 4. A *view graph* VG associated with a view V having n queryable attributes $a_1, \ldots, a_n$ is a couple {G, C} where G is an acyclic conceptual graph materializing the view and C={ $c_{a_1}, \ldots, c_{a_n}$} a set of distinct generic concept vertices corresponding to the queryable attributes of V. The type of each generic concept vertex c_{a_i} must correspond to the type associated with the attribute a_i.

4.2 The Queries

In the MIEL query language, a query is an instanciation of a given view by the end-user, by specifying, among the set of queryable attributes of the view, which are the selection attributes and their corresponding searched values, and which are the projection attributes. Since the CG base contains fuzzy values generated by the annotation process, the query processing has to deal with two new problems (compared with the query processing presented in [2]): taking into account the pertinence score associated with the semantic relations identified in Web data tables (see section 2.2) and comparing a fuzzy set expressing querying preferences to a fuzzy set, generated by the annotation process, having a semantic of similarity or imprecision. For the first problem, the end-user may specify a *threshold* which determines the minimum acceptable pertinence score to retrieve the data. The second problem is studied in section 4.3.

Definition 5. A MIEL *query* Q asked on a view V defined on n attributes $\{a_1, \ldots, a_n\}$ is defined by $Q = \{V, P, S, thresh\}$ where $P \subseteq \{a_1, \ldots, a_n\}$ is the set of projection attributes, $S = \{s_1, \ldots, s_m\}$ is the set of conjunctive selection attributes and $thresh$ is the minimum acceptable pertinence score for the relation represented by the view V. Each selection attribute s_i is associated with an approximative equality $(a_i \approx v_i)$ between an attribute $a_i \in \{a_1, \ldots, a_n\}$ and its searched value v_i which can be crisp or fuzzy and must be defined on a subset of the definition domain of a_i.

Example 1. *The query Q is expressed in the view GrowthParameterAw :* $Q = \{Microorganism, aw | (GrowthParameterAw(Microorganism, aw) \wedge (Micro-organism \approx MicroPreferences) \wedge (aw \approx awPreferences) \wedge (thresh \geq 0.5)\}$. *The fuzzy set MicroPreferences, which is equal to {1.0/Gram+, 0.5/Gram-}, means that the end-user is firstly interested in microorganisms which are Gram+ and secondly Gram-. The fuzzy set awPreferences, which is equal to [0.9, 0.94, 0.97, 0.99], means that the end-user is first interested in aw values in the interval [0.94, 0.97] which corresponds to the kernel of the fuzzy set. But he/she accepts to enlarge the querying till the interval [0.9, 0.99] which corresponds to the support of the fuzzy set. GrowthParameterAw relations having a pertinence score inferior to 0.5 are discarded.*

When a MIEL query Q is asked in the CG system, the view graph associated with the view V of Q is specialized by instantiating some of its concept vertices in order to take into account the selection attributes. The result is a *query graph* defined as follows:

Definition 6. Let $Q = \{V, P, S, thresh\}$ be a MIEL query and $VG = \{G, C\}$ the view graph associated with V. Let $(a \approx v)$ be a selection attribute of Q. Let

c_r be the generic concept vertex associated with the relation r represented by the view V in G. Let c_a be the generic concept vertex associated with a in G and c_v the concept vertex linked to c_a by the relation vertex *IsAnnotatedBy*. The *query graph* is obtained by a specialisation of G as follows: (i) if a is a numerical type, c_v is an individual concept vertex of type *NumVal* of which the crisp or fuzzy individual marker corresponds to v; (ii) if a is a symbolic type, c_v is a generic concept vertex of which the crisp or fuzzy type corresponds to v; (iii) moreover, an individual concept vertex of type *Numval* and of fuzzy individual marker $[thresh, 1]$ is linked to c_r by the relation vertex *HasForScore*.

Example 2. *A query graph corresponding to the query of example 1 is presented in figure 5.*

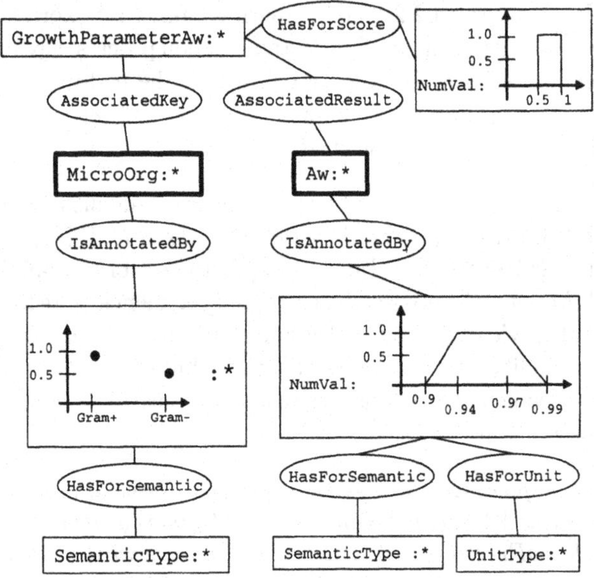

Fig. 5. The selection/projection attributes are framed in bold. One of the selection attribute values is expressed by a generic concept vertex with a fuzzy type, the second one is expressed by an individual concept vertex with a numerical fuzzy marker.

4.3 The Answers

An answer to a MIEL query Q must (1) satisfy the minimal acceptable pertinence score of Q; (2) satisfy all the selection attributes of Q in the meaning of definition 7 which is presented below and (3) associate a constant value with each projection attribute of Q. To measure the satisfaction of a selection attribute, we have to consider the two semantics -imprecision and similarity- associated with fuzzy values of the CG base. On the one hand, a fuzzy set having a semantic of imprecision is a normalized fuzzy set (its kernel is not empty) representing an ordered disjunction of exclusive possible values. In the framework of the possibility theory (see [9]), two classical measures have been proposed to compare a fuzzy set representing preferences to a fuzzy set representing an imprecise datum: a possibility degree of matching and a necessity degree of matching. On the other hand, a fuzzy set having a semantic of similarity represents an ordered conjunction of similar values. The comparison of a fuzzy set representing preferences to a fuzzy set representing similarity is not a classical problem studied in the bibliography. We propose to use the measure proposed in [10] to make such a comparison.

Remark 2. *When the fuzzy value of a selection attribute has a hierarchized symbolic definition domain, it is represented by a fuzzy set defined on a subset of its definition domain. Such a fuzzy set defines degrees implicitly on the whole definition domain of the selection attribute. In order to take those implicit degrees into account, the* **fuzzy set closure** *has been defined in [11].*

Definition 7. Let $(a \approx v)$ be a selection attribute, v' a value of the attribute a stored in the CG base, $sem_{v'}$ the semantic of v' (similarity or imprecision), μ_v and $\mu_{v'}$ being their respective membership functions defined on the domain *Dom* and *cl* the function which corresponds to the fuzzy set closure. The comparison result depends on the semantic of the fuzzy set: If $sem_{v'} = imprecision$, the comparison result is given by the **possibility degree of matching** between $cl(v)$ and $cl(v')$ noted $\Pi(cl(v), cl(v')) = sup_{x \in Dom}(min(\mu_{cl(v)}(x), \mu_{cl(v')}(x)))$ and the **necessity degree of matching** between $cl(v)$ and $cl(v')$ noted $N(cl(v), cl(v'))$ $= inf_{x \in Dom}(max(\mu_{cl(v)}(x), 1 - \mu_{cl(v')}(x)))$. If $sem_{v'} = similarity$, the comparison result is given by the **adequation degree** between $cl(v)$ and $cl(v')$ noted $ad(cl(v), cl(v')) = sup_{x \in Dom}(min(\mu_{cl(v)}(x), \mu_{cl(v')}(x)))$.

As the comparison results associated with the selection attributes are not commensurable, it is not possible to aggregate them in order to deliver a total order of the anwers. A partial order could be defined, but we consider that it is not easy to be interpreted by the end-user. So, we propose to aggregate, using the *min* operator (which is classically used to interpret the conjunction), the comparison results which correspond to fuzzy sets having the same semantic (similarity or imprecision). Therefore, each answer is associated with four scores: a pertinence score ps_r associated with the relation r, a global adequation score ad_g associated with the comparison results having a semantic of similarity and two global matching scores Π_g and N_g associated with the comparison results having a semantic of imprecision. Based on those scores, we propose to define a total order on the answers which gives greater importance to the most pertinent answers compared with the ontology. Thus, the answers are successively sorted according to: firstly, ps_r the pertinence score associated with the relation; secondly, ad_g which indicates in which extend the terms of the data cells are similar to the terms of the ontology (see section 2.2); and thirdly a total order defined on Π_g and N_g. In order to obtain a total order on the comparison results having a semantic of imprecision, we consider, as proposed in [12], that the necessity degree is of greater importance than the possibility degree (Π_g). Indeed, when the necessity degree (N_g) is positive, we are (more or less) certain that the item matches the requirement. So, the answers are first sorted on the values of their N_g and then on their Π_g in case the N_g values are equal.

Definition 8. An *answer* to a MIEL query $Q = \{V, P, S, thresh\}$ is a set of tuples, each of the form $\{ps_r, ad_g, \Pi_g, N_g, v_1, \ldots, v_l\}$, where ps_r is the pertinence score associated with the relation r represented by the view V; ad_g the global adequation score (see definition 9) associated with the comparison results having a semantic of similarity; Π_g and N_g the global matching scores (see definition 9) associated with the comparison results having a semantic of imprecision; and where $v_1, \ldots, v_l$ correspond to the crisp or fuzzy values associated with each projection attribute $a_i \in P$.

Example 3. *The answer to the query of example 1 (corresponding to the query graph of figure 5) considering the content of the Web data table of figure 2 is* : $\{\{ps_r = 1, ad_g = 0.5, N_g = 1, \Pi_g = 1,$ *Microorg=(0.5/Clostridium Perfringens+0.5/Clostridium Botulinum), aw=[0.943, 0.95, 0.96, 0.97]*$\}$, $\{ps_r = 1, ad_g = 0.5, N_g = 0.5, \Pi_g = 0.68,$ *Microorg=(0.5/Staphylococcus spp.+0.5/Staphylococcus aureus), aw=[0.88, 0.98, 0.98, 0.99]*$\}$, $\{ps_r = 1, ad_g = 0.5, N_g = 0, \Pi_g = 0.965,$ *Microorg=(1.0/Salmonella), aw=[0.94, 0.99, 0.99, 0.991]*$\}\}$.

4.4 The CG Query Processing

The query processing of a MIEL query Q consists in selecting the view graph associated with the view V of Q, building the corresponding query graph G, and "approximate"-projecting G into the CG base. As a matter of fact, the query processing in the CG system consists in searching for CGs which contain a more precise information than the information contained in the query (we search for specializations of the query graph) or, at least, for CGs which contain "approximate" answers. In order to find such CGs, we propose to use the "approximate"-projection operation which is a flexible mapping operation between two CGs. The "approximate"-projection is adapted from the classic projection operation using the comparison results presented in definition 7.

Definition 9. An *"approximate"-projection AP* from a conceptual graph G into a CG G' is a tuple (f, g, ad_g, N_g, Π_g), f (resp. g) being a mapping from the relation (resp. concept) vertices of G into the relation (resp. concept) vertices of G' such that: (i) the edges and their labels are preserved; (ii) the labels of the relation vertices can be specialized; (iii) each concept vertex c_i of G has an image $g(c_i)$ such that if $g(c_i)$ is linked to a generic concept vertex of type *Imprecision* (resp. *Similarity*) by a relation vertex of type *HasForSemantic*, then $g(c_i)$ satisfies c_i with the degrees $N(c_i, g(c_i))$ and $\Pi(c_i, g(c_i))$ (resp. the degree $ad(c_i, g(c_i))$) of Definition 7. The global matching scores (N_g, Π_g) and the global adequation score ad_g between G and G' are computed as follows: $N_g = min_i(N(c_i, g(c_i)))$ and $\Pi_g = min_i(\Pi(c_i, g(c_i)))$, with $1 \leq i \leq nb_{imp}$ (nb_{imp} being the number of concept vertices in G which have an image in G' having a semantic of imprecision); $ad_g = min_i(ad(c_i, g(c_i)))$, with $1 \leq i \leq nb_{sim}$ (nb_{sim} being the number of concept vertices in G which have an image in G' having a semantic of similarity).

The "approximate"-projection allows one to retrieve from the CG base a set of answer graphs which must satisfy the threshold *thresh* defined in the query.

Definition 10. Let G be a query graph associated with a MIEL query $Q=\{V, P, S, thresh\}$. Let AP be an "approximate"-projection from G into a CG base H and A_G a partial subgraph of H. Let c_r be the generic concept vertex associated with the relation r represented by the view V in G. A_G is an *answer graph* to the query graph G if there exists a surjective "approximate"-projection AP' from G into A_G such that $AP' \subseteq AP$, $N(c, g(c)) = 1$ and $\Pi(c, g(c)) = 1$ where c is the fuzzy individual concept vertex of type *Numval* associated with c_r by the relation vertex *HasForScore* and $g(c)$[2] its image by AP'.

An answer to a query Q is then a set of tuples, each tuple being built from an answer graph A_G of the query graph G associated with Q as defined below.

Definition 11. Let $Q = \{V, P, S, thresh\}$ be a MIEL query and G the query graph associated with Q. Let A_G be an answer graph of G. An *answer* to the query Q is a set of tuples, each of the form $\{ps_r, ad_g, \Pi_g, N_g, v_1, \ldots, v_l\}$ where:

- ps_r is the pertinence score associated with the relation r represented by the view V of Q. Let c_r be the generic concept vertex associated with the relation r in G and c'_r be the generic concept vertex image of c_r in A_G. The pertinence score is the individual marker of the individual concept vertex of type *Numval* linked to c'_r by the relation vertex *HasForScore* in A_G;
- ad_g, Π_g and N_g are the global scores associated with the "approximate"-projection AP from G into A_G as defined in definition 9.
- $v_1, \ldots, v_l$ correspond to the values associated with each projection attribute $a_i \in P$. Let c_{a_i} be the generic concept vertex associated with a_i in G. Let c'_{a_i} be the generic concept vertex image of c_{a_i} in A_G and c'_{v_i} the concept vertex linked to c'_{a_i} by the relation vertex *IsAnnotatedBy*. The value v_i associated with the projection attribute a_i is: (i) if a_i is a symbolic type, v_i is the fuzzy type of the generic concept vertex c'_{v_i}; (ii) if a_i is a numerical type, v_i is the marker of the individual concept vertex c'_{v_i} of type *NumVal*;

Example 4. *The answer to the query graph of figure 5 asked on the CG of figure 4 is: \{ ps_r=1.0, $ad_g = 0.5$, $\Pi_g = 1.0$, $N_g = 0.0$, Microorg=(0.5/Clostridium Perfringens+0.5/Clostridium Botulinum), aw=[0.943, 0.95, 0.96, 0.97] \}.*

5 Conclusion

In this paper, we have proposed a flexible querying system of fuzzy RDF annotations based on (i) translation rules of fuzzy RDF annotations into fuzzy CGs and (ii) an "approximate"-projection operation which is able to compare fuzzy CG queries with fuzzy CG annotations. A very next step will be to experiment

[2] $g(c)$ is considered as a particular case of a fuzzy individual concept vertex of type *Numval* having a semantic of imprecision.

the performances and to test the genericity of this system using different corpus of annotated Web data tables (risk in food and aeronautics). This work is done in the framework of the WebContent project financed by the French National Research Agency (ANR). In this project, the flexible querying system will be incorporated in a decision support system in the field of risk in food. Its added value will be to complement data retrieved from local databases unsufficient for statistical analysis with additional data retrieved from the Web.

References

1. Corby, O., Dieng, R., Hébert, C.: A conceptual graph model for w3c resource description framework. In: ICCS 2000. LNCS, vol. 1867, pp. 468–482. Springer, Heidelberg (2000)
2. Buche, P., Dibie-Barthélemy, J., Haemmerlé, O., Thomopoulos, R.: The MIEL++ Architecture When RDB, CGs and XML Meet for the Sake of Risk Assessment in Food Products. In: Schärfe, H., Hitzler, P., Øhrstrøm, P. (eds.) ICCS 2006. LNCS (LNAI), vol. 4068, pp. 158–171. Springer, Heidelberg (2006)
3. Hignette, G., Buche, P., Dibie-Barthélemy, J., Haemmerlé, O.: Semantic annotation of data tables using a domain ontology. In: Corruble, V., Takeda, M., Suzuki, E. (eds.) DS 2007. LNCS (LNAI), vol. 4755, pp. 253–258. Springer, Heidelberg (2007)
4. Zadeh, L.A.: Fuzzy sets. Information and Control 8, 338–353 (1965)
5. Zadeh, L.A.: Fuzzy sets as a basis for a theory of possibility. Fuzzy Sets and Systems 1, 3–28 (1978)
6. Dubois, D., Prade, H.: The three semantics of fuzzy sets. Fuzzy Sets and Systems 90(2), 141–150 (1997)
7. Chein, M., Mugnier, M.L.: Conceptual graphs, fundamental notions. Revue d'Intelligence Artificielle 6(4), 365–406 (1992)
8. Thomopoulos, R., Buche, P., Haemmerlé, O.: Different kinds of comparisons between fuzzy conceptual graphs. In: Ganter, B., de Moor, A., Lex, W. (eds.) ICCS 2003. LNCS, vol. 2746, pp. 54–68. Springer, Heidelberg (2003)
9. Dubois, D., Prade, H.: Possibility Theory - An Approach to Computerized Processing of Uncertainty. Plenum Press, New York (1988)
10. Baziz, M., Boughanem, M., Prade, H., Pasi, G.: A fuzzy logic approach to information retrieval using a ontology-based representation of documents. In: Fuzzy logic and the Semantic Web, pp. 363–377. Elsevier, Amsterdam (2006)
11. Thomopoulos, R., Buche, P., Haemmerlé, O.: Fuzzy sets defined on a hierarchical domain. IEEE Trans. Knowl. Data Eng. 18(10), 1397–1410 (2006)
12. Dubois, D., Prade, H.: Tolerant fuzzy pattern matching: an introduction. Fuzziness in Database Management Systems, 42–58 (1995)

Query-Answering CG Knowledge Bases*

Michel Leclère and Nicolas Moreau

LIRMM, Univ. Montpellier 2, CNRS
161, rue Ada
34392 Montpellier, France
{leclere,moreau}@lirmm.fr

Abstract. Conceptual graphs are a good choice for constructing and exploiting a knowledge base. In several of our projects (semantic portal for e-tourism, exploitation of digital object corpus, etc.), we have to query such bases. So it is natural to consider queries and bases as simple graphs and to compute the set of all projections from a query to a base. However, there is a problem of the return of this set of projections to the user. More generally, the main issue is about the definition of the notion of answers in an query-answering system made of knowledge bases formalized by graphs (Conceptual Graphs, RDF (Resource Description Framework) , Topic Maps, etc.). In this paper, we study several notions of answers and some of their characterizations. We distinguish between notions of answers by subgraphs of the base and answers by creation of result graphs. For the last type of answers, we define completeness, non-redundancy and minimality criteria of the answer sets and propose several notions of answers w.r.t these criteria.

1 Introduction

Many knowledge applications involve the elaboration and use of knowledge bases. Some examples are document management, digital object corpus management, enterprise knowledge repositories, construction of semantic portals, teaching aid management, the semantic web, etc. Two general contexts for using such bases can be noted, whereby use of these bases presupposes a query Q specifying the knowledge to be searched:

- Annotation context: resources are annotated by "descriptions" characterizing it; in this case, the exploitation is based on a search of resources whose annotations contain specific knowledge (e.g. [1,2,3]). This type of exploitation only requires a definition of a deduction notion allowing selection of descriptions D (and thus the resources R linked to these descriptions) which are deductions of the searched knowledge Q. The set of answers to a query Q on an annotation base B is $\{R \mid (R, D) \in B \ \wedge \ D \models Q\}$;
- Knowledge base context: some unstructured data that comply with a formal vocabulary defined by an ontology are stored in a base of assertions [4];

* This work is supported by the Eiffel project, funded by ANR-RNTL.

P. Eklund and O. Haemmerlé (Eds.): ICCS 2008, LNAI 5113, pp. 147–160, 2008.

a querying system allows extraction of knowledge from the base (moreover, this type of exploitation could be used in the annotation context, considering descriptions of a subset of resources). This type of exploitation also needs a deduction notion to characterize the existence of an answer, but it also requires a definition of what should be returned to a query Q on such a knowledge base.

In this paper, we propose a preliminary answer to this question in the framework of knowledge bases formalized by conceptual graphs [5]. The long-term goal of this work is to define a query language for conceptual graphs and to implant it as a knowledge server over the CoGITaNT framework [6]. This first approach is a study of different notions of answers constructed from projections of a query graph Q to a graph base B.

In a knowledge base querying system with a formal semantic, answers are built upon "pre-answers" that are logical consequences of the base proving the existence of answers. A query is often composed of two parts: an head which specifies how answers are constructed from the "pre-answers", and a body which details how to select these "pre-answers". In the case of a base only made of conjunctive assertions, these "pre-answers" are the "smallest parts" of the base which has the body of the query as a logical consequence.

In this work, we only consider such simple bases (*i.e.* without rules) formalized by conceptual graphs [7], although our results could be directly applied to other labeled graph formalisms, particularly to RDF/S knowledge bases of the semantic web [8] (*cf.* [9] for an equivalence of the two formalisms).

Unlike relational databases, these formalisms allow the introduction of an order relation over relations permitting the representation of a simple ontology and, more importantly, the use of variables in the base. In the querying mechanism, this leads to the problem of "pre-answer" equivalence and thus the problem of the definition of the answer notion.

We consider queries whose body is a conceptual graph and focus on the definition of several "pre-answer" notions. The main goal of this preliminary work is to study redundancy problems of these "pre-answers" which arise regardless of following operations applied to these "pre-answers" (e.g. specifyingthe concepts to keep by a lambda, or constructing a new graph with these answers).

As far as we know, this topic has not yet been studied. Several proposals have been made on querying relational databases with conceptual graphs [10,11,12]. Many studies have been conducted in an annotation context (e.g. [13,2]). An adaptation of relational algebra to the context of conceptual graph knowledge bases has been studied by S. Coulondre [14]. The relational bias limits answers to tuples of individuals over which several relational algebra operators are used (moreover, this author did not seem to consider variable-free knowledge bases). The most similar works were carried out by C. Gutierrez et al. [15], who studied querying of knowledge bases of the semantic web formalized by RDF/S, but the set of "pre-answers" is not fully detailed, particularly the problem of answer redundancies.

The following section briefly introduces the main notions of the conceptual graph formalism upon which our work is based. Section 3 defines the querying model. Section 4 proposes several notions of answers and answer criteria based on the answers redundancy problem. Finally, the conclusion proposes some ideas for other types of answers.

2 $\mathcal{SG}$ Formalism

The CG formalism we use in this paper has been developed at LIRMM over the last 15 years [7]. The main difference with respect to the initial general model of Sowa [5] is that only representation primitives allowing graph-based reasoning are accepted. Several extensions that preserve this link with graph theory have been introduced (rules, constraints, conjunctive types, nested graphs, etc.), however, for the sake of clarity and presentation, we only present the simplest model in this paper.

Simple graphs (SGs) are built upon a *support*, which is a structure $S = (T_C,\ T_R,\ I,\ \sigma)$ where T_C is the set of concept types, T_R is the set of relations with any arity (arity is the number of arguments of the relation). T_C and T_R are partially ordered sets. The partial order represents a specialization relation ($t' \leq t$ is read as "t' is a specialization of t"). I is a set of individual markers. The mapping σ assigns a signature to each relation specifying its arity and the maximal type for each of its arguments.

SGs are labeled bipartite graphs denoted by $G = (C_G, R_G, E_G, l_G)$ where C_G and R_G are the concept and relation node sets respectively, E_G is the set of edges and l_G is the mapping labeling nodes and edges. Concept nodes are labeled by a couple $t : m$ where t is a concept type and m is a marker. If the node represents an unspecified entity, its marker is the generic marker, denoted $*$, and the node is called a *generic* node, otherwise its marker is an element of I, and the node is called an *individual* node. Relation nodes are labeled by a relation r and, if n is the arity of r, it is incidental to n totally ordered edges.

A graph $G = (C_G, R_G, E_G, l_G)$ is *consistent* w.r.t. a support $S = (T_C, T_R, I, \sigma)$ if :

- the labels of the concept nodes (resp. relation nodes) belong to $(T_C \times (I \cup \{*\}))$ (resp. T_R);
- the relation nodes satisfy their signatures defined by σ.
- for each individual marker i of G, types of concept nodes with this marker have a greatest lower bound. This condition can differ if one considers a conformity relation in the support, if one imposes a lattice structure to the ordered set of concept types, if banned types are considered (a disjointness type axiom), etc.

A specialization/generalization relation corresponding to a deduction notion is defined over SGs and can be easily characterized by a graph homomorphism called *projection*. When there is a projection π from G to H, H is considered to be more specialized than G, denoted $H \leq G$. More specifically, a projection π from G to H is a mapping from C_G to C_H and from R_G to R_H, which preserves edges (if there is an edge numbered i between r and c in G then there is an edge numbered i between $\pi(r)$ and $\pi(c)$ in H) and may specialize labels (by observing type orders and allow substitution of a generic marker by an individual one).

Conceptual graphs are provided with a first-order-logic semantics, defined by a mapping denoted Φ.

The fundamental result of *projection soundness and completeness* establishes the equivalence between projection and deduction on formulas assigned to SGs: given two SGs G and H on a support S, there is a projection from G to H if and only if $\Phi(G)$ can be deduced from $\Phi(H)$ and $\Phi(S)$. Completeness is obtained up to a condition on H: H has to be in a normal form, so any individual marker must appear at most once in it (i.e. a specific entity cannot be represented by two nodes). An SG consistent w.r.t. a support can be easily normalized by joining concept nodes with a same individual marker. The normal form of a consistent graph G is denoted $norm(G)$.

Two notions of equivalence can be defined over SGs: a syntactic equivalence, and a semantic one. The syntactic equivalence is characterized by the existence of an isomorphism between two graphs (which is a bijective mapping from nodes of one of the graphs to nodes of the other preserving edges and without label specialization). The semantic equivalence is characterized by the existence of a projection from the first graph to the normal form of the second, and from the second to the normal form of the first and corresponds to a logic equivalence between formula associated with graphs: $\Phi(S) \models \Phi(G) \leftrightarrow \Phi(H)$ *iff there is a projection from G to* $norm(H)$ *and a projection from H to* $norm(G)$ (denoted $G \equiv H$).

This equivalence relation defines classes of equivalent SGs. SGs in figure 1 are from the same equivalence class. In each class, some graphs contain useless knowledge repetitions (redundancies) and there is a sole smallest graph with no redundancy, called the irredundant graph of the class (cf. [16]). A graph that is not in normal form contains redundancies (if two concept nodes have the same individual marker).

A subSG $H = (C_H, R_H, E_H, l_H)$ of an SG $G = (C_G, R_G, E_G, l_G)$ is an SG, where :

- $C_H \subseteq C_G$ and $R_H \subseteq R_G$
- E_H is a restriction of E_G to elements of $C_H \times R_H$
- l_H is a restriction of l_G to elements of H.

A strict sub-SG of G is a sub-SG with a strictly inferior number of nodes.

Characterization: An SG is said to be *redundant* if it is not in normal form or if it is equivalent to one of its strict sub-SGs. Otherwise, it is said to be *irredundant*,[1].

Property: [16] An equivalence class contains one and only one irredundant SG, which is the graph (single up to isomorphism) with the smallest set of nodes.

An algorithm to compute the irredundant form of an SG, whose complexity is polynomially related to the complexity of the projection algorithm, has been described in [17].

[1] Note that our irredundant definition is stricter than that given in [7].

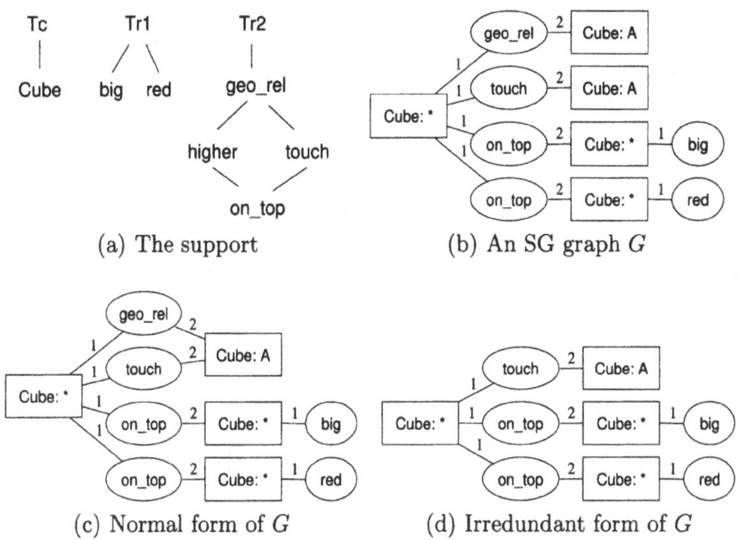

(a) The support (b) An SG graph G

(c) Normal form of G (d) Irredundant form of G

Fig. 1. Three equivalent SGs built on a support

3 Studied Querying Model

The chosen context is a knowledge base composed of assertions of entity existences and relations over these entities, called *facts*, and stored in a single graph (not necessarily connected) consistent w.r.t. a given support. A support can be seen as a basic ontology. This framework does not put forward hypotheses on how facts have been collected and does not prohibit the existence of metadata (date, etc.) on facts composing the base (e.g. a selection mechanism of facts to be queried based on metadata). The only important hypothesis put forward is that all of the facts of the base to be queried are consistent relative to each other (with respect to individual markers) and consistent w.r.t. the support. Figure 1(a) presents the support used in the following examples of the paper.

The SG *base* is assumed to be normalized. The base can be redundant because the irredundancy of the base does not solve the problem of answer redundancies, and computation of the irredundant form is expensive as the base can be large. Computation of the normal form is linear in the size of the base. Moreover, it is easy to write an incremental algorithm (called for each addition of knowledge in the base and whose complexity depends only on the size of the addition) for the normalization, whereas it seems difficult to find such an efficient algorithm for computation of the irredundant form.

From propositions of querying languages proposed for such knowledge base (e.g. SPARQL [18]), their definitions are clearly based on several mechanisms:

1. Adaptation of the base to the query, which consists of computing a base D' from a base D by application of a set of updating operations P (e.g. rule applications).

2. Selection of relevant parts of the base that respond to the query. This selection is made of "patterns" which specify selection criteria of base parts useful for contruction of the answer. These patterns have the same formalism as the base or indirectly of the same formalism because of the addition of variables. Thus these patterns are used like filters to select relevant base parts.
3. Verification of properties allowing to impose complementary selection criteria differing from a simple assertion of relations between entities : path existence, constraints (not present in the representation language) on the entity linked to a variable.
4. Construction of an answer from these parts, which is a specification of the type of answer to return (tuples of values associated with query variables, a graph built from all parts, the number of answers, etc.).

The second point is the core of the querying mechanism. Constructing a query boils down to making assertions with unknown values (variables) which is information to be retrieved. When the formalism allows the introduction of variables in the base, it is important to know what to do when these variables are linked with query variables.

In our formalism, the selection criterion is a given SG Q, called the *query* SG. There is no constraint on the query (in terms of relevance, normalization or irredundancy). However, it seems natural to verify the consistency of the query w.r.t. the support of the base to avoid queries with no links to the base. One can consider to compute the irredundant form of the query as the size of the query is generally small. One can consider an unconnected query as several queries.

Therefore as one considers formalisms provided with a formal semantic, one can define "relevant parts" as "the smallest subgraphs" of the base whose query graph is a logical consequence; such subgraphs are called *pre-answers* in [15]. In conceptual graph formalism, the existence of an answer is directly linked with the existence of a projection and a pre-answer is just the query image of a projection.

Definition 1 (Proofs of answers). *Let B be an SG base and Q an SG query, a proof of answer is a projection π from Q to B. The set of proofs of answers from Q to B is denoted $\Pi(Q,B)$: $\Pi(Q,B) = \{\pi_i \mid \pi_i : Q \longrightarrow B$ is a projection$\}$.*

Definition 2 (Images of proofs). *An image of a proof (or pre-answer), denoted $\pi(Q)$, is a sub-SG of B, image of the proof of answer π from Q to B. The images of proofs sequence from a query Q to a base B, denoted $IP(Q,B)$, is $IP(Q,B) = \langle \pi_1(Q), ..., \pi_n(Q) \rangle$, where n is the size of $\Pi(Q,B)$.*

All examples of the paper are from base and query of figure 2. Base is the SG of the figure 1(d). There are six projections of the query to the base, and thus six images of proofs. We have named some vertices (c_1, r_2, etc.) of graphs to refer directly to one vertex or to distinguish different subgraphs (see figure 3).

Therefore answers are based on images of proofs, one may have to return the same subgraph of B several times. In the example of figure 2, projections $\pi_i = \{(c_a, c_1), (r_b, r_2), (c_b, c_3), (r_c, r_3), (c_c, c_4)\}$ and $\pi_j = \{(c_a, c_1), (r_b, r_3), (c_b, c_4),$

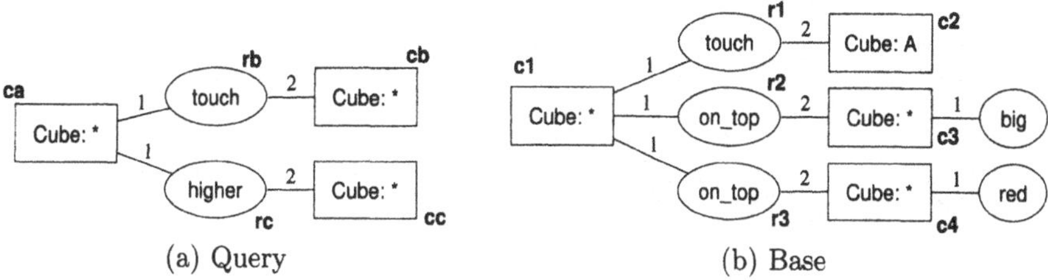

Fig. 2. Base and query (with named vertices)

$(r_c, r_2), (c_c, c_3)\}$ define the same subgraph (R_5 on figure 3). One can choose two ways to solve this problem:

- Considering it as the same answer (several proofs for the answer)
- Differentiating answers by representing the projection in answer graphs (e.g. by adding id of query vertices to their images vertices in answers).

We consider only the first case, as we want to express answers in the same language (structure and vocabulary) as the base and query.

4 Different Notions of Answers

In this section we study several kinds of answers. The first type of notion is to keep base subgraphs, similary to images of proofs.

4.1 Answering by Base Subgraphs

The most basic answer that can be returned is the set of images of proofs.

Definition 3 (Answer by image subgraphs). *The set of images of proofs of a query Q in a base B, noted $R_{IP}(Q, B)$, is $R_{IP}(Q, B) = \{\pi(Q) \mid \pi \in \Pi(Q, B)\}$.*

This answer notion can be used to select exploration start points of the base (by focusing a subgraph of B), to explore the base or to be a first step of base-updating queries. Figure 3 shows all of the five subgraphs answering the query.

4.2 Answering by Base-Independent Graphs

In many cases, the query language should allow knowledge extraction rather than "pinpointing" knowledge in the base. Answers are "copies" of base subgraphs. Thus, answers result from the construction of isomorphic graphs of images of proofs.

Since answer graphs are constructed up to an isomorphism, two isomorphic graphs should be considered equal. The set of answers is no longer a set of subgraphs of B, but rather a set of graphs isomorphic to images of proofs subgraphs of B. In such a set of answer graphs there are no two isomorphic graphs.

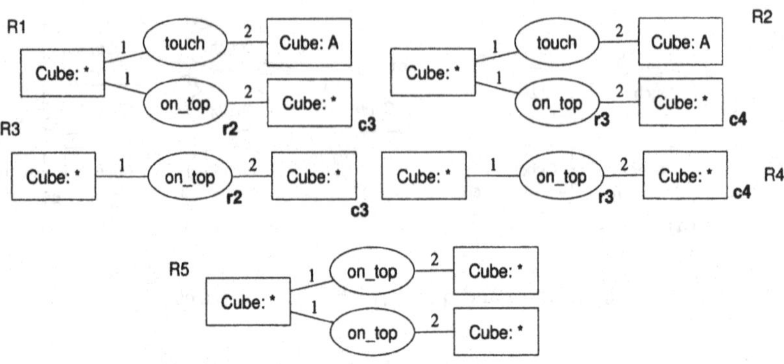

Fig. 3. The five subgraphs composing $R_{IP}(Q, B)$

Definition 4 (Graph set (iso-set)). *A graph (iso-)set* $\{G_1, ..., G_n\}$ *is a set of graphs in which, for all* i, j *with* $i \neq j$, G_i *is not isomorphic to* G_j.

Hereafter, the term "set of graphs" is short for the preceding iso-set notion (a set has no more two isomorphic graphs)[2].

An iso-answer is an answer notion corresponding to computation of all isomorphic graphs to images of proofs. On the example, $R_{ISO}(Q, B)$ is equal to graphs in figure 4.

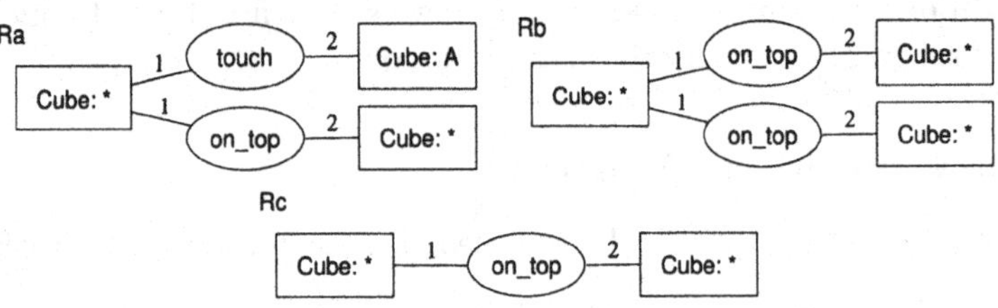

Fig. 4. Graphs composing $R_{ISO}(Q, B)$

Definition 5 (Iso-answer). *An iso-answer from* Q *to* B *is an isomorphic graph corresponding to an image of a proof from* Q *to* B. *The set of iso-answers is* $R_{ISO}(Q, B) = \{G \mid \text{where } G_i \in R_{IP}(Q, B) \text{ with } G_i \text{ isomorphic to } G\}$

With this answer notion, the link with the base is lost (particularly when the answer has no individual marker) since it is unknown which nodes of the base corresponds to a generic node of an answer, or even how many images of proofs correspond to the same answer graph[3]. Only the proof of the existence of a

[2] The images of proofs sets can contain isomorphic subgraphs (if they are not the same subgraph of the base).

[3] The open world assumption of knowledge bases does not indicate whether two different but isomorphic images of proofs represent the same "situation" of the world, or two similar "situations".

particular knowledge in the base is preserved with this notion. Thus, one can question the relevance of only considering isomorphism (i.e. a syntactic equivalence) as an equivalence criterion. In our example, graphs R_b and R_c in figure 4 both state that "there is a cube on top of a cube".

It seems advisable to detect equivalent answers (in terms of semantics) and to only keep one graph of each equivalent class. We define a criterion of such a set of answers :

Definition 6 (Without equivalence criterion). *A set of answers $R(Q,B) \subseteq R_{ISO}(Q,B)$ is "without equivalence" iff $\forall G_i \in R(Q,B)$, $\nexists G_j \in R(Q,B)$ with $G_j \neq G_i$ with $G_i \equiv G_j$.*

Two types of equivalences arise. The first is from the implicit relation of equality of two nodes having the same individual marker. This type of equivalence is avoided by the base normalization; therefore all images of proofs are in normal form (even if the query is not normalized). The second comes from the intrinsic redundancies of the language, and cannot be handled previously, because even if the base and query are irredundant, proof subgraphs are not necessarily irredundant (in the example, image of a proof R_5 is redundant despite that base and query are irredundant).

In each equivalence class, there is a single graph (up to isomorphism) that is the smallest graph of his class: it is the irredundant form. It seems natural to choose this graph to represent one equivalence class of images of proofs. In the example, $R_{IRR}(Q,B)$ is equal to graphs R_a and R_c in figure 4.

Definition 7 (Irredundant answers). *The set of irredundant forms of images of proofs from Q to B is $R_{IRR}(Q,B) = \{Irr(G) \mid G \in R_{ISO}(Q,B)\}$.*

This notion of irredundant answers may seem strange because one may think that an answer could not be isomorphic to an image of a proof. The following property holds that the irredundant form of each image of a proof is itself an answer (up to isomorphism) and that the set $R_{IRR}(Q,B)$ does not have equivalent answers.

Proposition 1. *$R_{IRR}(Q,B) \subseteq R_{ISO}(Q,B)$ is "without equivalence".*

Proof. For all $G \in R_{IRR}(Q,B)$, there is $G' \in R_{IP}(Q,B)$ such that G is isomorphic to $Irr(G')$. So there is a projection π' from Q to B such that $\pi'(Q) = G'$. There is also, by definition of the irredundant form, a projection π_r from G' to B such that $\pi_r(G') = Irr(G')$. Thus there is a projection $\pi = \pi_r \circ \pi'$ from Q to B such that $\pi(Q) = Irr(G')$ and so isomorphic copy of $Irr(G')$ is an answer of $R_{ISO}(Q,B)$. The unicity property of the irredundant form (up to isomorphism) in each equivalence class and our notion of graphs defined up to an isomorphism lead to the conclusion that $R_{IRR}(Q,B)$ is without equivalence.

The notion of answers without equivalence is not sufficient because, as one does not want to keep two equivalent answers, only answers that add new knowledge compared to other answers should be conserved. In the example, answer R_a states that "there is a cube that touches cube A and that is on top of a cube" and answer R_b states that "there is a cube on top of a cube", so knowledge of R_b is expressed by R_a. So we introduce a new incomparability criterion in the next section.

4.3 Sets of Incomparable Answers

Two answers are comparable if knowledge stated by the first is deducible[4] from the other. Once there are comparable answers in a set of answers, there is redundancy between them. The idea is to eliminate this redundancy by keeping only incomparable answers.

Definition 8 (Incomparability criterion). *A set of answers $R(Q,B) \subseteq R_{ISO}(Q,B)$ is "without redundancy" when all of its answers are incomparable: $\forall G_i, G_j \in R(Q,B)$ with $G_j \neq G_i$, $G_i \not\leq G_j$.*

Given a set of answers $R_{ISO}(Q,B)$ there is not a single subset "without redundancy" (e.g. $\{R_a\}$ or $\{R_b\}$). A natural constraint is to make such a subset to not remove too many answers, *i.e* it has to be maximal by inclusion.

Definition 9 (Maximality criterion). *A set of incomparable answers $R(Q, B) \subseteq R_{ISO}(Q,B)$ is "maximal" when no other answer can be added without adding redundancy: $\forall G \in R_{ISO}(Q,B) \setminus R(Q,B)$, $\exists G' \in R(Q,B)$ such that $G' \leq G$ or $G \leq G'$.*

This notion still does not ensure unicity of such a subset. One can want to constrain a bit more the notion of maximality by forcing the subset of answers R to be complete, as one wishes that all of the answers of $R_{ISO}(Q,B)$ can be deduced from answers of the subset. We thus extend the logical interpretation operator Φ from SGs to sets of SGs by taking the conjunction of logical formula associated with each graph of the set : so if $R = \{r_1, ...r_n\}$, $\Phi(R) = \Phi(r_1) \wedge ... \wedge \Phi(r_n)$.

Definition 10 (Completeness criterion). *A set of answers $R(Q,B) \subseteq R_{ISO}(Q,B)$ is complete iff $\Phi(S), \Phi(R(Q,B)) \models \Phi(R_{ISO}(Q,B))$.*

We call the *normalized disjoint union* (*NormalizedDisjointUnion*) of a set of graphs E, the normal form of the graph resulting from the join of nodes and edges (and labeling functions) of graphs of E.

Proposition 2. *A set of answers $R(Q,B) \subseteq R_{ISO}(Q,B)$ is complete iff the normalized disjoint union of $R(Q,B)$ is more specialized than all of the answers of $R_{ISO}(Q,B)$, i.e. $\forall G \in R_{ISO}(Q,B)$, $NormalizedDisjointUnion(R(Q,B)) \leq G$.*

The proof is trivial, note that $\Phi(NormalizedDisjointUnion(R(Q,B)) \equiv \Phi(R(Q,B))$.

[4] One could distinguish cases of a simple knowledge inclusion from cases of general deduction by using ontology knowledge, depending on the level of ontology knowledge appropriation by the user.

Corollary 1. *A set of complete incomparable answers is maximal.*

The completeness criterion still does not ensure unicity of a subset of answers (because of equivalent answers). A natural choice is to take the smallest set of answers. This leads to the definition of a minimality criterion based on a notion of answer size and equivalence of answers sets.

Definition 11 (Answer size). *The size of a set of answers $R(Q, B)$ is the sum of the number of nodes of all answers: $\Sigma_{g \in R(Q,B)} card(g)$.*

Definition 12 (Answers sets equivalence). *Two sets of answers $R(Q, B)$ and $R'(Q, B)$ of a query Q on base B and consistent w.r.t. a support S are equivalent iff $\Phi(S) \models \Phi(R(Q, B)) \leftrightarrow \Phi(R'(Q, B))$.*

Definition 13 (Minimality criterion). *A set of answers $R(Q, B) \subseteq R_{ISO}(Q, B)$ is minimal iff there is not an equivalent set of answers with a strictly smaller size.*

Proposition 3. *$R(Q, B)$ is minimal iff $R(Q, B) \subseteq R_{IRR}(Q, B)$.*

Corollary 2. *Each minimal set is unique (up to isomorphism).*

The previous proposal and its corollary can be easily deduced from irredundant graph properties.

Completeness and minimality constraints define a notion of answer that seems more appropriate when one searches to retrieve knowledge stored in a knowledge base.

Definition 14 (Most specific answers). *The set of the most specific irredundant answers is $R_{MIN}(Q, B) = \{G \in R_{IRR}(Q, B) \mid \nexists G' \neq G \in R_{IRR}(Q, B)$ with $G' \leq G\}$.*

On the example, $R_{MIN}(Q, B)$ is equal to graph R_a in figure 4.

Theorem 1. *$R_{MIN}(Q, B)$ is the only complete minimal and incomparable subset of answers of $R_{ISO}(Q, B)$.*

Proof. • Incomparability: $R_{MIN}(Q, B)$ is by definition composed of the most specific elements of $R_{IRR}(Q, B)$, and as two elements of $R_{IRR}(Q, B)$ are not equivalent, all of the answers of $R_{MIN}(Q, B)$ are incomparable. • Completeness: $R_{IRR}(Q, B)$ is complete because each answer of $R_{ISO}(Q, B)$ is equivalent to a graph of $R_{IRR}(Q, B)$. By the definition of $R_{MIN}(Q, B)$, one deletes in $R_{IRR}(Q, B)$ only G_i which is more general than another graph of $R_{IRR}(Q, B)$. Thus, $R_{MIN}(Q, B)$ is complete. • Minimality and unicity: $R_{MIN}(Q, B)$ is a subset of $R_{IRR}(Q, B)$, thus it is minimal and unique w.r.t. corollary 2. □

One can define an answer notion like that on the most specific element, but this time with the most general ones. This notion is like a "summary" of the set of answers: not all of the answers are returned, but a minimal subset generalizing all of the answers.

Definition 15 (Summary). *A set of answers $R(Q,B) \subsetneq R_{ISO}(Q,B)$ is a summary iff it is minimal and it generalizes all of the answers of $R_{ISO}(Q,B)$, that is $\forall G \in R_{ISO}(Q,B)$, $\exists G' \in R(Q,B)$ such that $G' \geq G$.*

Definition 16 (Maximal answer). *The set of all of the most general irredundant answers is $R_{MAX}(Q,B) = \{G \in R_{IRR}(Q,B) \mid \nexists G' \neq G \in R_{IRR}(Q,B)$ with $G' \geq G\}$.*

Property 1. $R_{MAX}(Q,B)$ is the sole minimal and maximal subset of incomparable answers of $R_{ISO}(Q,B)$.

The proof is similar to the proof of property 1. In the example, $R_{MAX}(Q,B)$ is equal to graph R_c in figure 4.

4.4 Case of Bases Composed of (Only) Individual Concepts

A special case is when a knowledge base does not contain any variables (corresponding to relational databases). In such bases, there is only a kind of redundancy from redundant relations between concepts (relations whose type is comparable and with the same ordered set of neighbors). Computation of the irredundant form is thus linear and incremental.

Proposition 4. *In an irredundant base whose concepts are all individual, all of the images of proofs of any query on this base are non-isomorphic and irredundant.*

Proof. A base in normal form and which only contains concepts that are individuals does not contain concept nodes with the same label (nor comparable concept nodes in terms of specialization/generalization). Moreover, as the base is irredundant, it does not contain redundant relations between concepts (all the more with the same label). So there is not any isomorphism (or projection) from a subgraph of the base to another (except identity relation). Images of proofs are base subgraphs so they are non-isomorphic. They are irredundant because none of the base subgraph can be projected in one of its strict subgraphs (since there is no projection from a base subgraph in another, except identity). □

Corollary 3. *If B is irredundant, for any query Q there is a bijection from the set of images of proofs $R_{IP}(Q,B)$ to their copies $R_{ISO}(Q,B)$, and $R_{ISO}(Q,B) = R_{IRR}(Q,B)$.*

5 Conclusion

In this paper we define two main notions of answers to a query in the knowledge base querying framework: the first is composed of base subgraphs that can allow browsing in the knowledge base or that can be used as a first step to update queries; the second consists of graphs independent of the base. We define several good criteria for this last notion: the non-equivalence of answers, the incomparability of answers, the completeness of a set of answers, and the minimality (in terms of size) of a set of answers. The most interesting notion of answer w.r.t. these criteria is the set of the most specific irredundant answers. However, in one of our projects, we need the notion of answers of the set of most general irredundant answers, as these answers are used as the body of new queries in another knowledge base.

An another answer notion, not developed in this paper is the construction of a graph resulting from the disjoint sum of all answers. the definition of such a notion gives a closed querying system (query, knowledge base and answer are in the same formalism) and allows us to reuse the result of a query as a knowledge base (nested queries).

Finally, in this paper, we overcome redundancies between answers by deleting answers. Another possibility is to overcome redundancies (when it is not a redundancy of the knowledge base itself) by completing answers by the addition of knowledge from the base allowing to differentiate redundant answers. We are currently working on a definition of such an answer contextualization mechanism. This kind of mechanism seems relevant for such knowledge bases because, contrary to relational databases in which a hypothesis is put forward that the creator of the request knows the schema of the database, they are by definition weekly structured and the only reasonable hypothesis put forward is that the creator can verify that his query is correct w.r.t. the ontology. A contextualization mechanism thus allows to respond to queries of the type: *"What knowledge can I have on animals owned by Mary?"* with a set of such answers : { *"Mary owns a pedigree animal"*, *"Mary owns a cat offered by her father"*, *"The preferred animal that Mary owns is a cat"*} rather than the only answer *"Mary owns a cat"*.

References

1. Hollink, L., Schreiber, A., Wielemaker, J., Wielinga, B.: Semantic annotation of image collections. Knowledge Capture, 41–48 (2003)
2. Moreau, N., Leclère, M., Chein, M., Gutierrez, A.: Formal and graphical annotations for digital objects. In: SADPI 2007: Proceedings of the 2007 international workshop on Semantically aware document processing and indexing, pp. 69–78. ACM, New York (2007)
3. Dieng-Kuntz, R., Corby, O.: Conceptual Graphs for Semantic Web Applications. In: Conceptual Structures: Common Semantics for Sharing Knowledge: Proceedings of 13th International Conference on Conceptual Structures, Kassel, Germany, July 17-22 (2005)
4. Fensel, D., Decker, S., Erdmann, M., Studer, R.: Ontobroker: Or How to Enable Intelligent Access to the WWW. In: Proceedings of the 11th Banff Knowledge Acquisition for Knowledge-Based Systems Workshop, Banff, Canada (1998)
5. Sowa, J.F.: Conceptual Structures: Information Processing in Mind and Machine. Addison-Wesley, Reading (1984)
6. Genest, D., Salvat, E.: A platform allowing typed nested graphes: How cogito became cogitant. In: Mugnier, M.-L., Chein, M. (eds.) ICCS 1998. LNCS (LNAI), vol. 1453, pp. 154–161. Springer, Heidelberg (1998)
7. Chein, M., Mugnier, M.L.: Conceptual Graphs: Fundamental Notions. Revue d'Intelligence Artificielle 6(4), 365–406 (1992)

8. Hayes, P.: RDF Semantics. Technical report, W3C (2004)
9. Baget, J.F.: Rdf entailment as a graph homomorphism, pp. 82–96 (2005)
10. Sowa, J.F.: Conceptual graphs for a data base interface. IBM Journal of Research and Development 20(4), 336–357 (1976)
11. Boksenbaum, C., Carbonneill, B., Haemmerlé, O., Libourel, T.: Conceptual graphs for relational databases. In: Mineau, G.W., Sowa, J.F., Moulin, B. (eds.) ICCS 1993. LNCS, vol. 699, pp. 142–161. Springer, Heidelberg (1993)
12. Haemmerlé, O., Carbonneill, B.: Interfacing a relational databases using conceptual graphs. In: Proceedings of the Seventh International Conference and Workshop on Database and Expert Systems Applications (DEXA 1996), pp. 499–505. IEEE-CS Press, Los Alamitos (1996)
13. Genest, D., Chein, M.: A content-search information retrieval process based on conceptual graphs. Knowl. Inf. Syst. 8(3), 292–309 (2005)
14. Coulondre, S.: Cg-sql: A front-end language for conceptual graph knowledge bases. Knowledge-Based Systems 12(5-6), 205–325 (1999)
15. Gutierrez, C., Hurtado, C., Mendelzon, A.O.: Foundations of semantic web databases. In: PODS 2004: Proceedings of the twenty-third ACM SIGMOD-SIGACT-SIGART symposium on Principles of database systems, pp. 95–106. ACM Press, New York (2004)
16. Mugnier, M.L., Chein, M.: Polynomial algorithms for projection and matching. In: Proceedings of the 7th Annual Workshop on Conceptual Structures: Theory and Implementation, London, UK, pp. 239–251. Springer, Heidelberg (1993)
17. Mugnier, M.: On generalization/specialization for conceptual graphs. Journal of Experimental & Theoretical Artificial Intelligence 7(3), 325–344 (1995)
18. Prud'hommeaux, E., Seaborne, A.: SPARQL Query Language for RDF. Technical report, W3C (2007)

Attribute Exploration Using Implications with Proper Premises

Heiko Reppe

Institut für Algebra
Technische Universität Dresden
D-01062 Dresden
Heiko.Reppe@mailbox.tu-dresden.de

Abstract. We propose a variation of the attribute exploration algorithm. Instead of implications with pseudo-intents as premises our approach uses implications with proper premises. It is known that the set of implications with proper premises is complete, but in general it is not minimal in size. This variation will allow us to calculate all implications of a formal context with premise size at most n, for some fixed $n \in \mathbb{N}$. This is of interest if the attribute set is large and the user requests valid implications with small premises. Other applications can be seen for formal contexts where the maximal premise size of an implication with proper premise is known, for example multivalued contexts scaled by multiordinal scales only.

1 Introduction

The procedure of attribute exploration is an algorithm developed by B. Ganter in [GW86, G87, G99] to build up an expert system on a certain domain. Implementations are available for instance in CONIMP[1] by P. Burmeister and in CONEXP[2] by S. Yevtushenko.

In the following, the elements of the domain G are called *objects*. These objects can be classified with the help of *attributes*. The set of all attributes will be called M. The natural relation "object g has attribute m" will be encoded by $g\ I\ m$. We use notions of formal concept analysis, the triple $\mathbb{K} = (G, M, I)$ will be called *formal context*. The following definitions can be found in [GW99].

For building up the expert system, the exploration algorithm gives questions to answer. These questions always have the form of implications, that is:

> Is it true, that if an object has attributes p_1 and p_2 ... and p_{k_1} then it also has attributes $c_1, c_2, \ldots,$ and c_{k_2} ?

We will focus on finite sets of attributes and hence k_1 and k_2 are natural numbers less or equal to the size of M.

---

[1] http://www.mathematik.tu-darmstadt.de/~burmeister/

[2] http://conexp.sourceforge.net/

P. Eklund and O. Haemmerlé (Eds.): ICCS 2008, LNAI 5113, pp. 161–174, 2008.

Formally, the questions are given in pairs of sets of attributes and often they are written as $P \to C$, with $P, C \subseteq M$. The set P is called *premise*, C is called *conclusion*, and the pair is called *implication*. A set A of attributes *respects* an implication $P \to C$ if $P \nsubseteq A$ or $C \subseteq A$. A set A of attributes *respects a set* $\mathcal{L}$ of implications if it respects every element of $\mathcal{L}$. The set of all implications respected by all all object intents $\{g' \mid g \in G\}$ is denoted by $\mathrm{Imp}(\mathbb{K})$. Additionally for $n \in \mathbb{N}$ we denote by

$$\mathrm{Imp}_n(\mathbb{K}) = \{A \to B \in \mathrm{Imp}(\mathbb{K}) \mid |A| \leq n\}$$

the set of implications with at most n-element premise.

An implication $A \to B$ *follows (semantically)* from a set of implications $\mathcal{L}$ if each subset of M respecting $\mathcal{L}$ also respects $A \to B$. A set $\mathcal{L}$ of implications is called *complete* if every implication of $\mathrm{Imp}(\mathbb{K})$ follows from $\mathcal{L}$.

If we consider a set of implications $\mathcal{L}$, then a larger set of implications $\langle\mathcal{L}\rangle \supseteq \mathcal{L}$ can be deduced from this set according to the three Armstrong rules [A74]:
for $X, Y, Z, W \subseteq M$

1. $X \to X \in \langle\mathcal{L}\rangle$,
2. if $X \to Y \in \langle\mathcal{L}\rangle$, then $X \cup Z \to Y \in \langle\mathcal{L}\rangle$, and
3. if $X \to Y \in \langle\mathcal{L}\rangle$ and $Y \cup Z \to W \in \langle\mathcal{L}\rangle$, then $X \cup Z \to W \in \langle\mathcal{L}\rangle$.

The set $\langle\mathcal{L}\rangle$ is then the smallest set containing $\mathcal{L}$ which is closed with respect to all three rules. It turns out that all implications of $\mathrm{Imp}(\mathbb{K})$ can be deduced from a complete set $\mathcal{L}$ in this way, and hence $\langle\mathcal{L}\rangle = \mathrm{Imp}(\mathbb{K})$. Moreover $\langle\cdot\rangle$ is a closure operator on $\mathrm{Imp}(\mathbb{K})$ [GW99].

Eventually, the role of an expert is to verify or reject an implication $P \to C$ questioned by the exploration algorithm. The verification formally needs a proof. The rejection is done by entering a counterexample. Hence, a *counterexample* is an object $g \in G$ with $P \subseteq g'$ but $C \nsubseteq g'$. The implemented algorithms of attribute exploration use implications, where the premise has a special form. They are pseudo-intents:

Definition 1. [GW99] *P is called* pseudo-intent *of $\mathbb{K} = (G, M, I)$ if and only if $P \subseteq M$, $P \neq P''$, and $Q'' \subseteq P$ holds for all pseudo-intents $Q \subsetneq P$. The set of implications $\{P \to P'' \setminus P \mid P \subseteq M$ pseudo-intent$\}$ is called* stem base *of $\mathbb{K}$.* $\Diamond$

The following theorem summarises Theorem 8 and Proposition 25 of [GW99].

Theorem 1. [GW99] *The stem base of a formal context $\mathbb{K} = (G, M, I)$ is non-redundant and complete. Moreover, every complete set of implications contains an implication $A \to B$ with $A \subseteq P$ and $A'' = P''$ for every pseudo-intent P.*

Consequently, we cannot find a complete set of implications containing fewer elements than the stem base. Beside the stem base other complete set of implications are known. One of them is the set of implications with proper premises.

Definition 2. [GW99] *For an attribute set $A \subseteq M$ of a context (G, M, I) we denote by*

$$A^\circ := A \cup \bigcup_{a \in A} (A \setminus \{a\})'' \quad \textit{and by}$$
$$A^\bullet := A'' \setminus A^\circ.$$

Hence, the latter is the set of those attributes contained in A'' but not in A or in the closure of any proper subset of A. We call A a proper premise *if $A^\bullet \neq \emptyset$, i.e., if $A'' \neq A^\circ$.* ◊

The notions $^\circ$ and $^\bullet$ are borrowed from convex geometry, where for a convex set C the set C° denotes the set of boundary points and $C^\bullet$ denotes the set of interior points. In our setting, we have:

$$\emptyset^\circ = \emptyset,$$
$$\forall m \in M: \ \{m\}^\circ = \{m\} \cup \emptyset'', \text{ and}$$
$$\forall A \subseteq M, \ |A| \geq 2: \ A^\circ = \bigcup_{B \subsetneq A} B''.$$

In particular, $\emptyset$ is a proper premise if $\emptyset'' \neq \emptyset$. Moreover, Definition 2 states that A° and $A^\bullet$ are disjoint sets with $A'' = A^\circ \cup A^\bullet$ holds for all $A \subseteq M$. Moreover, the following proposition is satisfied.

Proposition 1. [GW99] *If T is a finite subset of M, then*

$$T'' = T \cup \bigcup \{A^\bullet \mid A \textit{ is a proper premise with } A \subseteq T\}.$$

The set of all implications of the form

$$A \to A^\bullet, \quad A^\bullet \neq \emptyset,$$

of a formal context with finite attribute set is complete.

By Theorem 1 we already know that for every pseudo-intent P, there is a subset $A \subseteq P$ with $A \to A^\bullet$, which is an implication with proper premise and with $P'' = A'' = A^\circ \cup A^\bullet$. The following proposition shows the relation of pseudo-intents and the proper premises that they contain.

Proposition 2. *Let $\mathbb{K} = (G, M, I)$ be a formal context with finite attribute set. Let $P \subseteq M$ be a pseudo-intent, and $A \subseteq P$ be a proper premise for $A^\bullet$, then:*

1. *$A^\circ \subseteq P$ and*
2. *$A'' = P''$ implies $A^\circ = P$.*

Proof. For the first part consider $B \subsetneq A \subseteq P$. Then every implication $B \to B''$ follows from the stem base of $\mathbb{K}$. Hence there is a sequence of pseudo-intents Q_k, $1 \leq k \leq n$ with $Q_1 \subseteq B$, $Q_{l+1} \subseteq B \cup Q_l''$, for $1 \leq l \leq k-1$, such that $B \to B''$ follows from $\{Q_i \to Q_i'' \mid 1 \leq k \leq n\}$. Inductively we see that all pseudo-intents

Q_k are properly contained in P, thus their closures and B'' are contained in P and finally $A^\circ = \bigcup_{a \in A}(A \setminus \{a\})'' \subseteq P$.

For the second part it suffices to show $A^\bullet \subseteq P'' \setminus P$. Consider an attribute $x \in A^\bullet \cap P$. Then the implication $A \to \{x\}$ follows as in the proof above from elements of the stem base, and hence pseudo-intents properly contained in P. Since x is an element of $A^\bullet$ but not contained in any proper subset of A, we find a pseudo-intent $Q \subsetneq P$ such that $A \subseteq Q$ and $x \in Q'' \setminus Q$. This is a contradiction as $P'' = A'' \subseteq Q''$ and P pseudo-closed, hence $Q'' \subseteq P \neq P''$. □

Proposition 3. *Let $\mathbb{K} = (G, M, I)$ be a formal context, $n \in \mathbb{N}$ a nonnegative integer, and $\mathcal{L}_n = \{A \to A^\bullet \mid A \subseteq M, |A| \leq n, A^\bullet \neq \emptyset\}$. Then:*

$$\langle \mathcal{L}_n \rangle = \langle \mathrm{Imp}_n(\mathbb{K}) \rangle.$$

Proof. Because of $\mathcal{L}_n \subseteq \mathrm{Imp}_n(\mathbb{K})$, it follows $\langle \mathcal{L}_n \rangle \subseteq \langle \mathrm{Imp}_n \rangle$. By $\mathrm{Imp}_n(\mathbb{K}) \subseteq \langle \mathcal{L}_n \rangle$ we have the reverse direction. □

As one might expect, the set $\langle \mathrm{Imp}_n(\mathbb{K}) \rangle$ may contain implications with proper premises of size larger than n. An example for this is the Fano plane.

Example 1. The Fano plane is the projective geometry with seven points and seven lines, each line containing three points. Given two points there is exactly one line connecting them. We can understand this structure as a context, were the objects are the points and the attributes are the lines. The incidence relation is given by "point g belongs to line l".

The stem base of this structure consists of 21 implications. All of them have a two-element proper premise and 21 is exactly the number of all two-element subsets of the set of lines. Hence all implications can be deduced from $\mathrm{Imp}_2(\mathbb{K})$, but there are further implications with proper premises. Namely all three-element subsets of the set of lines which have no common point. These are all three-element subsets of the seven-element set of lines, except those meeting in a point. All together we receive 28 implications with a three-element proper premise.

The next example demonstrates that reducible implications with proper premises of size n occasionally follow only from implications with proper premises of larger size.

Example 2. Let us consider the formal context shown in Figure 1. The concept lattice of this context contains 24 elements. We are interested in the implications of this formal context. Figure 1 shows the stem base, containing 8 elements, the set of implications with proper premise contains three implications more. We see, that the two sets do not differ, if we restrict to the subsets with 1-element premise.

Of course, the sizes of proper premises are relatively small compared to the pseudo-intents. Actually, it is their characteristic to contain no redundant elements, i.e. elements with $a \in A \subseteq M$ and $A'' = (A \setminus \{a\})$.

Having a proper premise, the implication $\{a, b\} \to \{g\}$ is centre of the following consideration. Both complete sets of implications of Figure 1 contain

	a	b	c	d	e	f	g
1	×		×	×	×		
2	×		×	×		×	
3			×	×	×	×	×
4		×	×		×	×	
5		×		×	×	×	
6			×		×		×

stem base
$\{a\} \to \{c,d\}$,
$\{b\} \to \{e,f\}$,
$\{g\} \to \{c,e\}$,
$\{c,d,e,g\} \to \{f\}$,
$\{c,e,f,g\} \to \{d\}$,
$\{c,d,e,f\} \to \{g\}$,
$\{a,c,d,e,f,g\} \to \{b\}$,
$\{b,c,d,e,f,g\} \to \{a\}$

implications with proper premises
$\{a\} \to \{c,d\}$, $\{b\} \to \{e,f\}$,
$\{g\} \to \{c,e\}$,
$\{a,b\} \to \{g\}$, $\{a,g\} \to \{b,f\}$,
$\{b,g\} \to \{a,d\}$, $\{d,g\} \to \{f\}$,
$\{f,g\} \to \{d\}$,
$\{a,e,f\} \to \{g\}$,
$\{b,c,d\} \to \{g\}$,
$\{c,d,e,f\} \to \{g\}$

Fig. 1. We see a formal context on the left. Its set of concepts contains 24 elements. Additionally we listed the stem base as well as the set of implications with proper premises of this context. Both sets are ordered by the size of the premise.

$\{a\} \to \{c,d\}$, $\{b\} \to \{e,f\}$, and $\{c,d,e,f\} \to \{g\}$. Obviously, the implication $\{a,b\} \to \{g\}$ can be deduced from the latter mentioned and either of three would be eliminated $\{a,b\} \to \{g\}$ would become non-redundant. This explains that not alone the set of implications with proper premises is a redundant set, but also redundant implications with proper premises of size n need not follow from implications with proper premises with premise size smaller or equal to n.

Similar to $\{c,d,e,f\} \to \{g\}$, the redundant implications with proper premises of size 3 which are $\{a,e,f\} \to \{g\}$ and $\{b,c,d\} \to \{g\}$ are not contained in $\text{Imp}_2(\mathbb{K})$. The closure system respecting all implications of stem base with size 2 or smaller is distributive in this example, while the closure system respecting $\text{Imp}_2(\mathbb{K})$ is not.

Remark 1. The fact that all implications with pseudo-intent of size 0 and 1 are contained in every complete set of families (contingently split into several implications) was shown in [GN06].

2 Attribute Exploration with the Stem Base

The stem base of a formal context is a complete set of implications minimal in size. It can be computed via the next-closure algorithm. Without going into its details we explain its basic idea. In the beginning $\mathcal{L} = \emptyset$. The set $\mathcal{L}$ will gather all implications of the stem base that come up during the procedure.

The subsets of M are translated by the next closure algorithm into binary vectors of dimension $|M|$ and thereby ordered. The algorithm begins with the 0-vector encoding the empty set and increases the last bit in the next step. The algorithm computes all intents and pseudo-intents and asks the expert to verify that a set is a pseudo-intent. Every number corresponds to a subset A of M.

The empty set is either intent or pseudo-intent, hence the algorithm can start. For an intent or pseudo-intent A the next set B is computed such that in between the corresponding numbers no further number exits corresponding to an intent or pseudo-intent. The set B is additionally closed with respect to the implications of $\mathcal{L}$. If the set B is an intent of the context the algorithm proceeds otherwise the expert has to state the nature of this set.

This approach makes the algorithm fast when it comes to compute all intents and pseudo-intents of a formal context. Nevertheless, this procedure is time consuming especially when the set of attributes is large.

The algorithm is hardly adaptable to our task of computing all implications with small premises. The strategy of the algorithm does not pay attention to the premise size at all. Even though a pseudo-intent is very large indeed it may be reducible to a small proper premise, while certain implications of the stem base which are large in size already have a proper premise.

2.1 Attribute Exploration for Large Sets M

More precisely, we set up a scenario, when it is over-ambitious to compute a complete set of implications. We have in mind a formal context consisting of a very large set of objects G (it may be infinite) and a very large but finite set of attributes M.

Facing a very large attribute set, a data analyst is interested in general concepts, that means frequent closed itemset. The top part of the concept lattice, i.e. concepts with many objects on the extent side, reveals some dependencies between attributes. This is done in the analysis of *iceberg lattices*. When it comes to determine implications, the expert is interested in small premises since they are easier to understand. In contrast to the approach of TITANIC [S02], we want to determine all implications with restricted premise size, not depending on the support.

When it comes to build up an expert system that provides a formal context $(\widetilde{G}, M, \widetilde{I})$, with $\widetilde{G} \subseteq G$ and $\widetilde{I} = I \cap \widetilde{G} \times M$ one could ask for choosing $\widetilde{G}$ in a way that

$$\mathrm{Imp}_n(\widetilde{G}, M, \widetilde{I}) = \mathrm{Imp}_n(G, M, I)$$

is satisfied. In this case the number of concepts of $(\widetilde{G}, M, \widetilde{I})$ may be strictly smaller than the number of concepts of (G, M, I). This is due to the fact that every intent of $(\widetilde{G}, M, \widetilde{I})$ is an intent of (G, M, I). Thus, up to n-element subsets of M the closures are the same without regard to the context.

We draw attention to the well known fact that if $\mathcal{L}_1$, $\mathcal{L}_2$ are sets of implications and $\mathcal{H}_{\mathcal{L}_i}$ is the closure system containing all sets respecting $\mathcal{L}_i$, $i \in \{1,2\}$ then

$$\mathcal{L}_1 \subseteq \mathcal{L}_2 \implies \mathcal{H}_{\mathcal{L}_1} \supseteq \mathcal{H}_{\mathcal{L}_2}.$$

As we have $\mathrm{Imp}_1(\mathbb{K}) \subseteq \mathrm{Imp}_2(\mathbb{K}) \subseteq \cdots \subseteq \mathrm{Imp}_k(\mathbb{K})$ we conclude $\mathcal{H}_{\mathrm{Imp}_1(\mathbb{K})} \supseteq \mathcal{H}_{\mathrm{Imp}_2(\mathbb{K})} \supseteq \cdots \supseteq \mathcal{H}_{\mathrm{Imp}_k(\mathbb{K})}$. In [GR07] it was shown that if $\mathrm{Imp}_k(\mathbb{K}) \subsetneq \mathrm{Imp}_l(\mathbb{K})$ then $\mathcal{H}_{\mathrm{Imp}_l(\mathbb{K})}$ is not a sublattice of $\mathcal{H}_{\mathrm{Imp}_k(\mathbb{K})}$.

2.2 Multiordinal Scaled Multivalued Contexts

We have seen in Example 1 in some cases it is sufficient to know the implications of Imp_n, respectively those with proper premise, to compute every implication of

Imp($\mathbb{K}$). This in not only true for geometric incidence relations but for practical examples as well.

One way of transforming a multivalued data table into a formal context is scaling. Thereby a column of the original data table is mapped to a set of columns. The preferred way is by standardised scales, which are special formal contexts. One of them is the multiordinal scale which is defined as follows.

Definition 3. [GW99] *For a natural number* n *we define the set* $\mathbf{n} = \{1, 2, \ldots, n\}$ *and call the formal contexts of the form* $\mathbb{O}_n = (\mathbf{n}, \mathbf{n}, \leq)$ (one-dimensional) ordinal scales. *For natural numbers* $n_1, n_2, \ldots, n_k$ *we define the* multiordinal scale *as the formal context*

$$\mathbb{M}_{n_1,\ldots,n_k} = (\mathbf{n_1}, \mathbf{n_1}, \leq) \,\dot{\cup}\, \ldots \,\dot{\cup}\, (\mathbf{n_k}, \mathbf{n_k}, \leq)$$

in an analogous way. $\Diamond$

The *biordinal scale* is a special case of multiordinal scales. Here we face two opposing values of an attribute, which can appear in different graduations. For instance either of the attributes *warm* and *cold* appears, when describing todays temperature, but additionally we have graduations like {very warm, warm, cold, very cold}. If we describe colours and their intensity we have more values and at the same time graduation, this is an example of a multiordinal scale.

A corollary of [R07, Theorem 5] is the following proposition.

Proposition 4. *Let* (G, M, W, I) *a multivalued context with* $|M| = n$ *and by scaling of the multivalued attributes with multiordinal scales we receive the formal context* $\mathbb{K} = (G, N, J)$, *then* $A \subseteq N$, $A^\bullet \neq \emptyset$ *implies* $|A| \leq n$.

Proof. Let A be a k-element subset of N with $k > n$. Hence, A contains two values of one multivalued attribute $m \in M$. For those attributes a, b holds that either they are comparable in the scale belonging to m or they are incomparable.

In the first case we can choose the value of m with the larger intent and reduce A by the other. Still, the closure of the reduced set is the same. Thus A is no proper premise.

In the second case the closure of $A'' = N$, but N is the closure of the two-element set $\{a, b\}$. Thus A was no proper premise. $\square$

These contexts give a class of examples, where we only have to consider implications of a certain size. In other words, we have:

$$\mathrm{Imp}(\mathbb{K}) = \langle \mathrm{Imp}_n(\mathbb{K}) \rangle = \langle \{A \to A^\bullet \mid A \subseteq M, A^\bullet \neq \emptyset\} \rangle.$$

3 Naive Approach

The first idea is to do an attribute exploration in three steps. We begin with an empty set of implications $\mathcal{L}$. If an expert has verified an implication it will be added to this set. Firstly, the expert has to verify the implication $\emptyset \to \emptyset''$, if $\emptyset \neq \emptyset''$. This is done in the classical attribute exploration as the first step as well.

The second step is to verify the attribute order $\sqsubseteq$. This is the quasi order, which is defined as follows.

Definition 4. [GW99] *Let* $\mathbb{K} = (G, M, I)$ *be a formal context. We define a quasi order* $\sqsubseteq$ *on the set* M *by*

$$m \sqsubseteq n :\iff m \in \{n\}''.$$

By $\approx$ *we denote the largest subset of* $\sqsubseteq$ *which is an equivalence relation, i.e.*

$$m \approx n :\iff m' = n'$$

A formal context is called attribute clarified *if* $\sqsubseteq$ *is an order relation.* ◊

Thus, in the second step, the expert has to verify all implications of the form:

$$\{m\} \to \{m\}^{\bullet} \quad \text{if} \quad \{m\}^{\bullet} \neq \emptyset.$$

By [GN06] these implications belong to every complete set of implications. Hence the expert has to verify them in the classical attribute exploration. In our approach we bring this forward. On the contrary to the classical version, if the exploration has started with a small collection of examples of the domain, the expert has to provide a large amount of counterexamples during this step.

As M is a large set we advise to use a clarified attribute set and additionally disregard the elements of $\emptyset''$ for the forthcoming. However, by $\leq_l$ we denote a linear extension of $\sqsubseteq/_{\approx}$, that we will use from now on.

The third step works in a common manner for all k with $2 \leq k \leq \max$, where max is the limit of interest or the largest possible number, which is $|M| - 1$. In the latter case the expert would perform the whole exploration, but then the classical approach would be a better choice.

Suppose, $k = n$ which means the expert has to verify implications with proper premises with an n-element premise. We calculate the n-element subsets of M with respect to the order $\leq_l$. If A is such a subset of M with $A^{\bullet} \neq \emptyset$ and $A \to A^{\bullet}$ does not follow from $\mathcal{L}$, then verify $A \to A^{\bullet}$.

With regard to the order $\leq_l$ we reduce the number of questions, the expert has to answer. For example, if the expert accepts an implication $X \to X^{\bullet}$, and for some $x \in X$ there is an element $y \neq x$ with $x \in \{y\}''$ then implication $(X \setminus \{x\}) \cup \{y\} \to X^{\bullet}$ holds in the context as well. If $((X \setminus \{x\}) \cup \{y\})^{\bullet} = X^{\bullet}$ there is no need to verify this implication with proper premise and the algorithm can skip it automatically. If the questions came up in reverse order, the expert had to answer both question.

Instead of considering all n-element subsets of M we want to present a way to reduce the number of sets. Naturally, in the worst case all of them have to be considered. Our algorithm makes use of one of the arrow relations in the formal context.

4 Arrow Relations of a Formal Context

Provided that the attribute set of the formal context is finite we can use its arrow relation $\swarrow \subseteq (G \times M) \setminus I$ to recognise implications with proper premises. This relation denotes that an object g does not have the attribute m, but moreover for all concepts strictly below the object concept γg it holds that m is contained in their intent. Hence this arrow "points at" irreducible objects. Formally, the definition of this relation is as follows.

Definition 5. [GW99] *Let (G, M, I) be a formal context, $g \in G$ an object, and $m \in M$ ab attribute. We denote by*

$$g \swarrow m :\iff g \not{I} m \text{ and if } g' \subsetneq h' \text{ then } h\, I\, m.$$

Furthermore we denote by

$$\begin{aligned} m^{\swarrow} &:= \{g \in G \mid g \swarrow m\} \text{ and} \\ g^{\swarrow} &:= \{m \in M \mid g \swarrow m\}. \end{aligned} \qquad \Diamond$$

Definition 6. *Let $\mathbb{K} = (G, M, I)$ be a formal context. A non-empty set of attributes $A \subseteq M$ is called* independent *in B, for some $B \subseteq G$, if there exists an injective map $f : A \to B$ such that for all $g \in f(A)$ and all $m \in A$*

$$g\, I\, m \iff g \neq f(m). \qquad \Diamond$$

Of course, if A is independent in B, then it is independent in G as well. The closure system of intents of a formal context $\mathbb{K} = (G, M, I)$ equals the power-set of M if and only if M is independent in G.

With the help of the relation $\swarrow$, one can rephrase the definition of $A^{\bullet}$ for some set $A \subseteq M$ and thereby the definition of an implication with proper premise. The Proposition 5 restates [GW99, Proposition 23] for an n-element attribute set A.

Proposition 5. *Let (G, M, I) be a formal context with finite set of attributes, $n \in \mathbb{N}_0$, and $A = \{m_1, m_2, \ldots, m_n\} \subseteq M$ a n-element set then:*

$$\begin{aligned} &\text{for } n \neq 1 \\ A^{\bullet} &= \{m \in M \mid m^{\swarrow} \subseteq G \setminus A' \text{ and } A \text{ is independent in } m^{\swarrow}\}, \\ &\text{for } n = 1 \\ A^{\bullet} &= \{m \in M \mid \emptyset \neq m^{\swarrow} \subseteq G \setminus \{m_1\}'\} \setminus \{m_1\}. \end{aligned}$$

Corollary 1. *Let $\mathbb{K} = (G, M, I)$ be a formal context and $\widetilde{\mathbb{K}} = (\widetilde{G}, M, \widetilde{I})$ be a subcontext, i.e. $\widetilde{G} \subseteq G$ and $\widetilde{I} = I \cap (\widetilde{G} \times M)$. If $A \to A^{\bullet}$ is an implication with proper premise in $\widetilde{\mathbb{K}}$ and $g \in G \setminus \widetilde{G}$ is a counterexample, then*

$$A \cup A^{\bullet} \subseteq g' \cup g^{\swarrow} = B,$$

where B is determined in the context $\widetilde{\mathbb{K}}$ with additional row g.

Proof. Since g is a counterexample for $A \to A^{\bullet}$, we deduce that $A^{\circ} \subseteq g'$. By the definition of $\swarrow$ the set $g^{\swarrow}$ contains all $m \in A^{\bullet}$, which do not belong to g'. $\square$

	m	...	m_1	m_2	m_3	...
g_1	↙			×	×	
g_2	↙		×		×	
g_3	↙		×	×		
g_4	↙			×	×	
g_5	↙		×			
⋮						

	m	...	m_1	m_2	m_3	...
g_1	?			×	×	
g_2	?		×		×	
g_3	?		×	×		
g_4	?			×	×	
g_5	?		×			
⋮						
g	↙		×	×	×	

Fig. 2. The formal context on the left shows the implication with proper premise $\{m_1, m_2, m_3\} \to \{m\}$ if $m^{\swarrow} = \{g_1, g_2, g_3, g_4, g_5\}$. The premise is independent in $m^{\swarrow}$. On the right a counterexample g for this implication was added to the context.

Corollary 2. *Let $\mathbb{K} = (G, M, I)$ be a formal context and $\widetilde{\mathbb{K}} = (\widetilde{G}, M, \widetilde{I})$ be a subcontext, i.e. $\widetilde{G} \subseteq G$ and $\widetilde{I} = I \cap (\widetilde{G} \times M)$. If $A \to A^\bullet$ is an implication with proper premise in $\mathbb{K}$, then there is an implication with proper premise in $B \to B^\bullet$ in $\widetilde{\mathbb{K}}$ such that*

$$B \subseteq A \text{ and } A^\bullet \subseteq B^\bullet.$$

Proof. We denote $A^\bullet$ with C, as this in not determined in $\widetilde{\mathbb{K}}$. Observe, that $C \nsubseteq A$ and $A \to C \in \text{Imp}(\mathbb{K}) \subseteq \text{Imp}(\widetilde{\mathbb{K}})$. In the context $\widetilde{\mathbb{K}}$ the set A might not be a proper premise for the set C. By the finiteness of M it can be reduced to a set U such that $C \subseteq U''$ and U is a proper premise. □

5 Proposed Algorithm

Tracing implications with proper premises by arrows in the formal context offers two strategies. Suppose, we have checked all subsets of M with at most $n-1$ elements. To find an implication with an n-element proper premise, we have to make sure that there are attributes with at least n down-arrows per column. If such an attribute does not exist, then we have determined all implications with proper premises already. Hence we determined a complete set of implications in that case.

If on the other hand there is an implication with proper premise of size n or larger in the context (G, M, I), then by Corollary 2 there is an implication $A \to B$ with proper premise of the context $(\widetilde{G}, M, \widetilde{I})$, $|A| < n$, that we rejected in the $(n-1)$-th step. Therefore, we find attributes with n or more down-arrows. And of course we can make advantage of the implications that were rejected in the previous step. There, we already had determined the $(n-1)$-element set A of attributes which was independent in $m^{\swarrow}$, for some $m \in B$. Moreover, we added a counterexample g such that $g\,I\,a$ holds for all $a \in A$.

In the simplest case we face exactly n down-arrows. Not only all attributes contained in $(M \setminus g') \cap (m^{\swarrow} \setminus \{g\})'$ fulfil Proposition 5 and hence have to be considered now as a possible extension of the premise, but also, only these have to be checked.

In the case that more than n down-arrows appear in the column n an attribute a that may extent a set A has to be non-incident with all objects contained in $A' \cap m^{\swarrow}$. Additionally, $(m^{\swarrow} \setminus A') \cap \{a\}'$ is at least an $n-1$-element set. These conditions are necessary but not sufficient. Nevertheless they help to reduce the number of set that have to be considered in the succeeding step by a great amount.

Our algorithm will not consider all $(n-1)$-element subsets of M. Thus, we have an additional case to consider. As we did in the naive approach we will start with the verification of $\emptyset''$ and $\sqsubseteq$. Hereafter, all implications with proper premise valid in the actual context $\widetilde{\mathbb{K}}$ are calculated. Hence, there are implications calculated in the begining, but they will be checked in later steps of the procedure or if the premise size exceeds max they will not be check at all.

The expert has to verify $A \to A'' \setminus A^{\langle \mathcal{L} \rangle}$, where $(\cdot)''$ denotes the closure in the actual context $(\widetilde{G}, M, \widetilde{I})$ and $A^{\langle \mathcal{L} \rangle}$ is the closure of A with respect to all implications that were verified already. Again, he will only be consulted only if $A'' \setminus A^{\langle \mathcal{L} \rangle}$ is nonempty. By this we see that background knowledge can be added naturally to the algorithm to speed up the algorithm. In the following we represent this algorithm chronologically in a very short manner.

We included a reduction of the attribute set to each step which filters the reducible attributes and selects one element of each class if the case that the attribute set is not clarified. Hence the attribute set gets smaller and we mark this by an index. We can assume that the the object set is clarified. Reducible objects will not influence the procedure at all ($\swarrow$ occur only in irreducible rows).

Initialisation

If the set of attributes $M \neq \emptyset$ is entered the expert may provide additional information on the domain by entering some objects and all their attributes or by providing background knowledge, i.e. implications, which will be gathered in the set $\mathcal{L}$. If neither objects nor implications were added the sets G and $\mathcal{L}$ are initialised with $\emptyset$ and $\{\emptyset \to \emptyset\}$ respectively. As above we denote by $(\cdot)''$ the closure in the actual formal context and by $(\cdot)^{\langle \mathcal{L} \rangle}$ the closure with respect to the verified implications.

First step – Verification of $\emptyset''$ for $n = 0$

For $\emptyset \subseteq M$ compute the set $\emptyset'' \setminus \emptyset^{\langle \mathcal{L} \rangle}$. As long as this set is nonempty consult the expert to confirm $\emptyset \to \emptyset'' \setminus \emptyset^{\langle \mathcal{L} \rangle}$.

REJECTION: Add a counterexample and recalculate $\emptyset'' \setminus \emptyset^{\langle \mathcal{L} \rangle}$.

CONFIRMATION: Add $\emptyset \to \emptyset'' \setminus \emptyset^{\langle \mathcal{L} \rangle}$ to the set $\mathcal{L}$.

If $\emptyset'' \setminus \emptyset^{\langle \mathcal{L} \rangle} = \emptyset$ we define $M_1 := M \setminus \emptyset^{\langle \mathcal{L} \rangle}$ and proceed with the second step if $M_1 \neq \emptyset$ otherwise stop.

Second step – Verification of $\sqsubseteq$ for $n = 1$

For all $m \in M_1$ and as long as $\{m\}'' \setminus \{m\}^{\langle \mathcal{L} \rangle} \neq \emptyset$ consult the expert to confirm $\{m\} \to \{m\}'' \setminus \{m\}^{\langle \mathcal{L} \rangle}$.

REJECTION: Add a counterexample and recalculate $\{m\}'' \setminus \{m\}^{\langle \mathcal{L} \rangle}$.

CONFIRMATION: Add $\{m\} \to \{m\}'' \setminus \{m\}^{\langle \mathcal{L} \rangle}$ to the set $\mathcal{L}$.

If M_1 is exploited then check first that there is an attribute with at least two $\swarrow$

per column. If this condition holds proceed with the intermediate step otherwise stop.

Intermediate step – Rearranging M and Initialisation
By $g \sqsubseteq h :\iff g \in h''$ a quasi order is defined on the set M_1, which splits into an equivalence relation $\approx$ and an order relation on the equivalence classes. We can choose one element for each class. Thereby we define the set M_2. Additionally we arrange M_2 along a linear extension of $\leq$, which is called $\leq_l$, such that we are forced to begin with the smallest element in each step.

Hereafter, we calculate all implications with proper premises with size at least 2 in the smaller context $(\widetilde{G}, M_2, \widetilde{I})$ and split them into the set of implications with two-element premises $\mathcal{L}_+$ and everything else $\mathcal{L}_-$ in ascending order according to the premises with respect to size first and sets of the same size with respect to $\leq_l$.

Third step – for $2 \leq n \leq \max$
Until $\mathcal{L}_+ = \emptyset$ extract the implication with the smallest premise $A \to B$ and recalculate $A'' \setminus A^{\langle \mathcal{L} \rangle} =: B$. If the latter is nonempty consult the expert to confirm $A \to A'' \setminus A^{\langle \mathcal{L} \rangle}$.
REJECTION: Add a counterexample and recalculate $A'' \setminus A^{\langle \mathcal{L} \rangle}$.
CONFIRMATION: Add $A \to A'' \setminus A^{\langle \mathcal{L} \rangle}$ to the set $\mathcal{L}$.
If all counterexamples are added after some iterations and $A'' \setminus A^{\langle \mathcal{L} \rangle} = \emptyset$ holds then we add $A \to B \setminus A^{\langle \mathcal{L} \rangle}$ to the list $\mathcal{L}_-$. That is the list of implications considered in the following initialisation of $\mathcal{L}_+$. Afterwards we proceed with the next set contained in $\mathcal{L}_+$.
If on the other hand $A'' \setminus A^{\langle \mathcal{L} \rangle} = \emptyset$ and no counterexamples were added we just proceed with the next set.
If $\mathcal{L}_+ = \emptyset$ determine an attribute $m \in M_n$ with at least $n \swarrow$ per column and check that $\mathcal{L}_-$ is nonempty. If both conditions hold, then we determine M_{n+1} and initialise $\mathcal{L}_+$ again otherwise stop.

We determine the set M_{n+1} by testing all verified implications with proper premises of size n. If $A \to A^\bullet$ is such an implication and $A \subseteq m'$ holds for some $m \in A^\bullet$, then m is reducible. We can omit this attribute for further considerations and thus we exclude all implications of $\mathcal{L}_-$ containing m in their premise.

For initialising $\mathcal{L}_+$ determine for an implication $A \to B \in \mathcal{L}_-$ with $|A| = n$ the attributes $m \in M$ that fulfil: for all $g \in A'$ holds $g \not\!I\, m$ and there is an attribute $b \in B$ with

$$b^{\swarrow} \cap A' \neq \emptyset \text{ and } |(b^{\swarrow} \setminus A') \cap m'| \geq n - 1.$$

We add $A \cup \{m\} \to B$ to the set $\mathcal{L}_+$ and unify the conclusions if $A \cup \{m\}$ is a premise already contained in $\mathcal{L}_+$. At last fill $\mathcal{L}_+$ with all implications that are contained in $\mathcal{L}_-$ with premise size n.

6 Conclusion

We proposed an algorithm for attribute exploration using a different set of implications then the implemented versions of attribute exploration use. We focused on implications with proper premises. Their advantage is the reduced size of the premise and they can be read from a formal context including the arrow relation $\swarrow$.

We think that the algorithm is applicable if the attribute set is very large and especially, when the users are not interested in all implications, but only those with small premises. We concentrated on reducing the number of sets that were considered in this process. Therefore we rearranged the set of attributes and we eliminated all attributes that are reducible.

A disadvantage of this algorithm is that the expert has to state all attributes for every counterexample. This turns out to be a difficulty if the attribute set is large. Burmeister and Holzer described a method that overcomes this problem, see [BH00]. To combine these strategies is less promising since the relation $\swarrow$ relies on the incidence relation I of the formal context.

The complexity of our algorithm depends on the task. If the user determines a complete set of implication, then the usual attribute exploration should be preferred. There the expert has to answer less questions. Our attribute exploration is appropriate as a start off for an exploration with the stem base. All questions that arise in the first and second step will arise in the original attribute exploration as well.

References

[A74] Armstrong, W.W.: Dependency structures of data base relationships. In: IFIP Congress, Geneva, Switzerland, pp. 580–583 (1974)

[BH00] Burmeister, P., Holzer, R.: On the Treatment of Incomplete Knowledge in Formal Concept Analysis. In: Ganter, B., Mineau, G.W. (eds.) ICCS 2000. LNCS, vol. 1867, pp. 385–598. Springer, Heidelberg (2000)

[G87] Ganter, B.: Algorithmen zur formalen Begriffsanalyse. In: Ganter, B., Wille, R., Wolf, K.E. (eds.) Beiträge zur Begriffsanalyse, pp. 196–212. B.I. Wissenschaftsverlag, Mannheim (1987)

[G99] Ganter, B.: Attribute exploration with background knowledge. Theoretical Computer Science 217(2), 215–233 (1999)

[GR07] Ganter, B., Reppe, H.: Base Points, Non-unit Implications, and Convex Geometries. In: Kuznetsov, S.O., Schmidt, S. (eds.) ICFCA 2007. LNCS (LNAI), vol. 4390, pp. 210–220. Springer, Heidelberg (2007)

[GW86] Ganter, B., Wille, R.: Implikationen und Abhängigkeiten zwischen Merkmalen. Preprint 1017, November 1986 TH Darmstadt (1986)

[GW99] Ganter, B., Wille, R.: Formal Concept Analysis – Mathematical Foundations. Springer, Heidelberg (1999)

[GN06] Gély, A., Nourine, L.: About the Family of Closure Systems Preserving Non-unit Implications in the Guigues-Duquenne Base. In: Missaoui, R., Schmidt, J. (eds.) ICFCA 2006. LNCS (LNAI), vol. 3874, pp. 191–204. Springer, Heidelberg (2006)

[R07] Reppe, H.: An FCA Perspective on n-Distributivity. In: Priss, U., Polovina, S., Hill, R. (eds.) ICCS 2007. LNCS (LNAI), vol. 4604, pp. 255–268. Springer, Heidelberg (2007)

[S96] Stumme, G.: Attribute Exploration with Background Implications and Exceptions, Data Analysis and Information Systems. Statistical and Conceptual approaches. Proc. GfKl 1995. Studies in Classification, Data Analysis, and Knowledge Organization 7, 457–469 (1996)

[S02] Stumme, G., Taouil, R., Bastide, Y., Pasquier, N., Lakhal, L.: Computing iceberg concept lattices with TITANIC. Data Knowledge Engineering 42(2), 189–222 (2002)

Sorting Concepts by Priority Using the Theory of Monotone Systems

Ants Torim[1] and Karin Lindroos[2]

[1] Tallinn University of Technology, Ehitajate tee 5, 19086 Tallinn, Estonia
torim@staff.ttu.ee
[2] Tallinn University of Technology, Ehitajate tee 5, 19086 Tallinn, Estonia
karin.lindroos@tv.ttu.ee

Abstract. Formal concept analysis is a powerful tool for conceptual modeling and knowledge discovery. As size of a concept lattice can easily get very large, there is a need for presenting information in the lattice in a more compressed form. We propose a novel method MONOCLE for this task that is based on the theory of monotone systems. The result of our method is a sequence of concepts, sorted by "goodness" thus enabling us to select a subset and a corresponding sub-lattice of desired size. That is achieved by defining a weight function that is monotone, correlated with area of data table covered and inversely correlated to overlaps of concepts. We can also use monotone systems theory of "kernels" to detect good cut-off points in the concept sequence. We apply our method to social and economic data of two Estonian islands and show that results are compact and useful.

1 Introduction

Knowledge discovery is the search of patterns in potentially large volumes of data. Clustering methods group similar objects together without providing explicit definitions for clusters. Concepts in formal concept analysis combine the group of objects (extent of the concept) with the set of attributes shared by those objects (intent of the concept). The number of concepts generated even from a small data table, however, can easily overwhelm an analyst. Several methods already exist for reducing analytical overload like iceberg concept lattices (see [8]), blocks (see [4]) and nested line diagrams. In this article we propose a novel method called MONOCLE for this task and demonstrate that its results are useful and interesting by applying it to social and economic data of two Estonian islands.

2 Formal Concept Analysis

Here we provide a short introduction into formal concept analysis (FCA). A detailed exposition is given in "Formal Concept Analysis, Mathematical foundation" by Ganter and Wille [4] or "Formal Concept Analysis: Foundations and Applications" by Wille, Stumme and Ganter [11]. For the following definitions we use "Introduction to Lattices and Order" by Davey and Priestley [2].

P. Eklund and O. Haemmerlé (Eds.): ICCS 2008, LNAI 5113, pp. 175–188, 2008.

Definition 1. *A* ***context*** *is a triple (G, M, I) where G and M are sets and $I \subseteq G \times M$. The elements of G and M are called objects and attributes respectively.*

We can say less formally that a context is a binary data table.

Definition 2. *For $A \subseteq G$ and $B \subseteq M$, define*

$$A' = \{m \in M \mid (\forall g \in A), (g, m) \in I\} , \tag{1}$$

$$B' = \{g \in G \mid (\forall m \in B), (g, m) \in I\} ; \tag{2}$$

so A' is the set of attributes common to all the objects in A and B' is the set of objects possessing the attributes in B.

Definition 3. *A* ***formal concept*** *is any pair (A, B) where $A \subseteq G$ and $B \subseteq M$, $A' = B$ and $B' = A$. The* ***extent*** *of the concept (A, B) is A while its* ***intent*** *is B.*

We can say less formally that a concept is a set of objects together with the attributes these objects have in common under the restriction that we cannot add an additional attribute without removing an object and we cannot add an additional object without removing an attribute. The special concept $\top$ has the extent G and the special concept $\bot$ has the intent M.

Algorithms for efficiently generating concepts from the context are described in [4], [11] and [2].

Subset relations $A_1 \subseteq A_2$ and $B_2 \subseteq B_1$ define an order on the set of all formal concepts and it can be shown [2] that they form a complete lattice, known as the **concept lattice** of the context. Concept lattices are commonly visualized as line diagrams[1] where concepts are shown as nodes, and subset relations between their extents (and inverse subset relations between their intents) are shown by lines. More general concepts are drawn above less general concepts.

Let us consider the example from Figure 1 which describes sizes of various watercourses. Object set G and attribute set M are abbreviated as follows: $G =$ {**C**hannel, **B**rook, **S**tream, **R**iver}, $M =$ {**v**ery **s**mall, **s**mall, **l**arge, **v**ery **l**arge}. The set of concepts for the context is $\{x_1, x_2, x_3, x_4\}$ where $x_1 = (\{C, B\}, \{s\})$, $x_2 = (\{S, R\}, \{l\})$, $x_3 = (\{C\}, \{vs, s\})$ and $x_4 = (\{R\}, \{l, vl\})$. The corresponding concept lattice is then drawn. As the extent of x_1 contains that of x_3 and the extent of x_2 contains that of x_4 these concepts are connected with lines and the more general concepts are placed higher in the diagram.

Concept lattices can become large for quite a small contexts. For example, a 488×234 sparse binary data table with economic data about settlements in Estonian island Saaremaa contained 1823 concepts. It is obvious that such a number of concepts is too large for the unaided human analysis. Several methods try to mitigate that problem. A full comparative review could be a topic for another article, here we give only a short review.

[1] For this article we used GaLicia Platform [3], [9] for the generation of complex concept lattice diagrams.

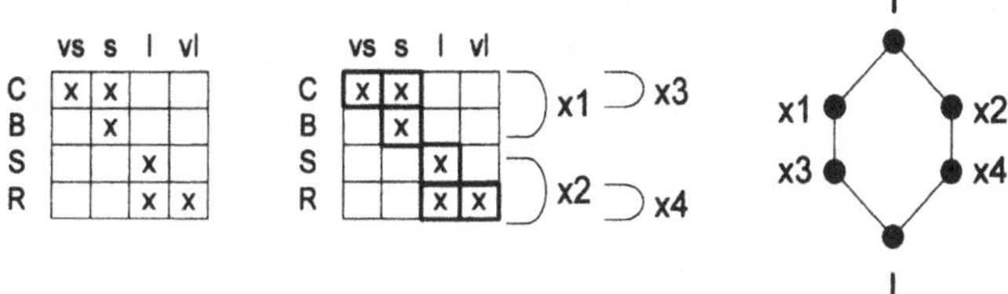

Fig. 1. A context as a binary data table, same context with the concepts marked inside the table by borders and labeled outside the table by their extents and the corresponding concept lattice. Taken from Davey and Priestley [2]. There is no requirement that attributes and objects in the concepts should be adjacent, we use such data tables only for the ease of illustration.

Blocks [4] introduce additional ones into the binary data table, generating bigger and fewer concepts. Our method sorts original concepts, without modifying them.

Nested line diagrams [4] summarize parallel lines and display them as just one line. Inner nodes contain sub-lattices. No concepts are removed, however, number of lines in the lattice is reduced.

Iceberg view, described by Stumme et al. [8], is based on selecting only the concepts that have extent of certain minimum size k, that is, cover at least k objects. Connecting this method with our theme, it can be described as sorting concepts by size of their extent and selecting those above some suitable cut-off point. Size of extent is intuitive and easy-to-calculate weight function. It does, however, eliminate concepts with few objects and many attributes. For some types of data, in our case economic data of settlements, these concepts are of great importance as they represent, for example, important regional centers. Our method takes into account both extent and intent sizes. But before describing our method, we need to give some background into the theory of monotone systems.

One measure for concepts goodness is **stability index**, proposed by Kuznetsov [6]. Stability measures independence of hypotheses on particular pieces of data that can be random, similar to the way scientific hypotheses are evaluated. Calculation of the stability index is, however, computationally more complex than our proposal.

Closure operators, described for example by Bělohlávek and Vychodil [1], represent a class of operators that constrain the lattice; retained concepts are guaranteed to form a complete lattice. Iceberg view method belongs into this general class.

3 Theory of Monotone Systems

The theory of monotone systems was developed in Tallinn University of Technology and introduced in 1976 in the article by Mullat [7]. A monotone system is a set of elements and a weight function. The weight function measures which elements are important for the system.

Definition 4. *A* ***monotone system*** *is a pair (W, w) where W is a finite set of elements, $w(x, H)$ is a weight for element $x \in H$ for any $H \subseteq W$ and the codomain of w is a linearly ordered set. Following property of monotonicity should hold for all $x \in H$ and for all $y \in H$ where $x \neq y$:*

$$w(x, H) \geq w(x, H \setminus \{y\}) \ . \tag{3}$$

That is, weights of the elements should decrease monotonically if any one element is removed from the system. There is a dual definition for monotonically increasing weights and a more general case where the removal of an element is replaced by an "operation" but for this article, these are not needed. Different weight functions and monotone systems algorithms are described in [10].

We want to measure the weight or "goodness" of subsytems of W. We use the weakest link principle and define the function F_{min} as:

$$F_{min}(H) = min\big(\{w(x, H) \mid \forall x \in H\}\big) \ . \tag{4}$$

We call the subsystems with the greatest value of F_{min} **kernels.**

Definition 5. *A subsystem $K \subseteq W$ is called the kernel of the system W if $F_{min}(K) \geq F_{min}(H)$ for any $H \subseteq W$.*

Minus technique means removing an element with the smallest weight from the monotone system and repeating this step until the system is empty. A minus technique sequence can therefore be found by a greedy algorithm, see [10]. Formal definition follows:

Definition 6. *We denote n-th element from the minus technique sequence for the system W by x_n. Let $H_1 = W$ and $H_n = (...((W \setminus \{x_1\}) \setminus \{x_2\})... \setminus \{x_{n-1}\})$.*

$$x_n = x \in H_n \text{ where } w(x, H_n) \leq w(y, H_n) \text{ for all } y \in H_n \tag{5}$$

The minus technique sorts the elements by their worth for the system. If we want to eliminate the k least interesting elements from the system we can apply minus technique and deal only with the set H_{k+1}. Thus we can use the minus technique to substitute arbitrary sized subset for the entire system. We can also use the kernels to suggest us good cut-off points. The following theorem deals with the relationship between the kernels and the minus technique.

Theorem 1 (Kernel as the global maximum). [2] *Let $w(x_k, H_k) = F_{min}(H_k)$ be the maximal weight in the minus technique sequence $x_1, x_2, .., x_n$ for the monotone system W. That is,*

$$F_{min}(H_k) \geq F_{min}(H_i) \text{ for all } i \in \{1...n\} \ . \tag{6}$$

Then the subsystem H_k is a kernel for the system W.

[2] This theorem was proven independently by A. Torim. Equivalent theorem, albeit with a longer proof, appeared in [7].

Proof. For all $A \subseteq H_1$ where $x_1 \in A$ we know that $F_{min}(A) \leq F_{min}(H_1)$ because of the property of monotonicity from the Equation 3. Therefore, either H_1 is a kernel or there is some kernel $K \subseteq H_2$.

If we know that $K \subseteq H_i$ for $i \in \{1...n\}$ then for all $A \subseteq H_i$ where $x_i \in A$ we know that $F_{min}(A) \leq F_{min}(H_i)$. Therefore, either H_i is the kernel K, or $K \subseteq H_{i+1}$.

By induction, there is some kernel $K \in \{H_1, H_2, ..., H_k, ..., H_n\}$. As $w(x_k, H_k) = F_{min}(H_k)$ is the maximal weight in the minus technique sequence, H_k is the kernel for the system W. □

The kernel as the global maximum provides a good cut-off point in the minus technique sequence. For practical purposes we often want more cut-off points to study either smaller or larger subsystems. Therefore we will also introduce the notion of **local kernels** that correspond to local maxima in the minus technique sequence.

Definition 7. *Let sequence $H_1, H_2, ..., H_n$ be the sequence of subsets corresponding to the minus technique sequence $x_1, x_2, .., x_n$. Then $H_k \in \{H_1, H_2, ..., H_n\}$ is a local kernel if $F_{min}(H_{k-1}) \leq F_{min}(H_k) \geq F_{min}(H_{k+1})$.*

Figure 2 shows an example of simple graph-based monotone system before and after the removal of an element.

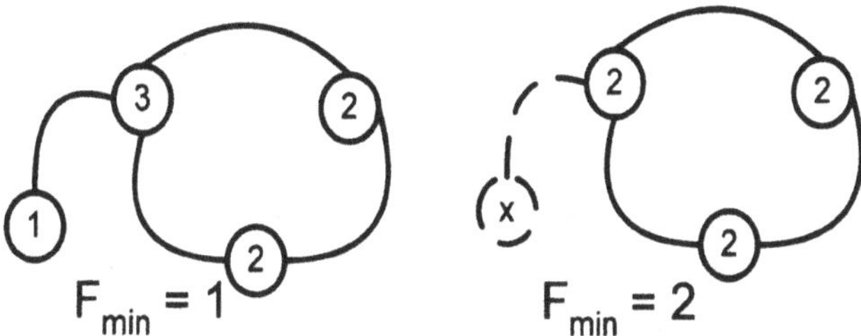

Fig. 2. A monotone system whose elements are vertices of the graph and the weight for the element is the number of adjacent vertices. Weights are shown inside the vertex circles. In this example, after removal of the element with the smallest weight, we have the kernel.

4 MONOCLE Method for Knowledge Discovery

We now introduce our **MONOCLE** (MONOtone Concept Lattice Elimination) method for knowledge discovery in binary data tables. We treat concepts as elements of the monotone system and we define an appropriate **MONOCLE weight function**. Generally, the MONOCLE data analysis process is as follows:

1. Concept generation.
2. Generation of minus technique sequence of concepts using MONOCLE weight function.

3. Data analysis using subsets of suitable size from the top of the minus technique sequence and possibly using global and local kernels to suggest good cut-off points.

The MONOCLE weight function is monotone and is correlated with the concept area. By concept area we mean the product of extent size and intent size $|A| \cdot |B|$ of a concept (A, B). We modify the weight of each attribute and object in our area calculation by its "rareness". Finally, we show that a certain invariance property holds for the MONOCLE weight function.

Definition 8. *Let W be the set of all concepts for some context and $H \subseteq W$. We denote the number of all concepts in H not containing the object g as $N_G(g, H)$ and define it formally as*

$$N_G(g, H) = \left|\{(A, B) \mid (A, B) \in H, g \notin A\}\right| . \tag{7}$$

We denote the number of all concepts in H not containing the attribute m as $N_M(m, H)$ and define it formally as

$$N_M(m, H) = \left|\{(A, B) \mid (A, B) \in H, m \notin B\}\right| . \tag{8}$$

Definition 9. *Let W be the set of all concepts for some context and $H \subseteq W$. Let the concept $x \in H$ have extent A and intent B. We define the MONOCLE weight function $w(x, H)$ as*

$$w(x, H) = \left(|A| + \sum_{g \in A} N_G(g, H)\right) \cdot \left(|B| + \sum_{m \in B} N_M(m, H)\right) . \tag{9}$$

We illustrate MONOCLE weight function by examples from the Figure 3.

Each object and attribute of the concepts in the set $H = \{a_1, a_2\}$ for the context (a) is not contained in exactly one concept, so $N_G(g, H) = 1$ and $N_M(m, H) = 1$ for any object g or attribute m in the context (a). Weights for the concepts in the context (a) are

$$\begin{aligned} &w(a_1, \{a_1, a_2\}) = w(a_2, \{a_1, a_2\}) \\ &= ((1 + 1) + (1 + 1)) \cdot ((1 + 1) + (1 + 1)) \\ &= 16 . \end{aligned} \tag{10}$$

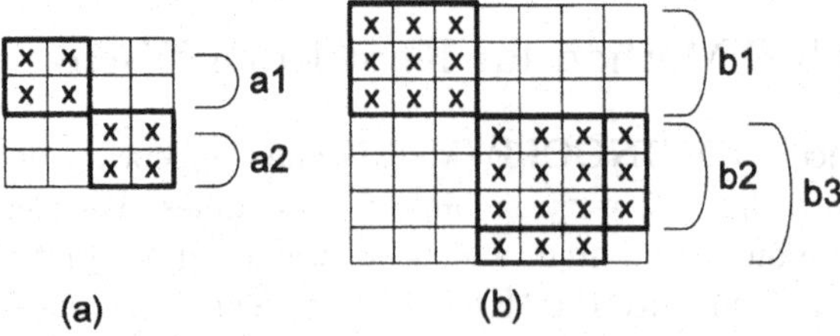

Fig. 3. Two sample contexts with the concepts marked inside the table by borders and labeled outside the table by their extents

Both a_1, a_2 and a_2, a_1 are correct minus technique sequences,

$$w(a_1, \{a_1\}) = w(a_2, \{a_2\}) = \big((0+1)+(0+1)\big)\cdot\big((0+1)+(0+1)\big) = 4\ . \quad (11)$$

so the corresponding sequence of F_{min} is 16, 4 ; $\{a_1, a_2\}$ is a kernel.

For the context (b):

$$w(b_1, \{b_1, b_2, b_3\}) = (3+3+3)\cdot(3+3+3) = 81\ ; \quad (12)$$

$$w(b_2, \{b_1, b_2, b_3\}) = w(b_3, \{b_1, b_2, b_3\}) = (2+2+2)\cdot(2+2+2+3) = 54\ . \quad (13)$$

Two minus technique sequences are b_2, b_1, b_3 and b_3, b_1, b_2 and the corresponding sequence of F_{min} is 54, 36, 12 ; thus kernel is $\{b_1, b_2, b_3\}$. Here, minus technique sequence is clearly different from the simple area calculation $|A|\cdot|B|$ where b_1 would be the first element removed from the system.

4.1 Invariance Property

We now demonstrate that we can change certain contexts in certain ways that preserve the weights of corresponding concepts in the old and new contexts.

Let us consider the three pairs of contexts, where concepts do not overlap, shown in Figure 4.

The set of objects G and the set of attributes M are unchanged for the pairs. For each concept in the upper contexts, we create r concepts in the lower contexts, leaving extent and intent size ratios between the concepts unchanged. For the pair (a) $r = 3/2$, for the pair (b) $r = 3$ and for the pair (c) $r = 2$. We can see that the weights of corresponding concepts are equal, for example:

$$w(a_1, \{a_1, a_2\}) = w(a'_1, \{a'_1, a'_2, a'_3\}) = 36 \quad (14)$$

$$w(b_1, \{b_1\}) = w(b'_1, \{b'_1, b'_2, b'_3\}) = 18 \quad (15)$$

$$w(c_1, \{c_1, c_2, c_3\}) = w(c'_1, \{c'_1, c'_2, c'_3, c'_4 . c'_5, c'_6\}) = 144 \quad (16)$$

$$w(c_2, \{c_1, c_2, c_3\}) = w(c'_3, \{c'_1, c'_2, c'_3, c'_4 . c'_5, c'_6\}) = 36 \quad (17)$$

$$F_{min}(\{a_1, a_2\}) = F_{min}(\{a'_1, a'_2, a'_3\}) = 36 \quad (18)$$

$$F_{min}(\{b_1\}) = F_{min}(\{b'_1, b'_2, b'_3\}) = 18 \quad (19)$$

$$F_{min}(\{c_1, c_2, c_3\}) = F_{min}(\{c'_1, c'_2, c'_3, c'_4 . c'_5, c'_6\}) = 36\ . \quad (20)$$

We now demonstrate that property formally.

Theorem 2 (Invariance property). *Let W be the system of non-overlapping concepts with set of objects G and set of attributes M. Let W' be another system of non-overlapping concepts with set of objects G' and set of attributes M' so that $|G| = |G'|$ and $|M| = |M'|$. Let r be a rational number so that for the sets of concepts defined by any pair of natural numbers n, m*

$$H = \big\{(A, B) \mid |A| = n, |B| = m, (A, B) \in W\big\} \quad (21)$$

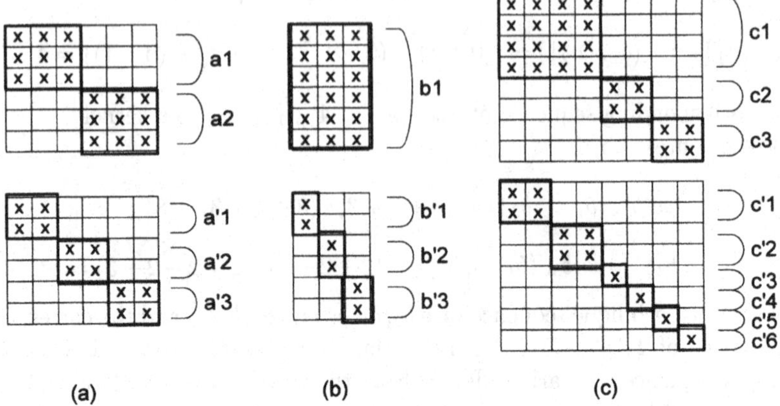

Fig. 4. Three pairs of contexts with non-overlapping concepts. The concepts are marked inside the table by borders and labeled outside the table by their extents.

$$H' = \{(A', B') \mid |A'| = \frac{n}{r}, |B'| = \frac{m}{r}, (A', B') \in W'\} \tag{22}$$

it holds that

$$|H'| = r \cdot |H|. \tag{23}$$

Then for any $c = (A, B) \in H$ *and* $c' = (A', B') \in H'$

$$w(c, W) = w(c', W'). \tag{24}$$

Proof. We can see that for non-overlapping concepts

$$|A| + \sum_{g \in A} N_G(g, W) = |W| \cdot |A| \tag{25}$$

$$|B| + \sum_{m \in B} N_M(m, W) = |W| \cdot |B| \ . \tag{26}$$

We also know that

$$|W'| \cdot |A'| = r \cdot |W| \cdot \frac{|A|}{r} = |W| \cdot |A| \tag{27}$$

$$|W'| \cdot |B'| = r \cdot |W| \cdot \frac{|B|}{r} = |W| \cdot |B| \ . \tag{28}$$

Thus

$$w(c', W') = (|W'| \cdot |A'|) \cdot (|W'| \cdot |B'|) = (|W| \cdot |A|) \cdot (|W| \cdot |B|) = w(c, W) \ . \tag{29}$$

□

5 Application for Analysis of Social and Economic Data

We now apply MONOCLE method to social and economic data of two largest Estonian islands: Saaremaa and Hiiumaa [3]. Our previous, non-FCA related results are documented in [5].

Our set of objects consists of settlements and the set of attributes consists of various social and economic characteristics like the presence of a school, a kindergarten, shops or certain types of industry. We excluded demographic attributes (number of children, workers and elderly) that were included in our research presented in [5] as these would tend to dominate the results and information provided by these attributes is somewhat less interesting than that from the more qualitative attributes. We have also applied the MONOCLE method to data with all the attributes present and results were generally consistent with those from [5] and in a more explicit and easier to interpret form. For Hiiumaa $|G| = 184$ and $|M| = 206$; for Saaremaa $|G| = 488$ and $|M| = 234$. The attribute sets are mostly similar, however some attributes are present for only one island, hence some differences.

The set of concepts for Hiiumaa contained 380 concepts, the set of concepts for Saaremaa contained 1823 concepts. The weights of minus technique sequences are presented in Figure 5.

As we can see from the Figure 5, the global kernels H_G and S_G are quite large. Smallest local kernel for Hiiumaa is H_L that is still pretty large. Smallest local kernel for Saaremaa S_1 contains 17 concepts, pretty good size for the general overview of the system. For Hiiumaa we select "almost" a local kernel H_2 that contains 10 concepts instead of the too large H_L. We also select subset H_2 that is equal in size to S_2 and S_1 that is equal in size to H_1 for comparison.

We present the concept lattices for Hiiumaa corresponding to H_1 and H_2 as the Figure 6. Note that concepts corresponding to intersections of extents and intents are also added. Lattices were generated from data tables that contained only concepts in H_1 or H_2 , using Galicia [3]. Markings for concepts in H_1 or H_2 were added later.

The following list is the tail of the minus technique sequence, numbered backwards: concepts in H_1 (all 17) and H_2 (first ten). Numbering corresponds to Figure 6. If extent or intent is large, we provide only its size. We use the format: Weight $w(x_n, H_n)$; {extent}, {intent}.

1. Weight 116; {Kärdla, Käina}, (58 attributes)
2. Weight 250; (68 settlements), {summer cabins}
3. Weight 382; {Käina}, (83 attributes)
4. Weight 574; {Kärdla}, (101 attributes)
5. Weight 625; {Emmaste}, (41 attributes)

[3] Saaremaa is the largest island (2,673 km^2) belonging to Estonia, Hiiumaa is the second largest (989 km^2). They are located in the Baltic Sea. The capital of Saaremaa is Kuressaare, which has about 15,000 inhabitants; the whole island has about 40,000 inhabitants. The capital of Hiiumaa is Kärdla, which has about 3,700 inhabitants; the whole island has about 10,000 inhabitants.

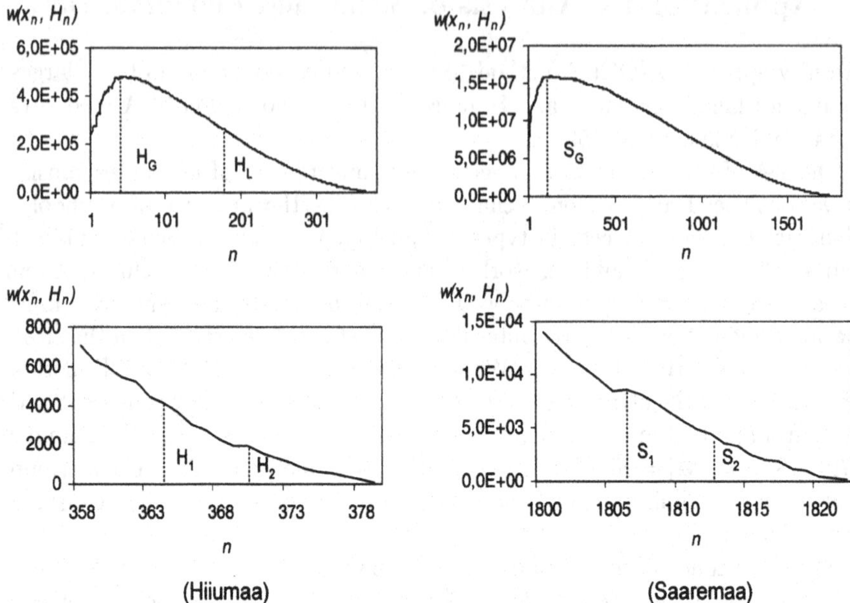

Fig. 5. Minus technique sequences for Hiiumaa and Saaremaa. Tails of sequences, containing most interesting concepts, are presented separately below main sequence. Several kernels and cut-off points used for following analysis $H_G, H_L, H_1, H_2, S_G, S_1, S_2$ are marked with dashed lines.

6. Weight 747; (17 settlements), {summer cabins, beach}
7. Weight 1050; {Kärdla, Käina, Emmaste}, (25 attributes)
8. Weight 1352; (22 settlements), {agriculture}
9. Weight 1536; (32 settlements), {housing}
10. Weight 1974; {Käina, Emmaste}, (27 attributes)
11. Weight 1976; {Kärdla, Emmaste}, (28 attributes)
12. Weight 2277; {Kärdla, Kõrgessaare, Käina}, (16 attributes)
13. Weight 2717; {Nõmme}, (22 attributes)
14. Weight 3000; {Kõrgessaare, Käina}, (19 attributes)
15. Weight 3620; (15 settlements), {summer cabins, housing}
16. Weight 4130; (22 settlements), {beach}
17. Weight 4444; {Kassari, Käina}, (17 attributes)

We present the concept lattices for Saaremaa corresponding to S_1 and S_2 as the Figure 7.

Following is the list of concepts for Saaremaa.

1. Weight 179; {Kuressaare}, (179 attributes)
2. Weight 272; (68 settlements), {landing places for fishing boats}
3. Weight 456; (87 settlements), {summer cabins}
4. Weight 936; {Kuressaare, Orissaare}, (52 attributes)
5. Weight 1204; {Kuressaare, Nasva}, (47 attributes)

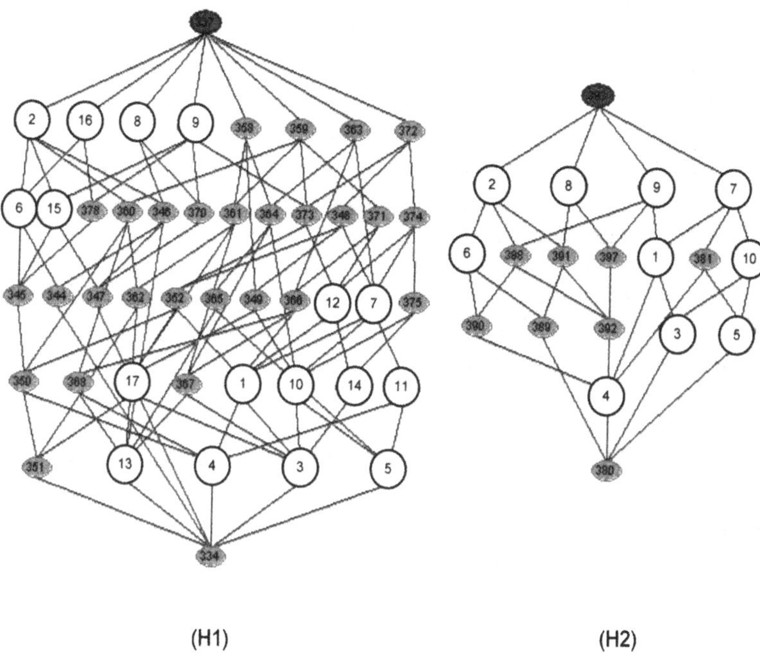

Fig. 6. Lattices H_1 and H_2 for Hiiumaa. Concepts in H_1 and H_2 are marked with big numbered circles.

6. Weight 1878; (55 settlements), {agriculture}
7. Weight 2007; {Kuressaare, Kärla}, (44 attributes)
8. Weight 2409; {Kuressaare, Valjala}, (39 attributes)
9. Weight 3330; (32 settlements), {landing places for fishing boats, summer cabins}
10. Weight 3582; {Nasva}, (54 attributes)
11. Weight 4448; {Kuressaare, Liiva}, (35 attributes)
12. Weight 4910; {Orissaare}, (58 attributes)
13. Weight 5681; {Kuressaare, Kudjape}, (30 attributes)
14. Weight 6534; (56 settlements), {housing}
15. Weight 7449; (41 settlements), {sights}
16. Weight 8085; {Kuressaare, Orissaare, Liiva}, (24 attributes)
17. Weight 8550; {Kuressaare, Valjala, Tornimäe, Kärla}, (17 attributes)

There is a clear division between concepts describing small monofunctional settlements (agriculture, summer cabins) and larger regional centres (Kärdla, Käina, Kuressaare). That divison is fundamental to the data and not the artifact of MONOCLE method - there are very few settlements that are neither monofunctional nor regional centres. The division seems to be clearer in the case of Saaremaa where larger centers, represented by the “artificial” concept in the upper right corner of lattices S_1 and S_2 do not have attributes common

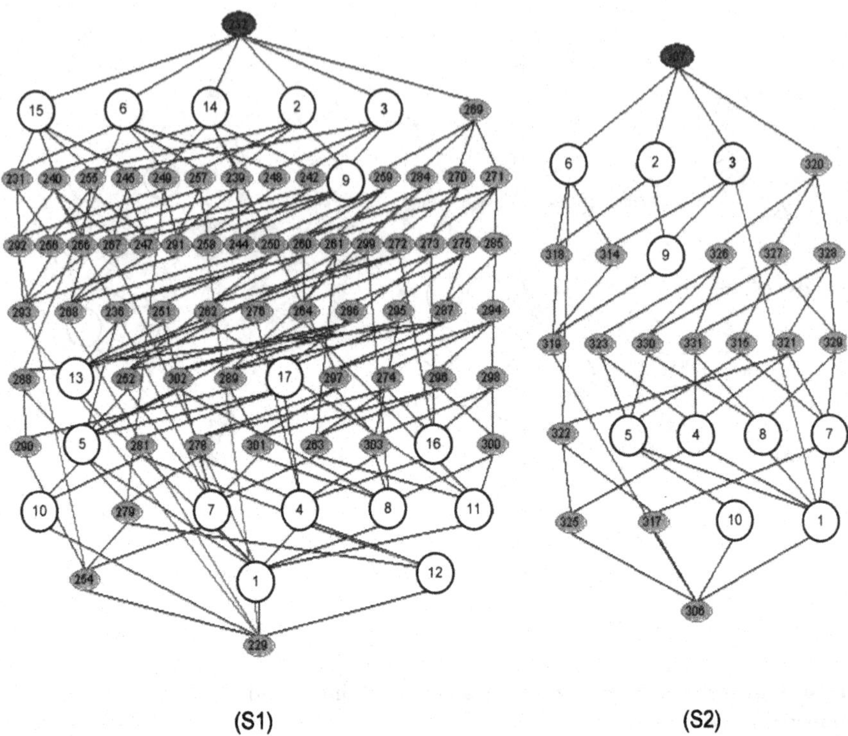

Fig. 7. Lattices S_1 and S_2 for Saaremaa. Concepts in S_1 and S_2 are marked with big numbered circles

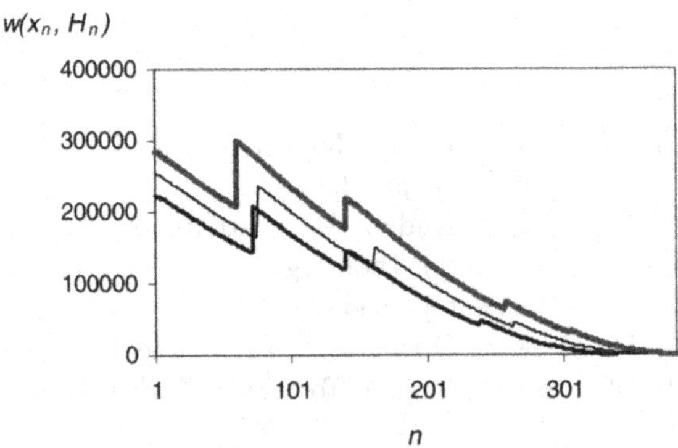

Fig. 8. Three minus technique sequences for random data tables where size and frequency of ones were same as that of Hiiumaa. Sawteeth correspond to different simple concept areas.

with concepts describing monofunctional settlements. Upper right "artificial" concept for S_1 has value {several enterprises with turnover over million crowns}, {Kudjape, Nasva, Kuressaare, Kärla, Liiva, Orissaare, Tornimäe, Valjala}. Role of Saaremaa's capital Kuressaare seems to be more important as that of Kärdla for Hiiumaa as {Kuressaare} is the extent of the last concept in the minus technique sequence.

We finally compare graphs presented in Figure 5 to that of random data, having same frequency of ones and size as the data table for Hiiumaa. Figure 8 shows graph for random data and we can see that graph for minus technique sequence is heavily influenced by the internal structure of data.

Running speeds for building the lattice and finding the minus technique sequence ranged from couple of seconds for Hiiumaa without demographic data to couple of minutes for Saaremaa with demographic data. Hardware was ordinary desktop computer and the program was written in Python. Detailed discussion of speed and complexity issues is outside the scope of this article.

6 Conclusions

We proposed and studied an interesting method for knowledge discovery that combines the elements from formal concept analysis and from the theory of monotone systems. We proved that we can find kernels from the minus technique sequence, we defined a monotone weight function that forms the heart of our MONOCLE method and showed that it has certain invariance property. We demonstrated the usefulness of our method by applying it to the social and economic data of two islands. Issues of speed and very large data tables were not dealt with in this article. Several methods like selecting only about 1000 concepts with greatest weights for the minus technique sequence calculation may be promising. Different weight functions, study of the stability of results with regard to weight function and with regard to selection of the minimal element when there are several candidates, are all fields that need further study.

Acknowledgements. We thank the Estonian Information Technology Foundation and Tiger University for funding this research. We thank the Galicia (Galois lattice interactive constructor) team from the University of Montreal for making their platform available as open source [3], [9]. We thank professor Rein Kuusik for his comments and suggestions.

References

1. Bělohlávek, R., Vychodil, V.: Formal Concept Analysis with Constraints by Closure Operators. In: Ganter, B., Godin, R. (eds.) ICFCA 2005. LNCS (LNAI), vol. 3403, pp. 176–191. Springer, Heidelberg (2005)
2. Davey, B.A., Priestley, H.A.: Introduction to Lattices and Order 2nd Revised edn. Cambridge University Press, Cambridge (2002)
3. Galicia Home Page [WWW] (December 30, 2007), `http://www.iro.umontreal.ca/~valtchev/galicia/`

4. Ganter, B., Wille, R.: Formal Concept Analysis, Mathematical Foundation. Springer, Heidelberg (1998)
5. Juurikas, K., Torim, A., Võhandu, L.: Multivariate Data Visualization in Social Space. In: Procs. of IADIS International Conf. on Applied Computing, pp. 427–432 (2006)
6. Kuznetsov, S.O.: On stability of a formal concept. Annals of Mathematics and Artificial Intelligence 49, 101–115 (2007)
7. Mullat, I.: Extremal Monotonic Systems (in Russian). Automation and Remote Control No 5, 130–139 (1976) (in English), [WWW] (December 31, 2007), `http://www.datalaundering.com/download/extrem01.pdf`
8. Stumme, G., Taouil, R., Bastide, Y., Lakhal, L.: Conceptual Clustering with Iceberg Concept Lattices. In: Proc. GI-Fachgruppentreffen Maschinelles Lernen 2001,Universität Dortmund 763 (2001)
9. Valtchev, P., Grosser, D., Roume, C., Hacene, R.H.: GALICIA: an open platform for lattices. In: Using Conceptual Structures: Contrib. to the 11th ICCS, pp. 241–254 (2003)
10. Võhandu, L., Kuusik, R., Torim, A., Aab, E., Lind, G.: Some Monotone Systems Algorithms for Data Mining. WSEAS Transactions on Information Science & Applications, 802–809 (2006)
11. Wille, R., Stumme, G., Ganter, B.: Formal Concept Analysis: Foundations and Applications. Springer, Heidelberg (2005)

Extending Attribute Dependencies for Lattice-Based Querying and Navigation

Nizar Messai, Marie-Dominique Devignes, Amedeo Napoli, and Malika Smaïl-Tabbone

INRIA Nancy - LORIA, BP 239, 54506 Vandœuvre-Lès-Nancy, France
{messai,devignes,napoli,malika}@loria.fr
http://www.loria.fr/~messai

Abstract. In this paper we study dependencies of attributes in the context of Formal Concept Analysis. These dependencies allow to define a hierarchy of attributes reflecting the importance or interest in attributes. A hierarchy of attributes is a set of attributes partially ordered with respect to their importance. It represents domain knowledge used to improve lattice-based querying and navigation. Actually, in lattice-based querying, hierarchies of attributes are used to define complex queries containing attributes with different levels of importance: more important attributes define the focus of the retrieval while less important ones define secondary information whose presence is desirable in the answers. Furthermore, the relation between attributes in a complex query represents implicit or explicit knowledge units that must be considered while computing answers. Similarly, in lattice-based navigation, the choice of moving to a particular concept rather than to another is influenced by the higher importance of the attributes in the concept intent. Hence, the design and use of a hierarchy of attributes leads to a navigation guided by domain knowledge.

1 Introduction

Formal Concept Analysis (FCA) [12] allows to build in a proper way the set of all *formal concepts* from a *formal context* i.e. a table with rows corresponding to objects and columns corresponding to attributes describing a relationship between objects and attributes. *Formal concepts* are maximal sets of objects sharing maximal sets of attributes. They are organized into a *concept lattice.* This particular way of reorganizing data makes FCA very useful for numerous data analysis and knowledge discovery tasks such as clustering, classification, ontology building, information retrieval (IR), etc.

In this paper we are particularly interested in the application of FCA in information retrieval. The basic ideas of lattice-based IR exist since the beginning of FCA i.e. improving information retrieval using a concept lattice [13,2,18,11]. They were mainly motivated by navigation (browsing) capabilities offered by concept lattices: formal concepts correspond to classes of relevant objects matching a given query, and moving to the upper-neighbors (respectively to the lower-neighbors) in the lattice hierarchy allows to consider more general (respectively

P. Eklund and O. Haemmerlé (Eds.): ICCS 2008, LNAI 5113, pp. 189–202, 2008.

more specific) queries [13]. Since then, many research works have successfully expanded lattice-based IR to a wide range of IR applications including Web document retrieval [5,4,19,14], and domain specific retrieval such as multimedia retrieval [9,17], file system retrieval [11], and biological database retrieval [16]. The success of these approaches is mainly due to two complementary factors. The first one is the use of concept lattices which naturally support multiple hybrid retrieval strategies including browsing, querying, query reformulation, etc. The second one is the effort being made in the FCA community to provide additional features such as lattice visualization and zooming –iceberg lattices and nested line diagrams–, semantic relationships highlighting –throughout merging thesaurus in concept lattices–, data and text mining techniques, etc. A detailed study of the the search functionalities allowed by FCA as well as references to their corresponding research works can be found in [6].

In lattice-based IR, a query is defined as a set of attributes considered together to characterize the objects to be retrieved. This representation does not allow the expression of semantic relationships between attributes. However, in many cases, there is a need for defining such relationships to allow a better understanding of the attribute meanings which improves the retrieval of relevant answers. Semantic relationships between query attributes also express the way a query must be interpreted and in which order attributes must be considered for the retrieval. Then it becomes possible to consider queries such as "*Italian restaurants with a "dehors" near the Louvre Museum*" as mentioned in [6]. The objects looked for here are restaurants and the query attributes are "*Italian*", "*with a dehors*", and "*near the Louvre Museum*". A restaurant fulfilling all the query attributes constitutes the ideal answer but if there is no such restaurant, it has to be possible for the user to express some preferences or priorities, i.e. which attributes must be considered in priority: for example [6], geographical proximity first (near the Louvre Museum), then the type of cuisine (Italian), and lastly possession of an open-air space (with a dehors).

In this paper, we propose a method for considering semantic relationships between attributes for lattice-based querying and navigation. This method is based on and extends the Attribute Dependency Formulas (ADFs) introduced in [1]. The extension is firstly due to the ability of defining arbitrary expressions representing the attribute dependencies and secondly to the application of extended ADFs to lattice-based querying and navigation. Moreover, taking into account semantic relationships between attributes leads to the definition of complex queries with attribute hierarchies reflecting preferences on attributes. In addition, a hierarchy of attributes can be seen as a guideline showing how attributes must be considered during retrieval sessions in adequation with domain knowledge. This extension of ADFs is quite general and can be applied to any lattice-based query answering or navigation.

The paper is organized as follows. Section 2 briefly presents FCA basics. Section 3 presents the formalization of hierarchically ordered attributes. Section 4 details the way hierarchies of attributes are taken into account in lattice-based IR and provides examples. Section 5 gives an overview of the implementation of hierarchies of attributes in the BR-Explorer system and shows preliminary results. Finally section 6 concludes the paper and gives some perspectives of the current work.

2 Background

2.1 Formal Concept Analysis

Formal Concept Analysis [12] starts from a given input data represented as a formal context to provide the set of all the formal concepts which form a concept lattice. A formal context is denoted by $\mathbb{K} = (G, M, I)$ where G is a set of objects, M is a set of attributes, and I is a binary relation between G and M ($I \subseteq G \times M$). $(g, m) \in I$ denotes the fact that the object $g \in G$ has the attribute $m \in M$ or that g is in relation with m through I.

Table 1 shows the running example of a formal context (taken from the BioRegistry corpus used for BR-Explorer testing [21,16]). The objects are biological databases ($DB1$, $DB2$, ... $DB8$) and the attributes are metadata of three types: organisms concerned by the information in the database (mammals, birds, amphibians and fish), quality of database content (updated, complete) and reference ontologies (Gene Ontology GO, NCBI taxonomy). The relation I expresses whether a database is annotated by a metadata (in which case there is a cross "×" in the corresponding row and column) or not. Consider for example the database $DB1$. The content of $DB1$ has the following characteristics: it deals with *amphibians* and *fish*, it is *complete* and uses *NCBI taxonomy* as reference ontology.

Table 1. The formal context $\mathbb{K} = (G, M, I)$ where Ma, Bi, Am and Fi are respectively the abbreviations of Mammals, Birds, Amphibians, and Fish

	Organisms				Content quality		Ontologies	
	Ma	Bi	Am	Fi	Updated	Complete	GO	NCBI
DB1			×	×		×		×
DB2			×	×	×	×		×
DB3	×		×					×
DB4	×	×			×			
DB5	×		×		×	×		
DB6	×	×				×	×	
DB7	×	×			×	×	×	×
DB8	×	×						×

The formal concepts (or simply *concepts*) are maximal sets of objects having in common maximal sets of attributes. Formally, a concept is represented by a pair (A, B) such that:

$$A = \{g \in G | \forall m \in B : (g, m) \in I\}$$
$$B = \{m \in M | \forall g \in A : (g, m) \in I\}$$

A is the *extent* and B is the *intent* of the concept. The concepts are computed by means of a Galois connection defined by two derivation operators:

$$' : G \rightarrow M \mid A' = \{m \in M | \forall g \in A : (g, m) \in I\}$$
$$' : M \rightarrow G \mid B' = \{g \in G | \forall m \in B : (g, m) \in I\}$$

For any concept (A, B), it holds that: $A' = B$, $B' = A$, $A'' = A$ and $B'' = B$. The set of all concepts in a formal context is denoted by $\mathfrak{B}(G, M, I)$. A concept $(A1, B1)$ is called a sub-concept of $(A2, B2)$ when $A1 \subseteq A2$ (or equivalently $B2 \subseteq B1$). In this case, $(A2, B2)$ is called super-concept of $(A1, B1)$ and we write $(A1, B1) \leq (A1, B1)$. The set of concepts in $\mathfrak{B}(G, M, I)$ ordered using the partial order "$\leq$" forms the concept lattice of the context (G, M, I) denoted by $\underline{\mathfrak{B}}(G, M, I)$. The concept lattice of the context given in table 1 is represented by the so-called line diagram (or Hasse diagram) shown in figure 1.

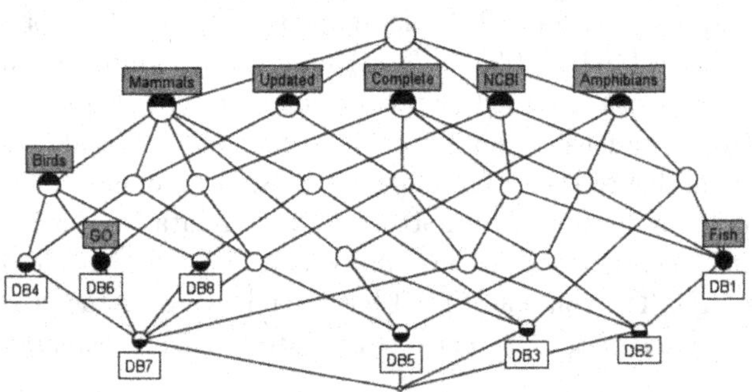

Fig. 1. The concept lattice $\underline{\mathfrak{B}}(G, M, I)$ corresponding to the formal context $\mathbb{K} = (G, M, I)$ shown in table 1

2.2 Queries in Lattice-Based IR

In lattice-based IR, queries are defined as sets of attributes describing the objects to be retrieved. We recall here the definition of query concept, introduced in [15], which will be used in the rest of this paper.

Definition 1 (Query concept). *A query concept Q is a pair (A, B) where B is the set of attributes describing the objects to be retrieved and A denotes the set of objects satisfying all the attributes in B. Initially, A contains only a dummy object x used to guarantee the existence of a concept having B as intent once Q is inserted in the concept lattice.*

A retrieval process starts by inserting the query concept into the concept lattice representing the data corpus. Then, depending on the approach, a particular exploration of the neighbors of the query concept provides a ranked set of relevant objects as an answer [3]. In BR-Explorer [15], only upper-neighbors of the query are considered. This choice is justified by the fact that the intents of these concepts are always part of the query intent.

In the current example of biological databases, consider a query for retrieving databases dealing with *Mammals*, having *Complete* and *updated* contents. The corresponding query concept is $Q = (\{x\},\{Mammals,\ Complete,\ Updated\})$. The concept lattice resulting from the insertion of Q into $\underline{\mathfrak{B}}(G, M, I)$ and the steps performed by BR-Explorer to compute the answer are shown in figure 2. The

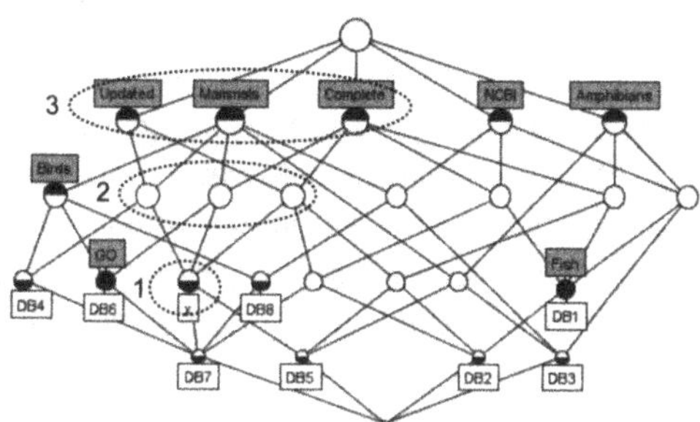

Fig. 2. Steps of BR-Explorer for the query $Q = (\{x\},\{Mammals,\ Complete,\ Updated\})$ on the concept lattice $\underline{\mathfrak{B}}(G, M, I)$

answer is the following:

1- $DB5, DB7$: *Mammals, Complete, Updated*
2- $DB4$: *Mammals, Updated*
$DB6$: *Mammals, Complete*
$DB2$: *Complete, Updated*
3- $DB3, DB8$: *Mammals*
$DB1$: *Complete*

In the answer, the database $DB2$ appears at rank 2 although being not relevant since it does not contain any information about *Mammals*. This irrelevancy is caused by the fact that nothing in the query Q asserts that the attribute *Mammals* is a main objective of the retrieval and should be present. In the following, we consider and solve this problem by defining the hierarchically ordered attributes in queries.

3 Hierarchically Ordered Attributes

3.1 Motivation

Considering Semantic Relationships in Queries. In lattice-based querying and navigation, semantic relationships between attributes are often needed in the queries as well as in the concept lattices. These relationships are not expressed by the lattice structure. Some research works were interested in extending concept lattices by integrating thesaurus ordering relation [2,19]. The main advantage of such approaches is to allow the definition of new meaningful queries (with respect to the thesaurus ordering relation) [6]. This concept reorganization improves browsing according to the thesaurus ordering relation.

However, in many cases, there is a need for local preferences on attributes (e.g. depending on a given point of view on attributes). In this case, the ordering of the lattice has not to be changed for integrating additional information. This is the purpose of the present research work to propose a method for extending the query and navigation capabilities of FCA in taking into account complex attribute dependencies.

Defining Priority Between Query Attributes. Consider again the example of biological databases. When looking for databases containing recent information about *mammals*, the query to be formulated must contain the attribute *mammals*. Then, among the set of databases in the answer, those that were recently *updated* will be preferred. Hence the attribute *mammals* is more important than the attribute *updated*.

Let us suppose now that the objective is to carry out statistical studies based on the recent contents of biological databases on the web. Thus the attribute *updated* has to be considered first. In the same way, secondary information such as the organisms (e.g. *mammals*) can also be useful. In this case, the attribute *updated* is more important than *mammals*.

These examples show that depending on the focus of a query, the attributes in a formal context may be considered at different levels of importance. Attributes chosen as most important define the main goal of the retrieval while less important attributes give secondary information. The definition of levels of importance reflects the priority of the attributes in a query.

3.2 Formalization

We recall the definition of Attribute Dependency Formulas introduced in [1] since the present formalization of attribute hierarchy is built upon this definition.

Definition 2 (Attribute Dependency Formulas). *Consider $m, m_1, \ldots, m_n \in M$. An Attribute Dependency Formula φ is in the form: $m \sqsubseteq m_1 \sqcup m_2 \sqcup \ldots \sqcup m_n$, where $m_1, \ldots, m_n$ are called primary attributes and m is called secondary attribute.*

In [1], ADFs are mainly used for category forming. Primary attributes are used to form large categories whereas secondary attributes are used to make a finer categorization within these categories. Applied to concept lattices, ADFs produce a reduced hierarchy which highlights the obtained categorization. In the following, we generalize the previous definition to deal with attribute priority in lattice-based querying.

Definition 3 (Attribute dependency). *(1) An attribute m_2 depends on an attribute m_1 whenever the presence of m_2 is not significant without the presence of m_1. We denote this dependency by $m_1 \succ m_2$.*
(2) More generally, an attribute dependency has the form $e_1 \succ e_2$ where e_1 and

e_2 may be atomic attributes, or "conjunction" of attributes, or "disjunction" of attributes.

In an attribute dependency $e_1 \succ e_2$, the attributes in e_2 are less important than the ones in e_1 and the presence of e_2 is meaningful only when associated with e_1. Figure 3 gives a graphical representation of basic examples of attribute dependencies.

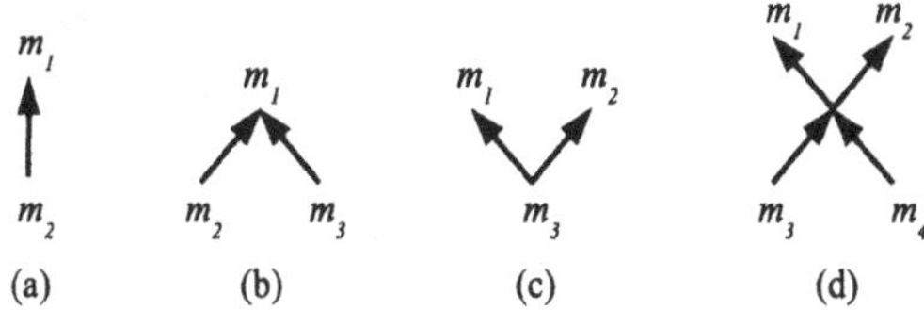

Fig. 3. Graphical representation of the following attribute dependency examples: (a)- $m_1 \succ m_2$, (b)- $m_1 \succ (m_2 \sqcup m_3)$ (equivalent to $(m_1 \succ m_2) \sqcap (m_1 \succ m_3)$), (c)- $(m_1 \sqcap m_2) \succ m_3$ (equivalent to $(m_1 \succ m_3) \sqcap (m_2 \succ m_3)$), (d)- $(m_1 \sqcap m_2) \succ (m_3 \sqcup m_4)$ (equivalent to $(m_1 \succ m_3) \sqcap (m_2 \succ m_3) \sqcap (m_1 \succ m_4) \sqcap (m_2 \succ m_4)$)

The attribute dependency operator "$\succ$" defines a partial order on given sets of attributes. Considering this partial order, attributes can be classified into hierarchies defined as follows.

Definition 4 (Hierarchy of attributes). *A hierarchy of attributes, denoted by $\mathcal{HA}$, is a set of attributes partially ordered according to a dependency relation.*

Figure 4 shows two examples of hierarchies of attributes corresponding to queries detailed in the previous sections.

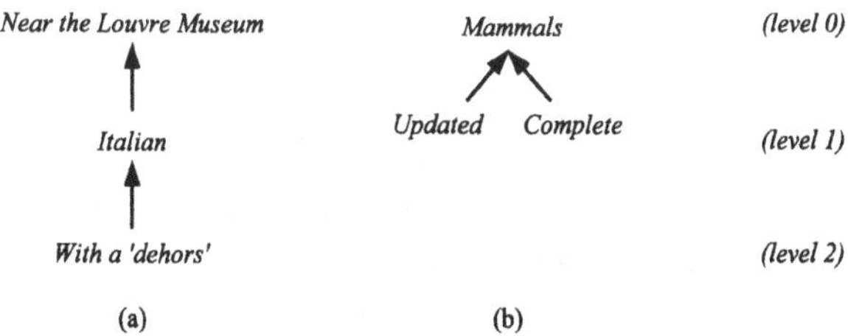

Fig. 4. A graphical representation of two examples of hierarchy of attributes. (a) corresponds to the hierarchy of attributes given with the query aiming to retrieve *Italian* restaurants *with a "dehors" near the Louvre Museum* [6]. (b) corresponds to the hierarchy of attributes given with the query aiming to retrieve biological databases which deal with *mammals*, and have *complete*, and *updated* contents.

3.3 Applying Hierarchies of Attributes to Concept Lattices

The application of a hierarchy of attributes to a concept lattice consists in removing the set of concepts which are incoherent with respect to the dependencies expressed in the attribute hierarchy. The coherence of a formal concept with respect to a hierarchy of attributes is defined as follows (in the same way as for ADFs [1]).

Definition 5 (Coherence of formal concepts). *(1) A formal concept (A, B) is said to be coherent with respect to an attribute dependency if there is no dependant attribute m in B without the attribute on which it depends.*
(2) A formal concept (A, B) is said to be coherent with respect to a hierarchy of attributes $\mathcal{HA}$ if and only if (A, B) is coherent with respect to all the dependencies in $\mathcal{HA}$.

Depending on the form of e_1 and e_2 in an attribute dependency "$e_1 \succ e_2$", we distinguish the following cases for concept coherence:

- case of dependency in the form "$m_1 \succ m_2$": (A, B) is coherent whenever $m_2 \in B$ then $m_1 \in B$.
- case of dependency in the form "$m_1 \sqcup ... \sqcup m_n \succ m_m$": (A, B) is coherent whenever $m_m \in B$ then $\exists i \in \{1, ..., n\}$ such that $m_i \in B$.
- case of dependency in the form "$m_1 \sqcap ... \sqcap m_n \succ m_m$": (A, B) is coherent whenever $m_m \in B$ then $\forall i \in \{1, ..., n\}$ $m_i \in B$.

These cases are easily generalized to dependencies where e_2 is a conjunction or a disjunction of attributes.

To illustrate the definition above, consider the concept lattice shown in figure 1 and suppose that we are interested in databases containing information about species living outside water. The attributes to be considered first (as focus) are *Mammals* and *Birds.* Attributes related to ontologies and content quality give secondary information on these species. This leads to the hierarchy of attributes:

$$\mathcal{HA}_1 : (Mammals \sqcup Birds) \succ (Updated \sqcup Complete \sqcup NCBI \sqcup GO)$$

An example of coherent concept with respect to $\mathcal{HA}_1$ is $C_1 = (\{DB5, DB6, DB7\},$ $\{$*Mammals, Complete*$\})$ because *Mammals* with *Complete* are both present in the intent of C_1. The concept $C_2 = (\{DB1, DB2, DB5, DB6, DB7\}, \{Complete\})$ is incoherent, due to the presence of *Complete* without any of the attributes in which it depends i.e. *Mammals* and *Birds.* The interpretation of the incoherence of C_2 may be the following: the attribute *Complete* means that the returned databases have complete information about the subjects they deal with without knowing whether such subjects are species living outside water (i.e. *Mammals* and *Birds*) or not. This makes the concept C_2 less informative.

The set of formal concepts coherent with respect to a hierarchy of attributes $\mathcal{HA}$ is denoted by $\mathfrak{B}_{\mathcal{HA}}(G, M, I)$ and the hierarchy of these concepts is denoted by $\underline{\mathfrak{B}}_{\mathcal{HA}}(G, M, I)$. The structure of a hierarchy of coherent concepts (with respect to ADFs) is discussed in [1]. There, it has been proved that such hierarchy is a

$\bigvee$-*sublattice*. This result is also valid for $\underline{\mathfrak{B}}_{\mathcal{HA}}(G, M, I)$. In addition, if the top of $\underline{\mathfrak{B}}(G, M, I)$ is coherent with respect to $\mathcal{HA}$ then $\underline{\mathfrak{B}}_{\mathcal{HA}}(G, M, I)$ is a complete lattice. This is the case of $\underline{\mathfrak{B}}_{\mathcal{HA}_1}(G, M, I)$ in the previous example. The concept lattice $\underline{\mathfrak{B}}_{\mathcal{HA}_1}(G, M, I)$ is shown in figure 5.

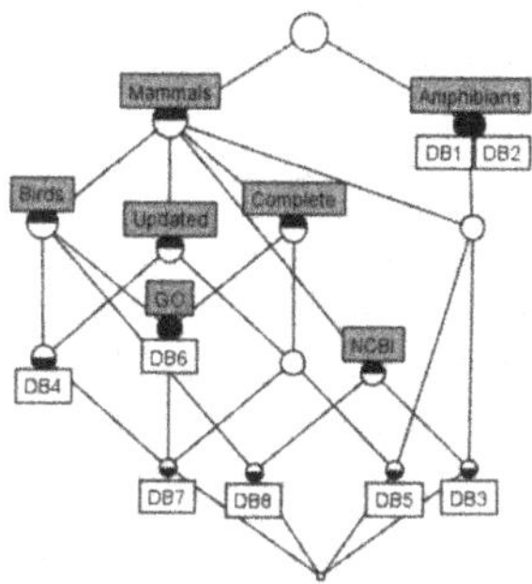

Fig. 5. The concept lattice $\underline{\mathfrak{B}}_{\mathcal{HA}_1}(G, M, I)$

There are two possible ways of computing $\underline{\mathfrak{B}}_{\mathcal{HA}}(G, M, I)$. The first one (the "naive" way) consists first in computing the whole concept lattice $\underline{\mathfrak{B}}(G, M, I)$ and then in deleting the incoherent concepts with respect to $\mathcal{HA}$. The second way consists in directly computing $\underline{\mathfrak{B}}_{\mathcal{HA}}(G, M, I)$. For that purpose, a way is to generalize the algorithm detailed in [1] (which is an adaptation of the *AddIntent* incremental algorithm [8]) and consider hierarchies of attributes instead of ADFs.

4 Lattice-Based Querying and Navigation with Respect to Hierarchies of Attributes

In this section, we detail the way hierarchies of attributes are used to improve lattice-based querying and navigation. In the case of lattice-based navigation, the reorganization of formal concepts in accordance with dependencies expressed in a hierarchy of attributes reduces the navigation space to the set of concepts which are coherent with respect to these dependencies. This reorganization can be seen as a preparation of the lattice for further navigation guided by domain knowledge expressed throughout the attribute dependencies.

In the case of lattice-based querying, the hierarchies of attributes are either *query-dependent* or *query-independent*. Query-dependent hierarchies of attributes are defined for a query and do not directly affect the concept lattice whereas query-independent hierarchies of attributes are applied to the concept lattice and can be reused for many other queries. In the following, we detail both ways of considering hierarchies of attributes.

4.1 Query-Dependent Hierarchies of Attributes

A query-dependent hierarchy of attributes represents a set of dependencies between the attributes of a given query. It gives the priority of each attribute in the query and, consequently, defines the way the query attributes must be considered during the retrieval process. The query is first inserted into the concept lattice. Then, at each step of the retrieval algorithm, the retrieved concepts are checked wether they are coherent or not with respect to the defined hierarchy of attributes. In the case of incoherence of a concept, there is no need to consider its super-concepts. In fact, as stated in the definition of coherence of formal concepts, the intent of an incoherent concept contains attributes without the attributes they depend on. And since the intents of super-concepts of a concept C are included in the intent of C, the super-concepts are either incoherent or irrelevant (i.e. their intents do not contain any query attribute) whenever C is incoherent.

For illustration, consider again the query concept $Q = (\{x\},\{$*Mammals, Complete, Updated*$\})$ in the current example (table 1). An example of a query-dependent hierarchy of attributes defined on this query is $\mathcal{HA}_Q$: $Mammals \succ (Complete \sqcup Updated)$. It corresponds to the following interpretation: "Q aims at retrieving databases dealing with *mammals* (the focus), and having a *complete* and *updated* content (secondary information)". Figure 6 shows the steps for retrieving relevant databases for Q with respect to $\mathcal{HA}_Q$. At the second step

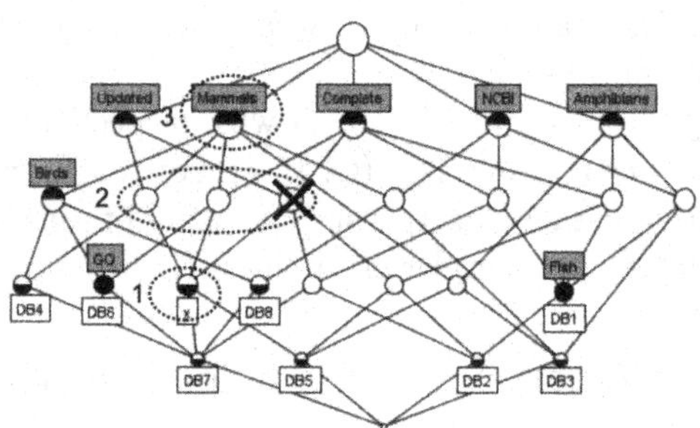

Fig. 6. Retrieval steps for the query $Q = (\{x\},\{$*Mammals, Complete, Updated*$\})$ with respect to the query-dependent hierarchy $\mathcal{HA}_Q$: $Mammals \succ (Complete \sqcup Updated)$

of the retrieval algorithm, an incoherent concept with respect to the dependencies in $\mathcal{HA}$ is reached (the concept corresponding to the node marked by a bold cross). This concept is ignored and its super-concepts are not considered in the following steps. The resulting answer is the following:

1- $DB5, DB7$: *Mammals, Complete, Updated*
2- $DB4$: *Mammals, Updated*
$DB6$: *Mammals, Complete*
3- $DB3, DB8$: *Mammals*

Compared to the answer produced for the same query without considering hierarchies of attributes (section 2.2), the present answer does not contain the databases $DB1$ and $DB2$. They do not appear since the formal concepts to which they belong are incoherent with respect to $\mathcal{HA}_Q$.

4.2 Query-Independent Hierarchies of Attributes

A query-independent hierarchy of attributes represents a set of attribute dependencies applied to the concept lattice before query sessions. These dependencies are not necessarily related to one specific query. They rather express global knowledge that must be considered when querying data in the concept lattice. Hence, querying a concept lattice with respect to a query-independent hierarchy of attributes $\mathcal{HA}$ consists firstly in applying $\mathcal{HA}$ to the concept lattice to reduce the retrieval space to coherent concepts. Then, once $\underline{\mathfrak{B}}_{\mathcal{HA}}(G, M, I)$ is obtained, query sessions can be performed. It can be noticed here that each query concept to be satisfied must be coherent with respect to $\mathcal{HA}$.

For illustration, consider the biological databases context (table 1) and suppose that we are interested in querying only databases dealing with species living outside water. Then, it is more efficient to query a reduced retrieval space (i.e. databases dealing with species living outside water) rather than querying the concept lattice $\underline{\mathfrak{B}}(G, M, I)$ representing all the biological databases. The query-independent hierarchy of attributes corresponding to this example is $\mathcal{HA}_1$: $(Mammals \sqcup Birds) \succ (Updated \sqcup Complete \sqcup NCBI \sqcup GO)$ detailed in section 3.3 and the corresponding concept lattice is $\underline{\mathfrak{B}}_{\mathcal{HA}_1}(G, M, I)$ shown in figure 5. Figure 7 shows the retrieval steps for the query $Q = (\{x\}, \{$*Mammals, Complete, Updated*$\})$ on the concept lattice $\underline{\mathfrak{B}}_{\mathcal{HA}_1}(G, M, I)$. As the dependencies expressed in $\mathcal{HA}_Q$ are also included in $\mathcal{HA}_1$, the returned answer is the same as in the previous section.

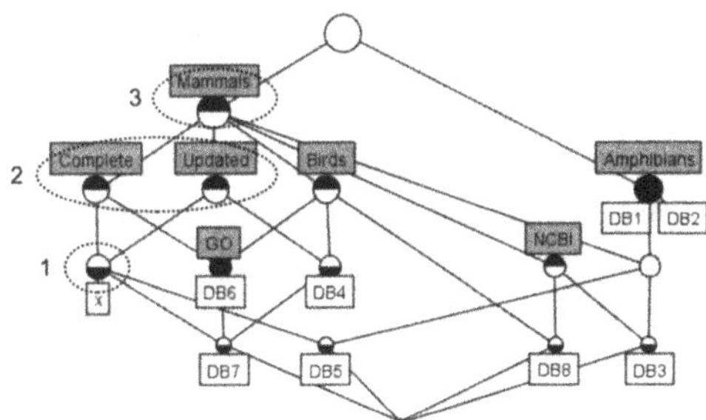

Fig. 7. Retrieval steps for the query $Q = (\{x\},\{$*Mammals, Complete, Updated*$\})$ on the concept lattice $\underline{\mathfrak{B}}_{\mathcal{HA}_1}(G, M, I)$

5 Implementation

The idea of lattice-based querying and navigation with respect to hierarchies of attributes detailed above is implemented in the BR-Explorer [16,15] system running on the BioRegistry corpus [21]. The general architecture of BR-Explorer is given in figure 8. The left part (*Annotation Collection* and *BR Construction*) corresponds to the data preparation task and is detailed in [21]. The extracted formal context contains 729 biological databases and 231 attributes. The corresponding concept lattice contains 638 formal concepts. The main part of BR-Explorer is based on the BR-Explorer retrieval algorithm detailed in [15]. The user interface of BR-Explorer has been designed to facilitate the definition of queries and hierarchies of attributes. In the current version of BR-Explorer, only query-dependant hierarchies of attributes module is implemented. The visualization of detailed answers as well as their neighborhood in the concept lattice have been designed for query refinement (from ontologies as detailed in [16] or simply by considering more adapted attributes), attribute dependency redefinition, and direct access to the Web sites of relevant databases. The screenshots of a detailed query session in BR-Explorer can be found at BR-Explorer Web page *http://www.loria.fr/~messai/BR-Explorer.*

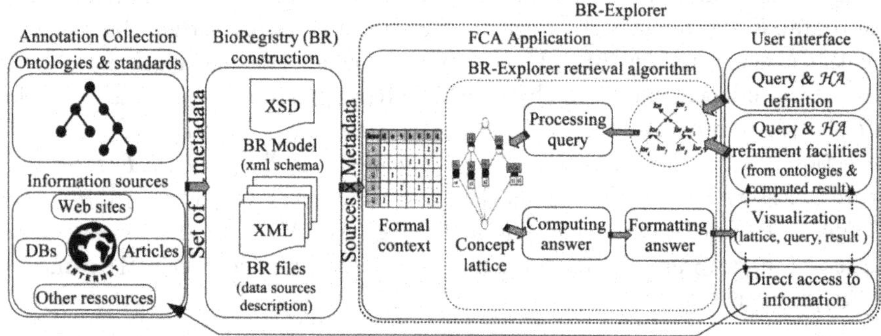

Fig. 8. Global architecture of BR-Explorer system

To experimentally evaluate the usefulness of hierarchies of attributes, a domain expert has defined a set of queries firstly without hierarchies of attributes and secondly with hierarchies of attributes, and submitted the queries to BR-Explorer. The comparison between answers of both cases has shown that the use of hierarchies of attributes considerably reduces the irrelevant databases especially in the case of queries containing “generic” attributes (indexing a large number of databases).

6 Conclusion and Perspectives

Taking into account the dependencies between attributes of a formal context allows to efficiently explore information contained in the lattice. Indeed, navigation in the corresponding concept lattice is done in accordance with a particular way of considering attributes. The definition of hierarchies of attributes expresses which attributes are to be preferred. The lattice-based IR extension consisting in combining a concept lattice and an attribute hierarchy significantly improves the precision and accuracy of lattice-based IR approaches by removing noisy answers. Furthermore, considering hierarchies of attributes allows to guide the navigation to relevant objects and can be seen as a knowledge-based navigation in the concept lattice.

In the future, the formalization of hierarchies of attributes will be extended to the definition of viewpoints [10,7] in concept lattices. This perspective is motivated by native common characteristics between both formalisms. Indeed considering the definition of viewpoints in [20], attributes in the top level of a hierarchy of attributes can be seen as *focus* and attributes below them can be seen as *view angles* refining the focus. Furthermore, coherent concepts with respect to a hierarchy of attributes match perfectly the *perspective viewpoints*.

References

1. Belohlávek, R., Sklenar, V.: Formal Concept Analysis Constrained by Attribute-Dependency Formulas. In: Ganter, B., Godin, R. (eds.) ICFCA 2005. LNCS (LNAI), vol. 3403, pp. 176–191. Springer, Heidelberg (2005)
2. Carpineto, C., Romano, G.: A lattice conceptual clustering system and its application to browsing retrieval. Machine Learning 24(2), 95–122 (1996)
3. Carpineto, C., Romano, G.: Order-theoretical ranking. Journal of the American Society for Information Science 51(7), 587–601 (2000)
4. Carpineto, C., Romano, G.: Concept Data Analysis: Theory and Applications. John Wiley & Sons, Chichester (2004)
5. Carpineto, C., Romano, G.: Exploiting the Potential of Concept Lattices for Information Retrieval with CREDO. Journal of Universal Computer Science 10(8), 985–1013 (2004)
6. Carpineto, C., Romano, G.: Using Concept Lattices for Text Retrieval and Mining. In: Ganter, B., Stumme, G., Wille, R. (eds.) Formal Concept Analysis. LNCS (LNAI), vol. 3626, pp. 161–179. Springer, Heidelberg (2005)
7. d'Aquin, M., Lieber, J., Napoli, A.: Decentralized case-based reasoning for the semantic web. In: Gil, Y., Motta, E., Benjamins, V.R., Musen, M.A. (eds.) ISWC 2005. LNCS, vol. 3729, pp. 142–155. Springer, Heidelberg (2005)
8. der Merwe, D.V., Obiedkov, S.A., Kourie, D.G.: AddIntent: A New Incremental Algorithm for Constructing Concept Lattices. In: Eklund, P.W. (ed.) ICFCA 2004. LNCS (LNAI), vol. 2961, pp. 372–385. Springer, Heidelberg (2004)
9. Ducrou, J., Vormbrock, B., Eklund, P.W.: FCA-Based Browsing and Searching of a Collection of Images. In: ICCS 2006. 14th International Conference on Conceptual Structures, Aalborg, Denmark, July 16-21, pp. 203–214 (2006)
10. Ferber, J., Volle, P.: Using coreference in object-oriented representations. In: 8th European Conference on Artificial Intelligence, ECAI 1988, Munich, Germany, August 1988, pp. 238–240 (1988)
11. Ferré, S., Ridoux, O.: Searching for objects and properties with logical concept analysis. In: Delugach, H.S., Stumme, G. (eds.) ICCS 2001. LNCS (LNAI), vol. 2120, pp. 187–201. Springer, Heidelberg (2001)
12. Ganter, B., Wille, R.: Formal Concept Analysis: Mathematical Foundations. Springer, Heidelberg (1999)
13. Godin, R., Mineau, G.W., Missaoui, R.: Méthodes de classification conceptuelle basées sur les treillis de Galois et applications. Revue d'intelligence artificielle 9(2), 105–137 (1995)

14. Koester, B.: Conceptual Knowledge Retrieval with FooCA: Improving Web Search Engine Results with Contexts and Concept Hierarchies. In: Perner, P. (ed.) ICDM 2006. LNCS (LNAI), vol. 4065, pp. 176–190. Springer, Heidelberg (2006)
15. Messai, N., Devignes, M.-D., Napoli, A., Smail-Tabbone, M.: BR-Explorer: An FCA-based algorithm for Information Retrieval. In: Fourth International Conference on Concept Lattices and their Applications - CLA 2006, Yasmine Hammamet, Tunisia, October 30 - November 1, 2006, pp. 285–290 (2006)
16. Messai, N., Devignes, M.-D., Napoli, A., Smal-Tabbone, M.: Querying a bioinformatic data sources registry with concept lattices. In: 13th International Conference on Conceptual Structures, ICCS 2005, Kassel, Germany, July 18-22, pp. 323–336 (2005)
17. Mimouni, N., Slimani, Y.: Indexing and Searching Video Sequences Using Concept Lattices. In: Fourth International Conference on Concept Lattices and their Applications - CLA 2006, Yasmine Hammamet, Tunisia, October 30 - November 1 2006, pp. 285–290 (2006)
18. Priss, U.: A Graphical Interface for Document Retrieval Based on Formal Concept Analysis. In: 8th Midwest Artificial Intelligence and Cognitive Science Conference, Dayton, Ohio, USA, pp. 66–70 (1997)
19. Priss, U.: Lattice-based Information Retrieval. Knowledge Organization 27(3), 132–142 (2000)
20. Ribière, M., Dieng-Kuntz, R.: A Viewpoint Model for Cooperative Building of an Ontology. In: Priss, U., Corbett, D., Angelova, G. (eds.) ICCS. LNCS, vol. 2393, pp. 220–234. Springer, Heidelberg (2002)
21. Smaïl-Tabbone, M., Osman, S., Messai, N., Napoli, A., Devignes, M.-D.: BioRegistry: A Structured Metadata Repository for Bioinformatic Databases. In: Computational Life Sciences, First International Symposium, CompLife 2005, Konstanz, Germany, September 25-27, 2005, pp. 46–56 (2005)

PACTOLE: A Methodology and a System for Semi-automatically Enriching an Ontology from a Collection of Texts

Rokia Bendaoud, Yannick Toussaint, and Amedeo Napoli

UMR 7503 LORIA, BP 239, 54506 Vandœuvre-lès-Nancy, France
{bendaoud,yannick,napoli}@loria.fr

Abstract. PACTOLE stands for "Property And Class characterization from Text for OntoLogy Enrichment" and is a semi-automatic methodology for enriching an initial ontology from a collection of texts in a given domain. PACTOLE is also the name of the associated system relying on Formal Concept Analysis (FCA). In this way, PACTOLE is able to derive a concept lattice from a formal context, consisting of a binary table describing a set of individuals with their properties. Given a domain ontology and a set of objects with their properties (extracted from a collection of texts), the PACTOLE system builds two concept lattices: the first corresponding to the restriction of the ontology schema to the considered objects and the second to the extracted pairs (object, property). As they are based on the same set of individuals, the two ontologies are merged using context apposition. The resulting final concept lattice is analyzed and a number of knowledge units can be extracted and furthermore used for enriching the initial ontology. Finally, the final concept lattice is mapped within the $\mathcal{FLE}$ KR formalism. The paper introduces and explains in details the PACTOLE methodology with the help of an example in the domain of astronomy.

1 Introduction

1.1 Motivation and Context

Ontologies are the backbone of Semantic Web. They help software and human agents to communicate by providing shared and common domain knowledge, and by supporting various tasks, e.g. problem-solving and information retrieval [11]. An ontology is usually based on a concept hierarchy and a set of relations between the concepts. In turn, a concept hierarchy structures domain knowledge into a set of hierarchically organized classes, making easier information search and reuse. However, the design and the enrichment of an ontology are hard and time-expensive tasks. Indeed, the knowledge acquisition bottleneck is one major factor slowing down ontology-driven applications [3]. This point is illustrated hereafter by an example taken from the domain of astronomy and used in the whole paper (this research work is carried out in the context of a project done in collaboration with researchers in astronomy). In this application domain, the

P. Eklund and O. Haemmerlé (Eds.): ICCS 2008, LNAI 5113, pp. 203–216, 2008.

design of a concept hierarchy and the identification/classification of celestial bodies, i.e. assigning a class to a given celestial body, are very difficult tasks, because of the growing number of discovered celestial bodies and the need of new classes to be defined. Traditionally, the classification task is performed "manually", according to the object properties appearing in the astronomy documents. The task consists in firstly reading scientific articles holding on the celestial object under study and secondly finding a possible class for that object. At present, more than three millions of celestial objects are classified in this way and made available in the SIMBAD database[1]. The SIMBAD database is one of the most important databases in astronomy memorizing the properties of celestial objects. But the SIMBAD database remains a database and has not the architecture of an ontology: no definition, no explicit representation of relations, no classification procedures built-in, and a considerable work has to be done for classifying the billion of remaining celestial objects. The task is tedious for human experts, who are not always confident with their own classification, mainly because classes lack precise and unambiguous definitions. Thus, the design of an ontology for guiding the classification of celestial bodies would be of great help for astronomy practitioners.

In this way, this paper presents a methodology and a system for designing an ontology from a collection of astronomical texts. One originality is that the resulting ontology is completed with the help of domain resources, e.g. domain ontology, database, or thesaurus. Accordingly, and this is the case in this paper, the methodology can be used for enriching the knowledge included in an existing resource, here a domain ontology based on the SIMBAD database. This approach can be used for partly solving the knowledge acquisition bottleneck. Moreover, it can be noticed that the methodology is not dependent on the domain and other experimentations have been carried out in the domain of biology. More precisely, the PACTOLE methodology –PACTOLE stands for "Property And Class Characterization from Text to OntoLogy Enrichment"– takes as input a collection of texts in astronomy and a domain resource, i.e. an ontology based on the SIMBAD database, and gives as output a set of new concepts and instances to be inserted in the initial ontology. The enrichment process is based on Formal Concept Analysis (FCA) [7]. In addition, for being inserted in the ontology, all knowledge units are represented within the Description Logics (DL) language $\mathcal{FLE}$ where the following constructors are available: conjunction ($\sqcap$), universal quantification ($\forall$), and existential quantification ($\exists$). The description logics $\mathcal{FLE}$ is used for representing concepts and relations in the ontology and has a sufficient power of representation for that task.

Actually, the PACTOLE system implements the PACTOLE methodology and builds two concept hierarchies using FCA: one concept hierarchy derives from the collection of texts and one concept hierarchy derives from the SIMBAD database (mentioned here before as the ontology based on the SIMBAD database). After that, the two concept hierarchies are merged by the operation of context apposition as introduced and discussed in [7].

[1] `http://simbad.u-strasbg.fr/simbad/sim-fid`

Applying in this way the FCA process for the enrichment of an ontology is an original design operation that brings forward two main benefits. Firstly, a FCA-based concept hierarchy provides a formal basis and specification for the resulting ontology. Moreover, many efficient FCA-based operations are designed for extending, maintaining, and managing a concept hierarchy, such as performing an incremental update of the hierarchy by adding either an object or an attribute (property), or assembling a concept lattice from parts. Secondly, as the concept hierarchy changes (because texts are changing for example), the ontology evolves in a correct and consistent way. The transformation of the concept lattice into a DL knowledge base (KB) allows then to query the KB with the help of a DL reasoner and to ask complex expert questions.

1.2 An Introductory Example

Let us consider the problem of detecting why two celestial objects are in the same class. To answer the question, the set of properties shared by both objects has to be characterized. The extraction of such set of common properties relies on a search in astronomical texts of elements that can be considered as properties for identifying the class of an object. For example, in a sentence such as "*We report the discovery of strong flaring of the object HR2517*", it is asserted that the object HR2517 can *flare*, i.e. showing an eruption of plasma at the surface of the object. The fact of flaring means for a celestial object, here HR2517, that the object is a particular type of star. In another sentence such as "*NGC 1818 contains almost as many Be stars as the slightly younger SMC cluster NGC 330*", it is asserted that the object NGC 1818 *contains* something. The fact of containing means that this celestial object is not a star.

In these sentences, the property of an object is given by a verb. A similar approach has been used in [6] and is based on Harris hypothesis [10], stating that terms in sentences are similar if they share similar linguistic contexts, here the similarity of verb-argument dependencies. In this way, individuals and their properties are extracted from a collection of texts using Natural language processing (NLP) tools. Then, the FCA process is used for building a concept hierarchy from a formal context, composed of a set of individuals, e.g. SMC, T, Tauri, a set of properties, e.g. contains, flaring, and a binary relation defined on the Cartesian product of both sets stating that an object has or has not a given property.

Given a concept hierarchy and the derived ontology represented in the $\mathcal{FLE}$ DL, complex expert questions can be answered. The questions are first given in natural language and then represented as DL queries. Such expert questions can be read as the following: *do the celestial objects 3C 273 and SMC belong to the same class?* or *What is the class of the celestial object V773 Tau?*.

1.3 Organization of the Paper

The following sections of this paper are organized as follows. The next section introduces the definitions of ontology enrichment and the basics of FCA. In the Section 3, the PACTOLE methodology is presented and the operations of knowledge extraction from texts, concept hierarchy design and representation (in $\mathcal{FLE}$), and ontology enrichment, are explained and illustrated. In Section 4, an evaluation of each step of the PACTOLE methodology system is given followed by a discussion and a synthesis of the present research work. Section 5 briefly presents related works on ontology design and enrichment. Finally, the Section 6 concludes the paper and shows future works.

2 Ontologies Enrichment and Formal Concept Analysis

In this section, the background definitions for the PACTOLE methodology are given. According to the general and commonly admitted statement in [9], an ontology is an explicit specification of a domain conceptualization. Moreover, an ontology is usually developed for the purposes of domain knowledge sharing and reuse. Following this way, the objective of the PACTOLE methodology is to enrich an existing domain ontology from a collection of texts, to solve a particular problem, e.g. expert question answering.

2.1 The Enrichment of an Ontology

The following definition of ontology enrichment is based on the work of Faatz and Steinmetz [5]. This enrichment operation is based on a so-called "set of formulas" for each concept of the initial ontology, including new concepts, new properties, and new instances.

Definition 1 (Ontology Enrichment). *Let Texts be a collection of written texts and Exp(Texts) a set of expressions that have been extracted from Texts by NLP tools. Expressions may be nouns or pairs (subject, verb). An algorithm for ontology enrichment from text denoted hereafter by AOET takes as input an ontology Ω and a set Exp(Texts), and returns as output an enriched ontology $\Omega \cup P$, where P is a set of formulas represented within the same representation formalism as Ω and obtained as follows. For each element $e \in$ Exp(Texts), AOET returns a formula f(e) that can be either an individual, a concept, or a role, involving e, and depending on the status of e in Exp(Texts), as explained in the following.*

2.2 Formal Concept Analysis

Formal concept analysis (FCA) [7] is a mathematical formalism allowing to derive a concept lattice (to be defined later) from a formal context $\mathbb{K}$ constituted of a set of objects G, a set of attributes M, and a binary relation I defined on the Cartesian product $G \times M$ (in the binary table representing $G \times M$, the rows correspond to objects and the columns to attributes or properties). FCA can be used for a number of purposes among which knowledge formalization and acquisition, ontology design, and data mining. The concept lattice is composed of *formal concepts*, or simply *concepts*, organized into a hierarchy by a partial ordering (a subsumption relation allowing to compare concepts). Intuitively, a concept is a pair (A, B) where $A \subseteq G$, $B \subseteq M$, and A is the maximal set of objects sharing the whole set of attributes in B and vice-versa. The concepts in a concept lattice are computed on the basis of a Galois connection defined by two derivation operators denoted by $'$:

$$' : G \rightarrow M; A' = \{m \in M; \forall g \in A : (g, m) \in I\}$$
$$' : M \rightarrow G; B' = \{g \in G; \forall m \in B : (g, m) \in I\}$$

Formally, a concept (A, B) verifies $A' = B$ and $B' = A$. The set A is called the *extent* and the set B the *intent* of the concept (A, B). The subsumption (or subconcept–superconcept) relation between concepts is defined as follows: $(A_1, B_1) \sqsubseteq (A_2, B_2) \Leftrightarrow A_1 \subseteq A_2$ (*or* $B_2 \subseteq B_1$). Relying on this subsumption relation $\sqsubseteq$, the set of all concepts extracted from a context $\mathbb{K} = (G, M, I)$ is organized within a complete lattice, that means that for any set of concepts there is a smallest superconcept and a largest subconcept, called the *concept lattice* of $\mathbb{K}$ and denoted by $\underline{\mathfrak{B}}(G, M, I)$.

3 The PACTOLE Methodology

PACTOLE is a methodology for enriching in a semi-automatic way an initial ontology based on a domain resources (thesaurus, database,...) with knowledge extracted from texts. PACTOLE is inspired from two methodologies, namely "Methontology" [8] and "SENSUS" [14]. From "Methontology", PACTOLE borrows the idea of keeping an expert in the loop to validate operations such as building from a set of terms extracted from resources defining a set of DL concepts. From "SENSUS", PACTOLE borrows the idea of being based on an existing ontology and enriching this initial ontology with resources such as texts. The PACTOLE process is based on five steps presented in Figure 1, each step in PACTOLE involves the experts validation.

The first step involves NLP processing for extracting from texts objects of the domain and their properties. The expressions that are considered are verb/subject,

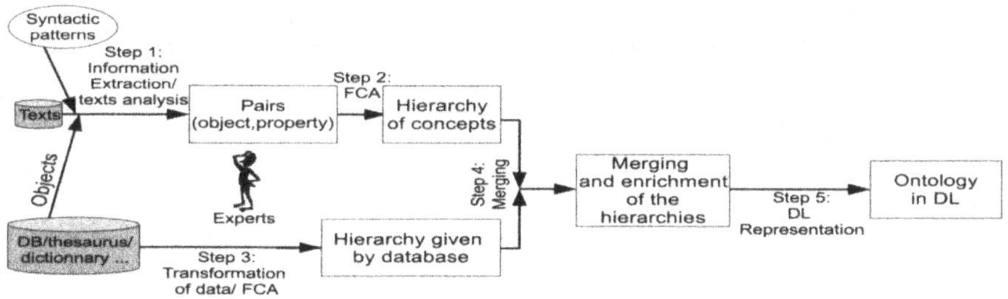

Fig. 1. PACTOLE Methodology

verb/object, verb/complement, and verb/prepositional phrase dependencies. They are good syntactic hints for assigning a property to an object. Each of these hints provides a pair (object, property). In the second step, FCA is used for building a concept lattice from the pairs (object, property). A concept in the hierarchy is composed of a maximal set of individuals sharing a maximal set of attributes (or properties) and vice-versa. The third step converts the existing knowledge resources into a lattice structure using FCA. During the fourth step, the two lattices are merged. The idea here is that the concept hierarchy from the initial knowledge resources can be partially enriched by the concept lattice resulting extracted from texts. During step five, the final (merged) lattice is represented with the $\mathcal{FLE}$ DL formalism. The following subsections give details on each step.

3.1 Text Analysis

This step aims at extracting from the texts a list of pairs (object, property). A preliminary task identifies celestial objects in the texts. Then, texts are parsed to extract syntactic dependencies, and some syntactic dependencies involving celestial objects are selected and translated into pairs of the form (celestial_object, property).

Detection of celestial objects. There is no normalization process for naming a celestial object in astronomy. Thus, identifying the names of the objects in the texts requires two complementary strategies which are suggested by the SIMBAD database: some names are already known (such as "`Orion`") and the string can be used to locate them in the texts. Some other names such as "`NGC 6994`" are described by a pattern "`NGC NNNN`" where NNNN is a number.

The system has extracted 1382 celestial objects from the collection of texts, this number representing 90% of the whole set of objects in the texts (as evaluated by the experts). Three new objects were identified: they were not in the SIMBAD database: `HH 24MMS`, `S140 IRS3`, `M33 X-9`. However, a few detected objects were not celestial objects. Three main failures in object identification have been pointed out:

- Underspecified patterns: some objects having the same pattern as celestial objects are not celestial objects: The `IRA X` pattern in SIMBAD covers `IRAS 16293` which is a celestial object but also `IRAM 30` which is a telescope,
- Abbreviations in texts: some authors use short ways to name objects in the texts, e.g. `S 180` instead of `Sand 180` as registered in SIMBAD,
- Typing errors in SIMBAD: some errors were made while typing the name of objects in SIMBAD, e.g. `Name Lupus 2` instead of `Lupus 2`.

Extraction of properties. The properties are extracted by parsing the texts with the shallow "Stanford Parser"[2] [4]. The Stanford Parser parses texts and extracts syntactic dependencies between a verb and its subjects, objects, complements, and preposition phrases. For example: *"NGC 1818 contains almost*

[2] http://nlp.stanford.edu/software/lex-parser.shtml

as many Be stars as the slightly younger SMC cluster NGC 330". The list of dependencies is the following:

- subject(contains-2,NGC 1818-1), direct_object(contains-2,Be stars-6)

Only verb dependencies are kept to build the pairs (celestial_object, property). The pair (contains, NGC 1818) is derived from the dependency subject(contains-2, NGC 1818-1), meaning that `NGC 1818` is able to `contain`. The pair (Be star, contained) is derived from dependency direct_object(contains-2, Be stars-6), meaning that Be stars can be `contained`.

Among the set of pairs (object, property), some are pure linguistic artefacts. They are not relevant to astronomy and should be filtered before the classification process. Firstly, properties which occurs only once are considered as noise and deleted. Secondly, the system deals with synonymy (`consists`, `contains` and `includes`...) for reducing dispersion. These properties are grouped and considered as the same property. Finally, for each remaining pair, an astronomer decides whether it is meaningful to keep the pair for the classification process. For example, properties such as `performing` or `oscillating` have been considered of low interest, while some others pairs such as `rotating` were considered as interesting.

This step allows the system to discover some properties which were previously unknown, in the sense that no correlation was known between celestial types of objects and properties. For example, the objects "`59 Aurigae, V1208 Aql`" can `pulse`, the object "`MM Herculis`" can `eclipse` or the objects "`AB Dor, OJ 287`" can `flare`.

3.2 Classifying Celestial Objects from the Texts Using FCA

The set of pairs extracted from the text are then transformed under the form of a binary table objects $\times$ properties leading to a formal context $\mathbb{K}_1 = (G, M_1, I_1)$ to which FCA method will be applied. Here G is a set of the celestial objects identified in the texts, M_1 is the set of properties extracted from texts and modified as described above, and I_1 is the relaion and $I_1(g, m_1)$ is a statement, that g has the property m_1. An example of such a lattice is given in Figure 2.

	observed	expanding	flaring	emits	includes
3C 273	X			X	X
TWA	X	X			
SMC	X			X	
T Tauri	X		X	X	
V773 Tau	X		X	X	

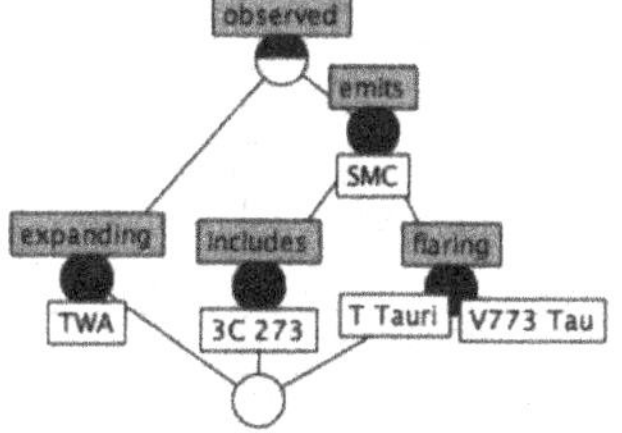

Fig. 2. The context $\mathbb{K}_1 = (G, M_1, I_1)$ and the lattice of this context

	Quasar	Association of Stars	Galaxy	Star	T Tau type Star
3C 273	X		X		
TWA		X			
SMC			X		
T Tauri				X	X
V773 Tau				X	X

Fig. 3. The context $\mathbb{K}_2$=(G, M_2, I_2) and the lattice of this context

3.3 Classifying Celestial Objects from Simbad Database Using FCA

The hierarchical structure defined in SIMBAD is encoded into a concept lattice so that both hierarchical structures – from SIMBAD and from texts – are expressed in the same formalism namely a concept lattice. The context related to SIMBAD is $\mathbb{K}_2$=(G, M_2, I_2) where G is a set of celestial objects identified in the texts, M_2 is the set of SIMBAD classes, and $I_2(g, m_2)$ is the relation stating that g has or has not the class m_2. An example of concept lattice extracted from SIMBAD is given on Figure 3.

3.4 Merging the Two Lattices

The PACTOLE system proposes to enrich the lattice resulting from SIMBAD with the concept lattice of celestial objects built from the texts. Merging these two concept lattices relies on the apposition operation as defined in [7]:

Definition 2. *Let* $\mathbb{K}_1$=(G_1, M_1, I_1), *and* $\mathbb{K}_2$=(G_2, M_2, I_2) *be formal contexts. If* $G = G_1 = G_2$ *and* $M_1 \cup M_2 = \emptyset$ *then:* $\mathbb{K} := \mathbb{K}_1|\mathbb{K}_2 := (G, M_1 \cup M_2, I_1 \cup I_2)$ *is the apposition of the two contexts* $\mathbb{K}_1$ *and* $\mathbb{K}_2$.

The two contexts are respectively $\mathbb{K}_1$=(G, M_1, I_1) (presented in Figure 2) and $\mathbb{K}_2$=(G, M_2, I_2) (presented in the Figure 3). The apposition context $\mathbb{K} = (G, M, I)$ is presented in the Table 1 where G is the same set of objects for $\mathbb{K}_1$ and $\mathbb{K}_2$, $M := M_1 \cup M_2$ where M_1 is the set of properties extracted from the texts and M_2 is a set of the classes of SIMBAD, and $I := I_1 \cup I_2$. The resulting concept lattice is presented in Figure 4.

Table 1. The context $\mathbb{K} = (G, M, I)$

	Quasar	Association of Stars	Galaxy	Star	T Tau type Star	observed	expanding	flaring	emits	includes
3C 273	X		X			X			X	X
TWA		X				X	X			
SMC			X			X			X	
T Tauri				X	X	X		X	X	
V773 Tau				X	X	X		X	X	

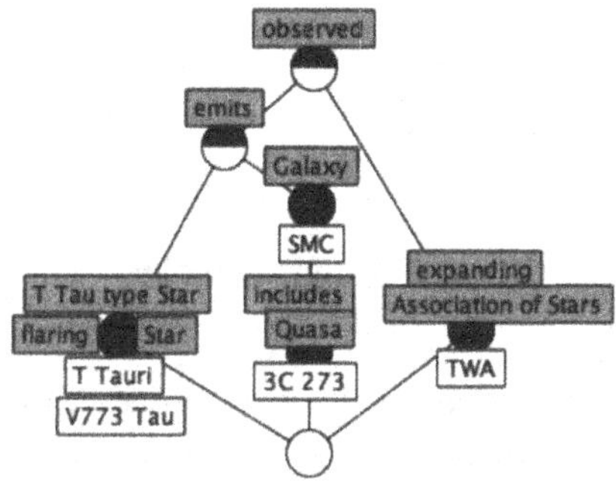

Fig. 4. Lattice of the context $\mathbb{K} = (G, M, I)$

3.5 Representing the Concepts with $\mathcal{FLE}$

The last step in PACTOLE is aimed at transforming the final lattice into an ontology represented in $\mathcal{FLE}$.

This transformation called α is based on a set of elementary transformations defined as follows: $\alpha : \mathbb{K} = (G, M, I) \rightarrow$ TBox $\sqcup$ ABox, where: $\mathbb{K}$ is a formal context, TBox and ABox being the bases of the ontology. The elementary transformations are the following:

1. A formal attribute $m_2 \in M_2$ is transformed in the TBox as an atomic concept c $\equiv \alpha(m_2) \equiv m_2$. A class in SIMBAD is represented as a concept, e.g. α(**quasar**)=**quasar**,
2. A formal attribute $m_1 \in M_1$ is transformed in the TBox as a defined concept c $\equiv \alpha(m_1) \equiv \exists m_1.\top$. Formal attributes are used as roles for defined concepts, e.g. α(**observed**) $\equiv$ $\exists$**observed.**$\top$,
3. A formal concept $c = (X, Y) \in \mathcal{C}$ is transformed in the TBox as defined concept $\alpha(c)$, i.e. $\alpha(c) \equiv \sqcap_{m \in Y} \alpha(m)$ where $\alpha(m)$ are either atomic or defined concepts, e.g. α(C$_4$) $\equiv$ **Star** $\sqcap$ **T_Tau_type_Star** $\sqcap$ $\exists$**observed.**$\top$ $\sqcap$ $\exists$**emits.**$\top$ $\sqcap$ $\exists$**flaring.**$\top$,
4. A subsumption relation between formal concepts C and D is transformed in the TBox as a general concept inclusion $\alpha(C) \sqsubseteq \alpha(\bar{D})$, e.g. α(C$_4$) $\sqsubseteq$ α(C$_1$),
5. A formal object $g \in G$ is transformed in the ABox as an instance $\alpha(g)$, e.g. α(**T Tauri**) = **T_Tauri** is an instance.

Table 2. Definition of each concept of the final lattice

N° in the lattice	Definition
C_0	$\exists$**observed.**$\top$
C_1	$\exists$**observed.**$\top$ $\sqcap$ $\exists$**emits.**$\top$
C_2	**Association_of_Stars** $\sqcap$ $\exists$**observed.**$\top$ $\sqcap$ $\exists$**expanding.**$\top$
C_3	**Galaxy** $\sqcap$ $\exists$**observed.**$\top$ $\sqcap$ $\exists$**emits.**$\top$
C_4	**Star** $\sqcap$ **T_Tau_type_Star** $\sqcap$ $\exists$**observed.**$\top$ $\sqcap$ $\exists$**emits.**$\top$ $\sqcap$ $\exists$**flaring.**$\top$
C_5	The bottom : $\bot$
C_6	**Galaxy** $\sqcap$ **Quasar** $\sqcap$ $\exists$**observed.**$\top$ $\sqcap$ $\exists$**emits.**$\top$ $\sqcap$ $\exists$**includes.**$\top$

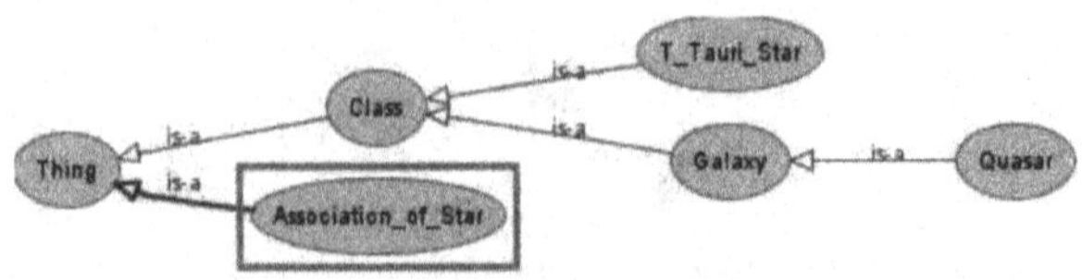

Fig. 5. Final ontology

The definition of each concept of the final lattice in Figure 4 is presented in Table 2. The resulting ontology shown in Figure 5 can be used for two kinds of tasks:

1. Instantiation of concepts. Let o_1 be a celestial object having the properties {a,b} and belonging to classes {C_1,C_2} in SIMBAD. A first task is instantiation, i.e. finding the class of an object such as o_1. The class of o_1 is a most general class X in the final ontology such that $\mathsf{X} \sqsubseteq \exists \mathsf{a}.\top \sqcap \exists \mathsf{b}.\top \sqcap \mathsf{C}_1 \sqcap \mathsf{C}_2$. When there exists more than one candidate class for being the class of an object o_1 say D_1 and D_2, the conjonction $D_1 \sqcap D_2$ becomes the class of o_1. For example, let us consider the question *"What is the class of the object V773 Tau, having the properties* {observed,flaring,emits} and belonging to the classes {Star,T Tau Star} in SIMBAD? The answer is the most general class $\mathsf{X} \sqsubseteq \exists \mathsf{observed}.\top \sqcap \exists \mathsf{flaring}.\top \sqcap \exists \mathsf{emits}.\top \sqcap \mathsf{Star} \sqcap \mathsf{T\_Tau\_Star}$, here the concept C_4 in the ontology.
2. Comparison of celestial objects. Let us consider two objects o_1 and o_2. A second task consists in comparing o_1 and o_2 and determining whether o_1 and o_2 are in the same class. One way for checking that is to find the class of o_1, then the class of o_2, and then to test whether the two classes are identical. For example, let us consider the two objects named 3C 273 and SMC. The object 3C 273 is an instance of the class C_6 and the object SMC is an instance of the class C_3. As $C_6 \sqcap C_3 = C_6$, the objects 3C 273 and SMC are not in the same class.

4 Evaluation

In this section, the PACTOLE methodology is evaluated, mainly by comparing the concept hierarchy associated to the resulting ontology and the initial existing hierarchy, here the SIMBAD database. The PACTOLE system has been applied on 11591 abstracts from the A&A "Astronomy and Astrophysics" journal for the years 1994 to 2002.

4.1 Evaluation of the Process

The Stanford Parser analyzes 68.5% of the sentences in the texts, where the maximum size of the parsed sentences is between 31 and 36 words. The system extracts three different sets of syntactic dependencies between verb and arguments, namely SO, SOC, and SOCP (detailed in Table 3) where:

Table 3. The results of the parser

	SO				SOC				SOCP			
	Pairs	Obj.	Prop.	Conc.	Pairs	Obj.	Prop.	Conc.	Pairs	Obj.	Prop.	Conc.
11591 abstracts	384	209	14	30	401	211	14	30	1709	470	23	70

- SO: subject(object,verb) + object(object,verb),
- SOC: SO + complement(object,verb),
- SOCP: SOC + preposition_X(object,verb), where X can be (in, of, ...).

A concept lattice with 94 concepts has been built from the SIMBAD database, where 470 objects and 92 properties have been considered in the formal context.

The lattice resulting from apposition was presented to the astronomers. Actually, new concepts have been discovered such as the concept ({`Orion, TWA`}, {`Association_of_stars, expanding, observed`}). This concept represents the `Association_of_stars` than can `expand`. The concept is considered as interesting by domain experts, and labelled as the `Association_of_Young_Stars`.

4.2 Evaluation of Hierarchy Correspondence

The correspondence between the concept hierarchy extracted from the collection of texts and the concept hierarchy extracted from SIMBAD database has to be checked. Here the objective is to check whether the PACTOLE system has defined each class of the concept hierarchy resulting from SIMBAD (validation classes) as a class with properties extracted from the collection of texts (experimentation classes). This correspondence relies on similarity between sets of instances. In order to do so, the measures of precision and recall have been used. The precision and the recall are calculated for each experimentation classes with respect to one of the closest class in verification class using the Euclidean distance. The global precision (Precision_F) and the global recall (Recall_F) are the average of all precisions (respectively of all recalls).

Calculate the global precision and recall. The precision is the number of common instances between C_{E_i} (experimentation class i) and C_{V_j} (validation class j) divided by the number of instances in C_{E_i}. The recall is the number of common instances between C_{E_i} and C_{V_j} divided by the number of instances in C_{V_j}. N is the number of classes in C_E.

$$Precision_i = \frac{C_{E_i} \cap C_{V_j}}{C_{E_i}}, \quad Recall_i = \frac{C_{E_i} \cap C_{V_j}}{C_{V_j}}$$

$$Precision\_F = \frac{\sum_{i=1..N}(Precision_i)}{N}, \quad Recall\_F = \frac{\sum_{i=1..N}(Recall_i)}{N}$$

Detection of the closest class. For each class has been searched for one of the closest class in the classes of SIMBAD using the Euclidian distance, if we find

two closest classes, one of them is taken. Let G be the set of objects, E the set of experimentation classes, and V a set of validation classes. For each class C_{E_i} $\in E$, and for each class $C_{V_j} \in V$, vector V_{E_i} and V_{V_j} are defined as:
$\forall g \in G$: if g is an instance of C_{E_i} then $V_{E_i}[g] = 1$ else $V_{E_i}[g] = 0$
$\forall g \in G$: if g is an instance of C_{V_i} then $V_{V_i}[g] = 1$ else $V_{V_i}[g] = 0$,
then:

$$Distance(V_{E_i}, V_{V_j}) = (\sum_{k=0}^{N}(V_{E_i}[g] - V_{V_j}[g])^2)^{1/2}$$

C_{V_j} is one of the closest class of C_{E_i} iff $\forall V_{V_p} \in V - \{V_{V_j}\}$ Distance $(V_{E_i}, V_{V_p}) \geqslant$ Distance (V_{E_i}, V_{V_j}).

For example, let G be the set of objects $G = \{$`3C 273, TWA, SMC, T Tauri, V773 Tau`$\}$ (see the Figures 2 and 3). One of the closest class for C_{E_1} with instances $\{$`3C 273, SMC, T Tauri, V773 Tau`$\}$ (Figure 2) is class C_{V_1} in SIMBAD with instances $\{$`3C 273, SMC`$\}$ (Figure 3). The distance between the vector associated to C_{E_1} that is V_{E_1} = [1,0,1,1,1] and the vector associated to C_{V_1} that is V_{V_1} = [1,0,1,0,0] is the minimal distance.

$Distance(V_{E_1}, V_{V_1}) = \sqrt{2}$, $Precision_1 = \frac{C_{E_1} \cap C_{V_1}}{C_{E_1}}$=0.5, $Recall_1 = \frac{C_{E_1} \cap C_{V_1}}{C_{V_1}}$=1.

4.3 Discussion

The PACTOLE system allows to extract new knowledge units in the astronomy domain and to enrich an ontology associated to the SIMBAD database. These knowledge units can be divided in three kinds. The first kind is related to the identification of new celestial objects (see the subsection 3.1). The second kind is related to the discovery of new correlations between celestial objects and their properties (see the subsection 3.1). The third kind is related to the proposition of new classes in SIMBAD (see the subsection 4.1). The experiment in astronomy shows also that using all syntactic dependencies (SOCP) leads to better results.

The SOCP set allows the extraction of more pairs, more properties and more classes (see Table 3). This set also offers a better precision and a better recall (see Table 4). The score of precision is high (74.71%) meaning that objects are classified in adequate classes. The score of recall is low for several reasons. The first reason is that the number of properties associated with objects is not sufficient. Sometimes, the system extracts only one or two properties for an object and this is too small for classification. The second reason is that verbs are not the sole properties for defining a class, considering for example adjectives, adverbs, measures, etc. The third reason is that some properties are implicit and they cannot be extracted by any analyzer.

Table 4. Resulting measures of precision and recall for differents set of dependencies

	SO		SOC		SOCP	
	Final Precision	Final Recall	Final Precision	Final Recall	Final Precision	Final Recall
FCA	58.33%	05.03%	58.91%	05.94%	74.71%	30.22%

5 Related Work

Buitelaar et al. [1] is a reference book on ontologies extracted from texts. The different aspects of ontology development are presented: methods, evaluation, and applications. Some approaches aim at building ontologies starting from scratch. For example, Faure et al. [6] use a syntactic structure to describe an object by the verb with which it appears and then statistic measures are used to build a concept hierarchy. Cimiano in [3] use a similar approach but use FCA for building a concept hierarchy. With respect to Cimiano our method proposes a formalization for the resulting ontology, adding defined concepts and we involve knowledge expert.

In the scientific domain, it is important to integrate expert knowledge because some knowledge units are implicit in texts. Stumme et al. [13] merge two ontologies for building a new one. The proposed method takes as input a set of natural language documents. NLP techniques are used to capture two formal contexts encoding the relationships between documents and concepts in each ontology. This method combines the knowledge of the collection of texts and the expert knowledge. In comparaison with our approach, the approach of Stumme et al. uses the texts for merging and not for enriching the two ontologies. Navigli et al. [12] propose to enrich an existing ontology using on-line glossaries. They use natural language definitions of each class and convert them into formal (OWL) definitions, compliant with the core ontology property specifications. Castano et al. [2] also propose to enrich an existing ontology by matching the existing ontology and new knowledge extracted from data. Regarding this methodology, PACTOLE uses a similar idea for evaluating the resulting ontology by similarity between existing and new concepts. This method is called "shallow similarity" by the authors. A difference is that they compare the set of properties while we compare the set of instances.

6 Conclusion and Future Work

In this paper, we have presented a methodology for semi-automatically enriching an ontology from a collection of texts. This methodology merges a concept hierarchy extracted from a collection of texts with text mining method and a concept hierarchy representing domain knowledge. We have shown how the resulting concept hierarchy can be represented within the DL language $\mathcal{FLE}$. The proposed methodology was applied to astronomy for extracting knowledge units about celestial objects for problem-solving purposes such as celestial object classification and comparison. We also evaluated the PACTOLE methodology in this context and proposed a definition for precision and recall for evaluating the hierarchy correspondence.

One future work consists in improving the PACTOLE system for the classification of objects annotated “Object of unknown nature” in SIMBAD and suggestion of classes for these objects. Another work consists in integrating relations between the celestial objects in the definition of classes. It is also planned to test the PACTOLE methodology and system in the domain of microbiology domain for the classification of bacteria.

References

1. Buitelaar, P., Cimiano, P., Magnini, B.: Ontology Learning from Text: Methods, Evaluation and Applications. IOS Press, Amsterdam (2005)
2. Castano, S., Ferrara, A., Hess, G.N.: Discovery-driven ontology evolution. In: Tummarello, G., Bouquet, P., Signore, O. (eds.) 3rd Italian Semantic Web Workshop, Pisa, Italy (2006)
3. Cimiano, P.: Ontology Learning and Population from Text: Algorithms, Evaluation and Applications. Springer, Heidelberg (2006)
4. de Marneffe, M.C., MacCartney, B., Manning, C.D.: Generating typed dependency parses from phrase structure parses. In: 5th International conference on Language Resources and Evaluation (LREC 2006), GENOA, ITALY (2006)
5. Faatz, A., Steinmetz, R.: Ontology enrichment evaluation. In: Motta, E., Shadbolt, N.R., Stutt, A., Gibbins, N. (eds.) EKAW 2004. LNCS (LNAI), vol. 3257, pp. 497–498. Springer, Heidelberg (2004)
6. Faure, D., Nedellec, C.: Knowledge acquisition of predicate argument structures from technical texts using machine learning: The system asium. In: Fensel, D., Studer, R. (eds.) EKAW 1999. LNCS (LNAI), vol. 1621, pp. 329–334. Springer, Heidelberg (1999)
7. Ganter, B., Wille, R.: Formal Concept Analysis, Mathematical Foundations. Springer, Heidelberg (1999)
8. Gómez-Pérez, A., Fernandez-Lopez, M., Corcho, O.: Ontological Engineering. Springer, Heidelberg (2004)
9. Gruber, T.R.: A translation approach to portable ontology specification. Knowledge Acquisition 5, 199–220 (1993)
10. Harris, Z.: Mathematical Structure of Language. Wiley, J. and Sons, Chichester (1968)
11. Maedche, A.: Ontology Learning for the Semantic Web. Springer, Heidelberg (2002)
12. Navigli, R., Velardi, P.: Ontology enrichment through automatic semantic annotation of on-line glossaries. In: Staab, S., Svátek, V. (eds.) EKAW 2006. LNCS (LNAI), vol. 4248, pp. 126–140. Springer, Heidelberg (2006)
13. Stumme, G., Maedche, A.: Fca-merge: Bottom-up merging of ontologies. In: 17th International Joint Conferences on Artificial Intelligence (IJCAI 2001), pp. 225–234. Morgan Kaufmann Publishers, Inc., San Francisco (2001)
14. Valente, A., Russ, T., MacGregor, R., Swartout, W.: Building and (re)using an ontology of air campaign planning. IEEE Intelligent Systems 14(1), 27–36 (1999)

Fair(er) and (Almost) Serene Committee Meetings with Logical and Formal Concept Analysis

Mireille Ducassé[1] and Sébastien Ferré[2]

[1] IRISA/INSA, Campus Universitaire de Beaulieu, 35042 Rennes Cedex, France
[2] IRISA/Univ. Rennes 1, Campus Universitaire de Beaulieu, 35042 Rennes Cedex, France
Mireille.Ducasse@irisa.fr, Sebastien.Ferre@irisa.fr

Abstract. In academia, many decisions are taken in committee, for example to hire people or to allocate resources. Genuine people often leave such meetings quite frustrated. Indeed, it is intrinsically hard to make multi-criteria decisions, selection criteria are hard to express and the global picture is too large for participants to embrace it fully. In this article, we describe a recruiting process where logical concept analysis and formal concept analysis are used to address the above problems. We do not pretend to totally eliminate the arbitrary side of the decision. We claim, however, that, thanks to concept analysis, genuine people have the possibility to 1) be fair with the candidates, 2) make a decision adapted to the circumstances, 3) smoothly express the rationales of decisions, 4) be consistent in their judgements during the whole meeting, 5) vote (or be arbitrary) only when all possibilities for consensus have been exhausted, and 6) make sure that the result, in general a total order, is consistent with the partial orders resulting from the multiple criteria.

1 Introduction

There are numerous situations in academic life where decisions are taken in committee, for example to hire people or to allocate resources. The problem is to put a total order in partially ordered sets. For example, the applicants for a job have different qualities that are not necessarily comparable. Assume that a committee has to decide between two persons, if one is systematically better than the other one for all the criteria, the decision is easy to take. In general, however, the candidates are numerous (more than 100 in some cases), and some are the best with respect to some criteria and only average with respect to other criteria.

The final decisions of such committee meetings are necessarily arbitrary, at least partially. While some people may enjoy the opportunity to intrigue, our experience is that most participants have a genuine approach and try to be as honest as possible. This article is dedicated to such honest participants who want the process to be as rational as possible.

We conjecture that the frustrations felt by genuine people come mainly from the fact that the selection criteria are hard to express and that the global picture is too large for participants to embrace it fully.

P. Eklund and O. Haemmerlé (Eds.): ICCS 2008, LNAI 5113, pp. 217–230, 2008.

In this article, we propose a decision process where Logical Concept Analysis (LCA) [4] and Formal Concept Analysis (FCA) [5] are used to address the above two problems. We do not pretend to totally eliminate the arbitrary side of the decision. After all, a committee is in general set up when it has been recognised that there is no obvious best solution. The role of the committee is therefore to make a decision and collectively take the responsibilities for it. We claim, however, that, with our approach, genuine people have the possibility to : 1) be fair with the candidates, 2) make a decision adapted to the circumstances, 3) smoothly express the rationales of a decision, 4) be consistent in their judgements during the whole meeting, 5) vote (or be arbitrary) only when all possibilities for consensus have been exhausted, and 6) make sure that the result, in general a total order, is consistent with the partial orders resulting from the multiple criteria.

In the following we illustrate our approach with an example which reconstitutes a committee meeting which had to choose among 43 job applicants. At the original meeting, the only tool which had been used was a spreadsheet. The actual arguments which had been put forward during the discussions are explicited *a posteriori* in this article.

Two tools are used, Camelis and Conexp. Camelis[1] [3] is a concept-based information system. It relies on concept analysis to support the organizing and browsing of a collection of objects. One specificity w.r.t other (pure) FCA based systems [7,1,2] is the use of logics to represent and reason on object descriptions, queries and navigation links. This allows typed attributes to be used, for instance, date intervals, string patterns, and Boolean connectives *and, or, not.* Conexp[2], developed by Sergey A. Yevtushenko, enables, among other features, to edit a Formal Concept Analysis context and to display concept lattices.

The example illustrates how two formal concept analysis tools can help alleviate frustrations and explain a decision. We show that the taken decision has a rationale behind it and argue that had the tools been used the discussions would have been much more serene. Regarding FCA and LCA, this case study also shows that both local navigation, such as advocated by Camelis, and global formal concept lattices are needed.

2 Running Example

The example which is used throughout the article reconstitutes a committee meeting which had to produce a sorted list of five candidates in order to fulfil a two-year position[3]. The application was open either to PhD or to PhD students about to defend. There were 43 candidates.

In the French academic system, before hiring people, reports must be written. At the computer science department of the INSA of Rennes, besides a qualitative free style report, referees also fill in a spreadsheet file where a number of objective criteria are assessed. This is somehow a many valued context.

[1] `http://www.irisa.fr/LIS/ferre/camelis`

[2] `http://conexp.sourceforge.net/`

[3] called “Attaché Temporaire d’Enseignement et de Recherche”

For this article we have straightforwardly used the actual spreadsheet file used in the recollected meeting as a formal context for Camelis. We have only replaced the names of the candidates (resp. the names of the research teams) by three (resp two) letter codes. The attributes are : the number of publications in international and national conferences as well as in journals, the location of the thesis, the expected date of the end of the thesis, whether the candidates have a computer science education, whether they have teaching experience, whether they have practical (programming) experience, whether there is a pedagogical project in the application file, whether they could integrate a research team of the laboratory. Two attributes, "bonus" and "malus", were meant to capture information which had not been anticipated.

3 Advocated Decision Process

The decision process that we advocate has three stages : firstly an analysis of the context driven by the attributes/criteria which eliminates obviously out of scope candidates, secondly an analysis of the context driven by the candidates, thirdly a discussion to make partial orders into a total order, this includes votes.

Indeed, it is not tractable to examine all the attributes of each candidate in a detailed way. This would require at least 5mn per candidate. With more than 40 candidates, doing this for all candidates means several hours of analysis, people are not ready to do that for candidates who are obviously out of scope. Starting by an analysis driven by the attributes helps to speed up the process in a fair way and to spend time on valuable candidates.

3.1 Context Analysis Driven by the Attributes

In the first stage the analysis of the context is driven by the attributes. The attributes are investigated in turn. For each one the committee decides how relevant the attribute is for this particular decision. In particular it is decided whether a given attribute is

selective: the committee decides that it is mandatory. The candidates who do not fulfil them are eliminated.

selective but counterbalanced: the committee decides that, in the absolute, the attribute would be mandatory, but in this context another attribute exhibited by some candidates could counterbalance the lack of this attribute. The counterbalancing attributes are specified. The candidates who do not fulfil either the selective or the counterbalancing attribute are eliminated.

relevant: the committee decides that the attribute is relevant but not mandatory. It is kept to later differentiate the candidates.

irrelevant: the committee decides that the attribute is not relevant for this particular decision.

Furthermore, new attributes may be identified, and the context is subsequently updated on the fly. Attributes that contain interesting information but which are not totally accurate can be restated.

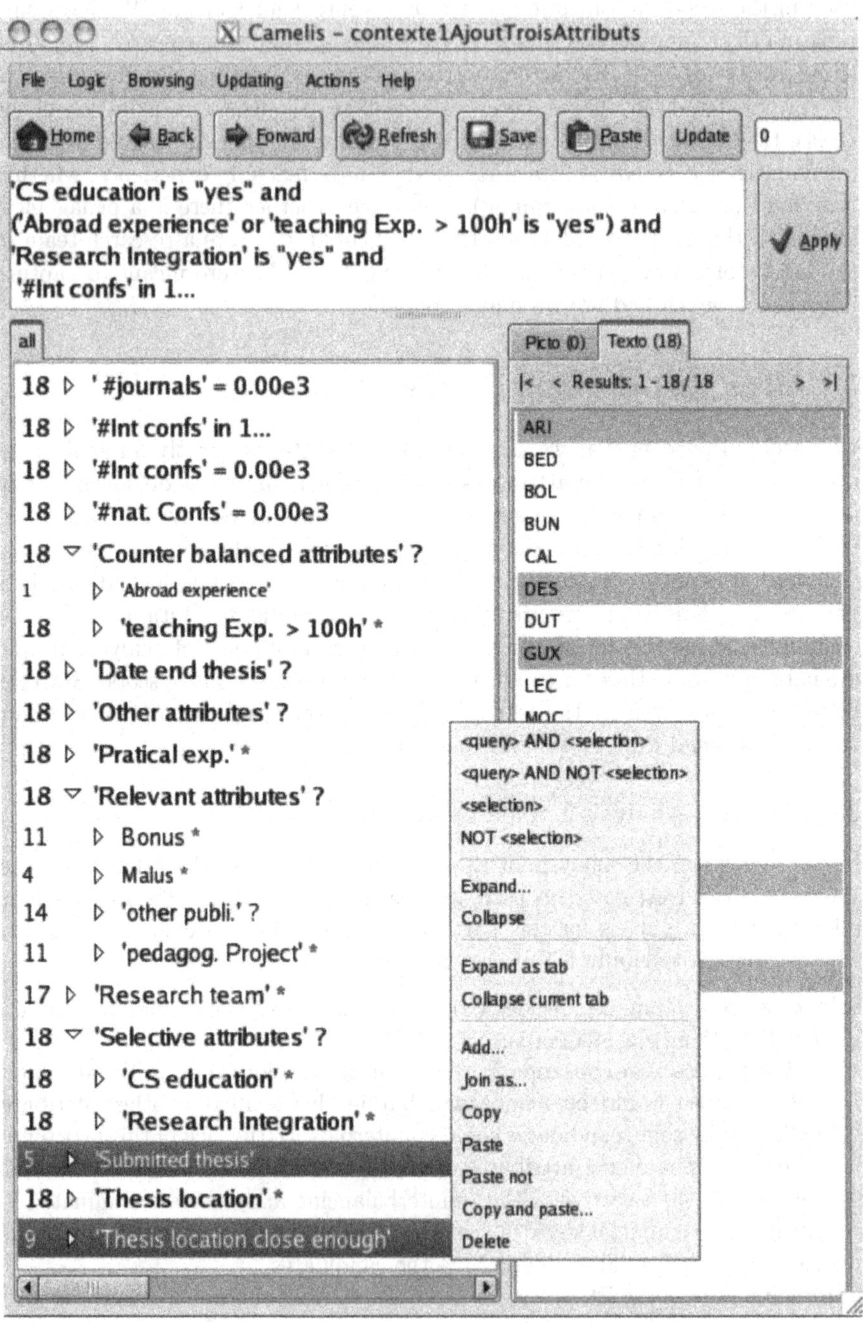

Fig. 1. Snapshot of Camelis during the attribute driven context analysis

In our example, this first stage is done under Camelis. A snapshot taken during the attribute driven stage is given Figure 1. The upper area contains the list of buttons and menus which are not detailed here.

The query area contains the selection criteria that the committee has specified so far. Namely on the figure this can be read as *the selected candidates must have a Computer science education AND (they should have an experience abroad OR a teaching experience) AND their integration in the research laboratory should be granted AND they should have at least a publication in an international conference.*

The bottom right-hand side window is the "object" window. It contains the name codes of the candidates who fulfil the criteria of the query area. One can see that, out of the initial 43 candidates, only 18 are left with the above query.

The bottom left-hand side window contains the taxonomy of all the attributes. The number on the left tells how many of the selected candidates have the attribute. One can see, for example, that only 11 candidates, out of the remaining 18, have a bonus. Starting from the top of the window, here is the information contained in the attribute window. One can read that the number of publications in journals, international conferences and national conferences are numbers (they can match "`0.00e3`").

The committee has decided that having a "teaching experience" was a selective but counterbalanced attribute. Indeed, while everybody agreed that it is a very important criterion, somebody pointed out that it would not be fair that one of the candidates was eliminated because he had not enough teaching experience. He was doing a PhD partly in the USA and has not been able to teach. The group decided that having an "experience abroad" is very interesting for the department and that it can counterbalance not enough teaching experience. This is the echo of the second line of the query area. At that moment of the meeting, the context did not yet contain the information about an experience abroad. It had been easily updated on the fly. All candidates with an experience abroad were identified by their respective referee. A new attribute was added and associated to them.

The "date of the end of the thesis" has not been considered yet. The attribute is therefore not yet sorted. This also applies to whether candidates have "practical experience", and to which "research team" the candidates might join. "Other attributes" is folded, it contains attributes that the committee has already assessed as irrelevant for this decision. The committee has decided that "Bonus", "Malus", "other publications" and the presence of a "pedagogical project" in the file were interesting properties but that they should not yet be used for the selection. The four attributes have therefore been put under "Relevant attributes". The "Computer science education" and the "research integration" have been used in the query. The two attributes have therefore been put under "Selective attributes".

The snapshot has been taken when the committee had just realized that "date end thesis" and "thesis location" could not be used as such. It had been decided that the candidates should either have submitted their thesis or that their thesis location should be close enough to Rennes to give them a better chance to complete their thesis. Therefore, instead of the precise thesis location, it is more accurate to know whether the thesis is done in a laboratory close enough to Rennes. With Camelis, it is easy to fix that: select all the thesis locations that are close enough to Rennes, then select the candidates associated with these locations and add a new attribute to those candidates. Similarly, instead of the estimated date of the PhD defence, it is better to know whether the thesis is already submitted. The context had therefore been extended on the fly by two new attributes, "Submitted thesis" and "Thesis location close enough", as well as their associations to candidates.

The figure illustrates how the two new attributes will be taken into account in the query. Namely, the committee would like to select only candidates with a "submitted thesis" or a "thesis location close enough" to Rennes. The two attributes have been clicked. Camelis has greyed them. It has also greyed the candidates not fulfilling one or the other. The committee can therefore see who is going to be eliminated if the disjunction of the two attributes is judged selective. "ARI", "DES", "GUX" and two other candidates hidden by the pop-up window might disappear. Each referee has a chance to tell if a candidate that he considers valuable might be lost. At the actual meeting, the committee had decided that the selection was fine. A right-click opens the pop-up window. The user is about to click on "<query> AND <selection>" which will add "AND (Submitted thesis' OR 'Thesis location close enough')" to the query.

3.2 Context Analysis Driven by the Objects/Candidates

At some point, the number of remaining candidates becomes small enough so that it becomes tractable to examine candidates in a detailed way. The committee analyzes all the attributes of each candidate in turn. In so doing, the committee can, of course, still decide that an attribute should be "selective" or "selective but counterbalanced".

During the first, attribute-driven, stage the committee checks that the candidates who are about to be eliminated indeed miss a required selective attribute. During the second, candidate-driven, stage the committee checks that the remaining candidates indeed have the attributes that their referees associated to them, in particular the selective ones. The committee also checks that no important attribute association is missing. It is most likely that new attributes emerge.

Figure 2 illustrates the investigation of a candidate assessment. The "CAL" candidate has been clicked in the right-hand side "object" window. His attribute values are shown, in two different ways, in the left-hand side attribute window as well as in the query window. The committee has just detected that the research team associated to this candidate is not "ic+tx" but "ic". It is about to select "Paste not" in the pop-up window to remove the attribute from the intent of "CAL". The next step will then be to "Paste" candidate "CAL" to attribute "Research team is "ic"".

In the actual session, the remaining of the second, candidate-driven, stage went as follows. Fourteen candidates were still in competition at the beginning of the second stage. While examining each candidate in turn, the committee questioned the potential integration into the research laboratory of two

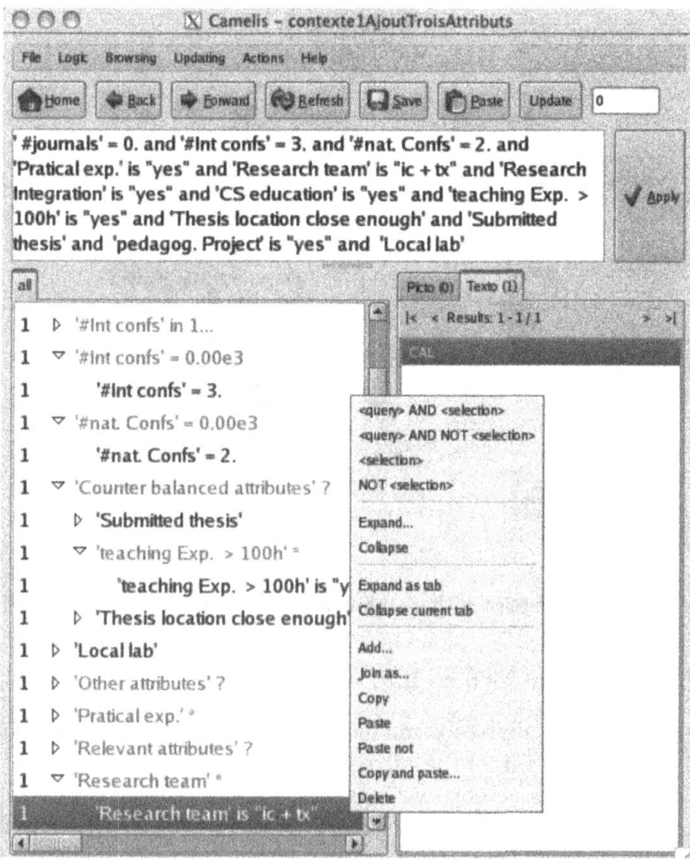

Fig. 2. Snapshot of Camelis during the candidate driven context analysis

candidates. After discussion, it was agreed that the referees may have been a bit overoptimistic. The two candidates were said not to be easily integrated in the lab. The context was therefore updated and there were 12 candidates left.

Investigating the "malus" attribute, the committee decided that one of the candidates, currently working in the laboratory, might never complete his thesis. His long term integration into the research laboratory was therefore questioned and the related attribute negated.

Among the remaining candidates, somebody pointed out that one of them was having a "major contribution in teaching". This would be interesting to keep. An attribute was added.

As there were still numerous good candidates in the list, the committee tested whether there would be enough good candidates to reinforce research teams already present at INSA. A new attribute had been introduced and six good candidates fulfil it. The committee decided that the attribute could be selective.

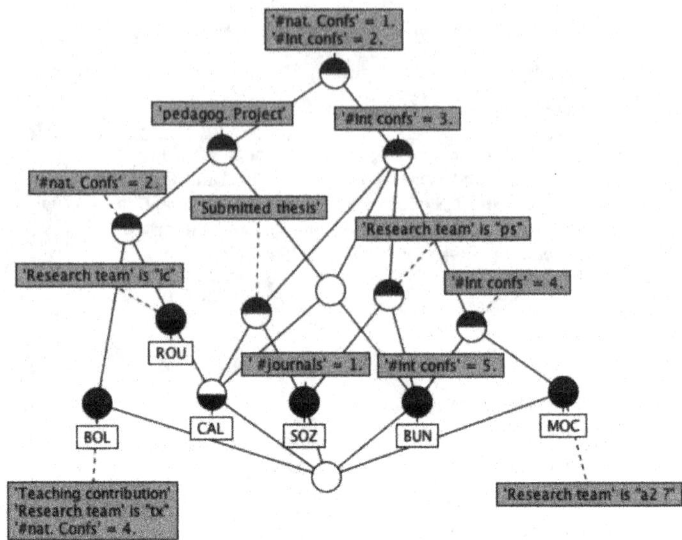

Fig. 3. The concept lattice with six remaining candidates and relevant attributes

3.3 Making Partial Orders into a Total Order

At some point, the committee is confident that the context is valid and that the selection query is relevant for the current decision process. Furthermore, no more objective selection can be done with general consensus. We conjecture that, at this point, the formal concept lattice could be useful as it gives a global picture of all the partial orders. The task of the committee is to rank 5 candidates, hence making the partial orders into a total order.

Figure 3 shows the lattice related to the six candidates coming out of the second stage. From Camelis, we exported into Conexp a context containing the remaining 6 candidates and attributes which had been identified "relevant". In order to see the partial orders on numerical attributes, the context had been completed. Namely a candidate exhibiting "#int. confs=3" has also be credited by "#int. confs=2" and by "#int. confs=1". We can see that research teams "ic" and "ps" still have two candidates. From the informal information given by the two teams, the committee decides that, for "ic", "ROU" is better than "CAL", and it, therefore, keeps "ROU" and eliminates "CAL". Similarly, for "ps", "BUN" is kept and "SOZ" is eliminated. These decisions use information not yet in the formal context. At this stage of the process, there are few remaining candidates, the decisions start to become arbitrary, it is not so crucial to update the context. The eliminated candidates are simply removed from the display.

Figure 4 shows the lattice related to the remaining four candidates on the left-hand side. Using, again, informal information, the committee reckoned that "MOC" and "BUN" were stronger with respect to research and that "ROU" and "BOL" were stronger with respect to teaching. Furthermore, it had been decided

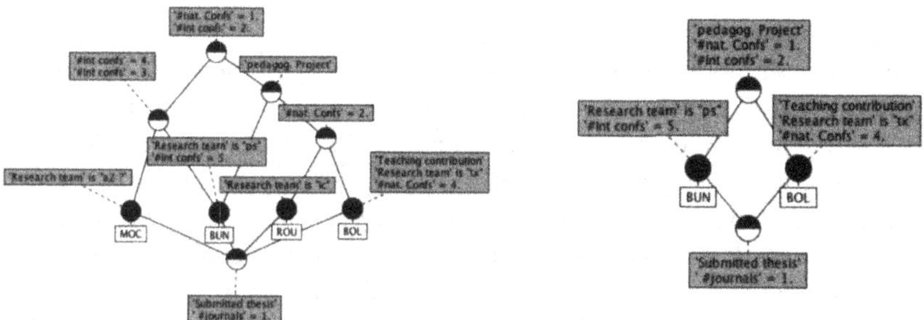

Fig. 4. The concept lattices with four remaining candidates and with the two finalists

that "BUN" was stronger than "MOC" considering the number and the quality of the publications. It had also been decided that "BOL" was stronger than "ROU" considering the teaching contribution. This resulted in the lattice shown on the right-hand side of Figure 4. The lattice shows the attributes common to the two candidates at the top. The attributes that neither of them has are displayed at the bottom. The specific attributes are attached to each candidate. At that point, the committee can vote.

4 Discussion

In this section we discuss the benefits of our method to make decisions. We argue 1) that the process is fair to the candidates, 2) that the decision is adapted to the circumstances, 3) that committee members can be (relatively) serene, 4) that the process requires LCA/FCA tools and 5) that using a fully automatic tools would be unwise.

4.1 The Process is Fair to the Candidates

This process is fair to the candidates. Until the last "political" stage, no candidate can be eliminated without an explicit reason and the reason is applied to all candidates. The selection criteria are explicitly specified. All candidates fulfiling a given criterion are treated equally. When the committee thinks that a candidate quality could counterbalance a required criterion, *all* candidates exhibiting this very quality will be considered equally. In general, every time the context is updated, all candidates are concerned. Reports of candidates eliminated by the current selection are also updated. Indeed, if the committee has second thoughts and relaxes some of the selective attributes, some candidates may no longer be eliminated by the relaxed attributes. They may therefore re-appear in the selection. As their attributes have been updated, they will benefit from all the decisions that have been taken after their initial elimination.

4.2 The Decision is Adapted to the Circumstances

The process allows the decision to be adapted to the circumstances. Indeed, even if a committee uses more or less the same criteria for different meetings, the context is every time different, even if only slightly. For example, for a given recruitment there may be very few candidates, the committee can decide either not to fulfil a position or to adapt the selective criteria. It can also happen that the set of candidates is especially strong and that for a given recruitment the selective criteria can be tightened. It can also happen that some candidates exhibit special qualities not yet identified. For all the cases, the formal context can be easily updated, the selection query can be easily refined, constraints can be easily relaxed. The approach has all the required flexibility to adapt to the situation.

4.3 Committee Members Can be (Relatively) Serene

It is always hard to take multi-criteria decisions. The discussion below takes the point of view of honest committee members who sincerely want that the decision benefits to the institution.

The process is consistent, flexible and backtrackable. The underlying concept lattice ensures that the process is consistent through out the meeting and that the result, in general a total order, is consistent with the, easy to express, partial orders. Flexibility and consistency are the basis of the fairness discussed in the above section. For example, with our approach, it cannot happen that a candidate is eliminated whereas he has an attribute that enabled another candidate to be given a second chance. Furthermore, without support it is easy to be inconsistent even genuinely. With our approach, if a criterion is said crucial at the beginning of the discussion and if the chosen candidates do not fulfil it, then at the very least it will be visible and the committee can discuss whether this is acceptable on the spot. Last but not least the result of the selection is independent of the order in which the atomic decisions are taken. As a result, every partial decision can always be questioned, the process is backtrackable. There is no need to be always on edge, no fatal decision is taken until the last minute.

The process is transparent and traceable. Some committee meetings sometimes feel like a "Three card trick" game. Our process is transparent and traceable. The context can be easily extended on the fly with new attributes and their associations to candidates. Every member of the committee can follow the updates and can raise an objection at any time. The new context is easy to check. The selection decisions are visible in the query area. It expresses the rationale of the final decision.

The end result comes from many small and (relatively) easy decisions. Anybody can propose a new attribute or a new association, or suggest that an attribute should be selective. If the committee agrees by consensus, it is fine. If no consensus emerges the committee can vote to decide whether a new attribute is relevant or selective. It can also vote to decide whether a given candidate has an attribute. Those are small decisions, relatively easy to take. Furthermore, if the committee decides that the attribute is not selective but only relevant, it is fine that the context is anyway updated. It will be taken into account later when people will vote. It is also fine that a criterion is labeled relevant even if only one person in the committee judges it so. Consensus is only mandatory for selective criteria. This can save a lot of fruitless discussions.

Intriguing is not so easy. At such meetings, there are often people who are only there to "push" their own candidate regardless of the means. For example, we have seen situations where a candidate who had never been discussed comes out of the votes because a sub-part of the committee had plotted beforehand. With our approach the candidates who are voted upon must have been examined in depth, their attributes must have been validated. They must satisfy a number of required properties. It is not so easy to manipulate a group on small concrete decisions. It is therefore most likely that candidates who do not fulfil the required properties will have been eliminated. Sometimes it also happens that a new criterion comes out at the end of the meeting and that it gives a decisive advantage to the very candidate who is supported by the referee who expresses the criterion. As this is bad practice, genuine referees can refrain from specifying an important criterion only because it is somehow too late. With our approach if somebody wants to defend a candidate, it is fine at any time because all candidates with the same attributes will be treated equally. Last but not least, voting occurs only when all possibilities for consensus have been exhausted. At that moment, any result is fair as the whole rational is fulfilled.

4.4 LCA/FCA Tools Are Relevant

The overall process is very hard to achieve without appropriate tools. When we started using a spreadsheet it was already a big improvement over oral or written reports even if structured. However, selecting criteria and candidates in the spreadsheet is very error prone. It is hard to be 100% sure that the process is consistent. Furthermore adding attributes and stating in the table who has them is very tedious and again error prone. Even worse, keeping track of the selection process is almost impossible, especially when it is a combination of a logical AND, OR and NOT connectors. In Camelis, everything that was so hard to do with a spreadsheet can be done naturally. Furthermore the global consistency is guaranteed and the query keeps the history of decisions.

At the actual meeting, neither Camelis nor the lattices were available. The decisions were, nevertheless, taken mostly with respect to the criteria displayed in the previous section. People had to keep the picture in their mind and it took a lot of time. The same arguments had to be repeated again and again, every time we needed them somebody had forgotten them.

It is not straightforward to get the initial context, especially when no formal reports are required. However, our experience shows that people are more keen to write them once they understand the potential gain. We are actually considering to use Camelis to fill in the initial reports, skipping the spreadsheet altogether. Indeed, updating the associations attributes-candidates is very easy. If the attributes already exist it is just a matter of drag and drop. Adding a new attribute is also straightforward. A verbose report would be easy to generate automatically from the final context. Furthermore, the criteria resulting from a meeting could be used to initialize the next recruitment.

4.5 Using a Fully Automatic Tools Would be Unwise

The context must be updated and validated. It is important to note that even with a large experience of committee sessions there is no way the formal context can be filled and not been questioned. The formal context has to be updated and validated. Firstly, it is impossible to guarantee that all the important criteria have been foreseen and anticipated. Secondly, the context depends of the set of candidates: it is possible that not enough candidates fulfil the selection criteria that have been used in previous meetings. Thirdly, some of the candidates can show interesting features that had not been previously identified. Fourthly, referees may make mistake while filling in the context. Lastly, referees make judgements that the remaining of the committee may want to question. Some people are too kind and other are too strict, it is also easy to miss an important feature in a CV. As a consequence, it is out of question that the decision is taken automatically using the context as it is at the beginning of the meeting. One of the objectives of the process described so far is, on the contrary, that the members of committee collectively agree both on a set of attributes/criteria and on who satisfy them. Namely the context is revised and updated during the process and is as much a result of the meeting as the resulting ranked list of candidates.

No magical number. Once the context is agreed upon by the committee, we still do not advocate to build a program that would compute magical numbers. Numbers have the nice feature to be naturally ordered. However, the committee has to take full responsibility for the final decision. With our process, all the attributes are identified. All the selective attributes have been agreed upon. The weights and priority among the attributes which have been labelled relevant depend of each committee member who takes them into account while voting. This makes it easier for committee members to shoulder the final decision.

5 Related Work

Concept analysis has been applied to numerous social contexts, such as social networks [8] and computer-mediated communication [6]. Most of those applications are intended to be applied *a posteriori*, in order to get some understanding of the studied social phenomena. On the contrary, we propose to use LCA and FCA in the course and as a support of the phenomena itself. In our case, the purpose is to support a social/committee decision process. Our approach is to other social applications, what information retrieval is to data-mining. Whereas data-mining automatically computes a global and static view on *a posteriori* data, information

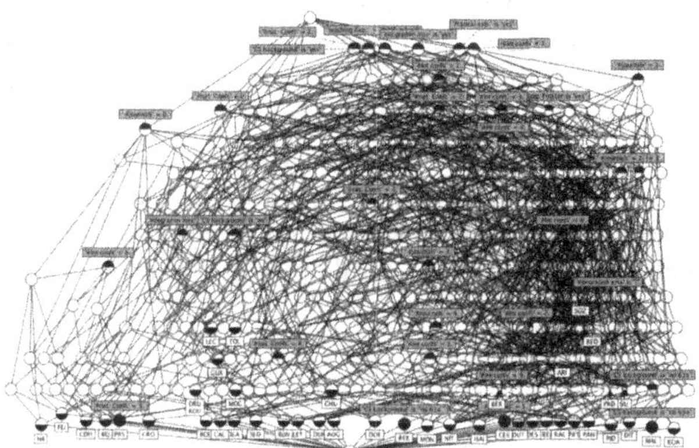

Fig. 5. Concept lattice for the 43 candidates and 40 attributes out of 62

retrieval (i.e. navigation in the concept lattice) presents the user with a local and dynamic view on *live* data, and only guides users in their choice.

A reason for not showing the global concept lattice is that it is too large to be managed by hand. Figure 5 shows the resulting concept lattice. Even in the case of our small context reduced to the 43 candidates and 40 attributes (out of 62), the number of concepts is 1239. Indeed, that formal context is dense. All candidates have many attributes, hence the large number of concepts. Local views such as proposed by Camelis or other FCA-based tools in the domain of information retrieval [2] are better suited for the first two stages described in this article than global lattices.

A specificity of Camelis is the use of logics. This has consequences both on the queries that can be expressed, i.e. on the set of candidates that can be selected, and on the attribute taxonomy, i.e. on the presentation of the criteria occurring in the selected candidates. The use of logics allows to express inequalities on numerical attributes (e.g., number of publications), disjunctions and negations in queries. In pure FCA, only conjunctions of Boolean attributes can be expressed. Previous sections have shown how disjunction is important to take into account counterbalanced selection criteria. In the taxonomy, criteria are organized according to the logical subsumption relation between them (e.g., "nb. papers = 2" is placed under "nb. papers 1.."). In pure FCA, criteria would be presented as a long flat list. Logics helps to make the taxonomy more concise and readable by grouping and hierarchizing together similar criteria. The taxonomy can be updated dynamically, making it possible to group together all irrelevant attributes. In this way, irrelevant attributes are displayed in one line, instead of many, but they are still accessible, and can be moved back as relevant attributes.

6 Conclusion

There are numerous situations similar to the one recollected in this article. Every time a scarce resource has to be assigned by a group which wants to put rationale into its decision, our approach could be used. We have illustrated that with our approach genuine people have the possibility to smoothly express the rationales of a decision. The resulting query gives an explanation of the selection. The committee can take the responsibility of its decision. The committee can be consistent in their judgements during the whole meeting and can be fair with the candidates. It can make a decision adapted to the circumstances. Voting can be postponed to the moment when all possibilities for consensus have been exhausted. The result, in general a total order, is consistent with the expressed partial orders.

Acknowledgements. The authors thank Daniel Herman for fruitful comments.

References

1. Cole, R., Stumme, G.: CEM - a conceptual email manager. In: Ganter, B., Mineau, G.W. (eds.) ICCS 2000. LNCS, vol. 1867. Springer, Heidelberg (2000)
2. Ducrou, J., Vormbrock, B., Eklund, P.W.: FCA-based browsing and searching of a collection of images. In: Schärfe, H., Hitzler, P., Øhrstrøm, P. (eds.) ICCS 2006. LNCS (LNAI), vol. 4068, pp. 203–214. Springer, Heidelberg (2006)
3. Ferré, S.: CAMELIS: Organizing and browsing a personal photo collection with a logical information system. In: Diatta, J., Eklund, P., Liquière, M. (eds.) Int. Conf. Concept Lattices and Their Applications, pp. 112–123 (2007)
4. Ferré, S., Ridoux, O.: An introduction to logical information systems. Information Processing & Management 40(3), 383–419 (2004)
5. Ganter, B., Wille, R.: Formal Concept Analysis: Mathematical Foundations. Springer, Heidelberg (1999)
6. Hara, N.: Analysis of computer-mediated communication: Using formal concept analysis as a visualizing methodology. Journal of Educational Computing Research 26(1), 25–49 (2002)
7. Lindig, C.: Concept-based component retrieval. In: IJCAI 1995 Workshop on Formal Approaches to the Reuse of Plans, Proofs, and Programs (1995)
8. Roth, C.: Binding social and semantic networks. In: European Conf. on Complex Systems, ECCS (September 2006), `http://ecss.csregistry.org/`

Contextual Cognitive Map

L. Chauvin, D. Genest, and S. Loiseau

LERIA - Université d'Angers
2 boulevard Lavoisier 49045 Angers Cedex 01 France

Abstract. The model of cognitive maps introduced by Tolman [1] provides a representation of an influence network between notions. A cognitive map can contain a lot of influences that makes difficult its exploitation. Moreover these influences are not always relevant for different uses of a map. This paper extends the cognitive map model by describing the validity context of each influence with a conceptual graph. A filtering mechanism of the influences according to a use context is provided so as to obtain a simpler and more adjusted map for a user. A prototype that implements this model of contextual cognitive map has been developed.

1 Introduction

A *cognitive map* is a graphical representation of an influence network between notions. A *notion* is described by a text. An *influence* is a causality relation from a notion to another. The effect of the influence can be represented by a numeric or a symbolic value often + or − [2][3][4]. A cognitive map provides a communication medium for humans making the analysis of a complex system. Cognitive maps have been used in many fields such as in biology [1][5], ecology [6][7], management [2][8][4]. Some systems [2][9] associate an inference mechanism to cognitive maps which uses the influences to compute new influences, called *propagated influences* between any pair of notions.

Large cognitive maps are difficult to understand and exploit. Lot of notions and influences are not always appropriate for a specific use. Irrelevant influences decrease the quality of inferences made with a cognitive map because the propagated influences are often incoherent or ambiguous.

The model of contextual cognitive maps presented in this paper provides a mean to express the *validity context* of each influence of a map. The validity context of an influence represents the different cases in which this influence is relevant. For each category of user, a *use context* is defined. Using the validity contexts, a *filtering mechanism* extracts the notions and the influences that are relevant for any use context. So for a user, using its use context, the obtained cognitive map is simpler than the initial cognitive map and allows to compute propagated influences that are more adjusted to him.

Conceptual graphs are graphical representations of knowledge like cognitive maps. In this paper, we propose to use them for representing contexts. A conceptual graph is associated to each influence of a map. A use context of a cognitive

P. Eklund and O. Haemmerlé (Eds.): ICCS 2008, LNAI 5113, pp. 231–241, 2008.

map is also described by a conceptual graph. The projection from the validity context of an influence to the use context is used in the filtering mechanism.

In the section 2, conceptual graph basic definitions are reminded. Section 3 presents our new model of contextual cognitive map. Section 4 presents the filtering mechanism according to a use context. The influence propagation mechanism is presented in the section 5. Section 6 presents the prototype that we have developed using this model.

2 Conceptual Graphs Model

The conceptual graph model presented here is a simplified version of the model defined in [10]. Any conceptual graph is defined on a support which organizes, using the relation "a kind of", a vocabulary composed of concept types and relation types.

Definition 1 (Support). *A* support *S is a pair (T_C, T_R) where T_C is a set of concept types, T_R is a set of relation types. T_C and T_R are partially ordered by a "is a kind of" relation noted $\leq$.*

Example 1. The support described in figure 1 defines concepts types like *pedestrian* (which is a sort of *person*) and relations types like *agent*.

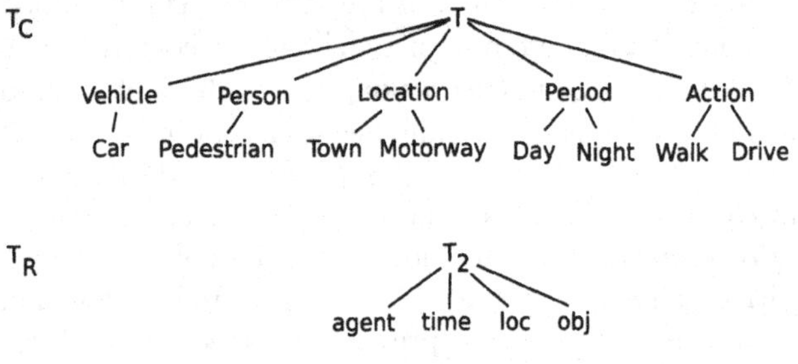

Fig. 1. A support

A conceptual graph serves as a graphic representation of a fact.

Definition 2 (Conceptual graph). *A* conceptual graph *$G = (C, R, E, label)$ defined on a support (T_C, T_R), is a non oriented bipartite multigraph where:*

- *C is a set of concept nodes and R is a set of relation nodes.*
- *$E \subseteq C \times R$ is a set of edges.*
- *Each node has a type given by the* label *function: if $r \in R$, $label(r) \in T_R$, if $c \in C$, $label(c) \in T_C$. Edges adjacent to a relation node r are totally ordered. They are numbered by the* label *function from 1 to the degree of r.*

The main operation of this model is the projection from a conceptual graph to another.

Definition 3 (Projection). *A* projection *from a conceptual graph $G = (C, R, E, label)$ to a conceptual graph $H = (C', R', E', label')$ defined to the same support, is a mapping $\sqcap$ from C to C' and from R to R' such that:*

- *the edges and labels are preserved :* $\forall(r,c) \in E, (\sqcap(r), \sqcap(c)) \in E'$ *and* $label((r,c)) = label'((\sqcap(r), \sqcap(c)))$.
- *the nodes labels can be decreased :* $\forall x \in C \cup R$, $label(\sqcap(x)) \leq label(x)$.

We consider that there is a projection from an empty conceptual graph to any conceptual graph.

A logical semantics has been proposed for conceptual graphs in first order logic [11].

3 Contextual Cognitive Maps Model

A contextual cognitive map is an oriented multigraph where nodes are labeled by notions. Arcs represent influences. An influence is a relation of possible causality between notions. The effect of an influence is represented by a symbol. A context is associated to each influence by the designer. All contexts are described by conceptual graphs defined on a same support.

Definition 4 (Context). *A* context *is a conceptual graph defined on a support* T_K.

Definition 5 (Contextual cognitive map). *Let N be a set of notions. Let S be a set of symbols. Let K be a set of contexts defined on a support T_K.*
A contextual cognitive map *defined on N, S, T_K, K, is an oriented labelled multigraph $(V, label_V, I, label_I)$ where:*

- *V is a set of nodes.*
- *$label_V : V \mapsto N$ is a labeling function. It associates to each notion of N a node of V.*
- *$I \subseteq V \times V \times K$ is a set of arcs. An arc (v_1, v_2, k) of I means that a notion $label_V(v_1)$ influences a notion $label_V(v_2)$ and k is the validity context of this influence.*
- *$label_I : I \mapsto S$ is a labeling function. It associates to each arc of I a symbol of S.*

Example 2. The contextual cognitive map of figure 2 is inspired from road safety problems. This map is defined on the symbol set $\{+, -\}$. The effect of an influence can be positive or negative. For example, if we consider the notions *Human errors* and *fatal accident,* the fact of doing *human error* increases the risks of having a *fatal accident.* Each influence of figure 2 is associated to a context of

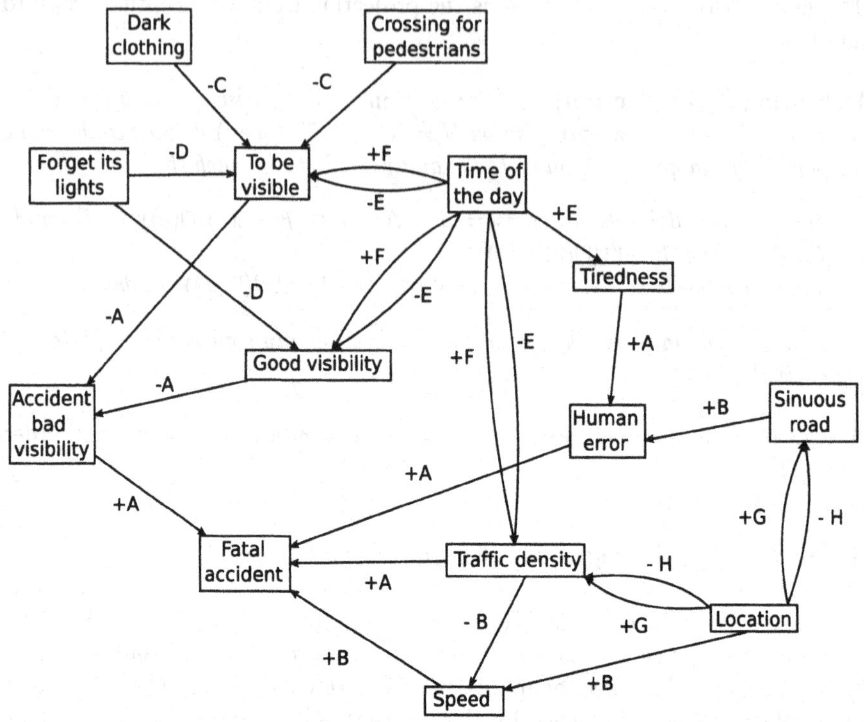

Fig. 2. Cognitive map

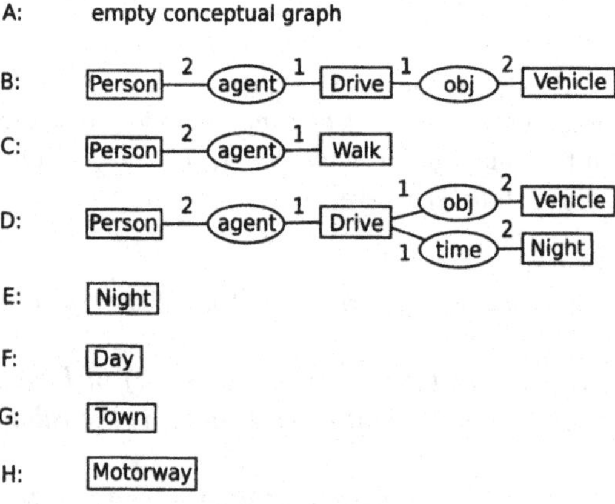

Fig. 3. Set of contexts of the map

figure 3. These contexts are defined on the support of figure 1. The letters near arcs of the figure 2 are not part of the contextual cognitive map model. The selection of an arc in a user interface makes appear the associated context. Some influences like the *Tiredness* which influences *Human errors* are always relevant whatever the use context of the map. In this case the empty conceptual graph (A) is associated to these influences. The influence of the notion *Forget its lights* on the *Good visibility* is relevant for a vehicle driver (B). For a pedestrian (C), the notion *Dark clothing* negatively influences *To be visible.* Some notions are connected by two influences that are associated to different contexts. For example, *Time of the day* negatively (resp. positively) influences *Good visibility* if the context is *Night* (E) (resp. *Day* (F)).

4 Filtering Mechanism According to a Use Context

Once the map is built, the map can be used. To do that the user specifies the context in which he uses it.

Definition 6 (Use context). *Let N be a set of notions. Let S be a set of symbols. Let K be a set of contexts defined on a support T_K. Let $M = (V, label_V, I, label_I)$ be a contextual cognitive map defined on N, S, T_K, K. A* use context *use_cont of M is a conceptual graph defined on T_K describing the context in which M is used.*

A set of use contexts can be provided in which the user can select one or he can build a new conceptual graph to describe its use context.

Example 3. The figure 4 shows a set of use contexts provided to the user.

To obtain a map that fits with the use context, influences that are valid for this context are determined.

Definition 7 (Valid arcs). *Let N be a set of notions. Let S be a set of symbols. Let K be a set of contexts defined on a support T_K. Let M =*

A pedestrian in a town during the night:

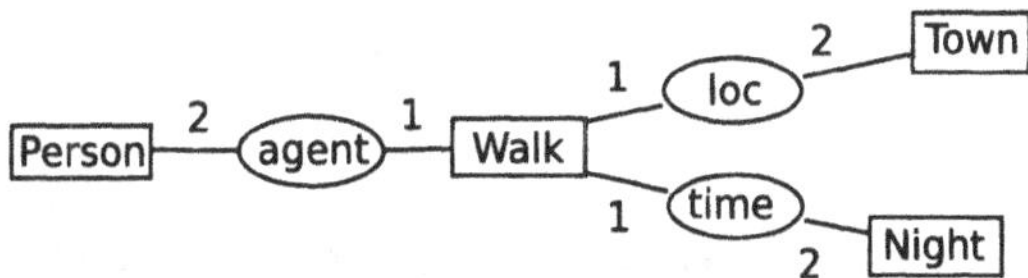

A motorist on a motorway during the day:

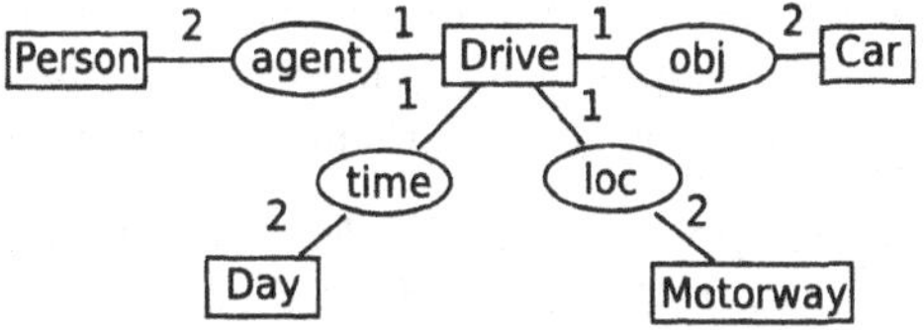

Fig. 4. Use contexts

$(V, label_V, I, label_I)$ be a contextual cognitive map defined on N, S, T_K, K. We define $ValidArcs(I, use\_cont)$ the subset of I as :
$ValidArcs(I, use\_cont) = \{(v_1, v_2, k) \in I \mid \exists \text{ a projection from } k \text{ to } use\_cont\}$

A notion is considered interesting for the user if it is connected to a valid influence.

Definition 8 (Valid nodes). *Let N be a set of notions. Let S be a set of symbols. Let K be a set of contexts defined on a support T_K. Let $M = (V, label_V, I, label_I)$ be a contextual cognitive map defined on N, S, T_K, K. We define $ValidNodes(V, use\_cont)$ the subset of V as :*
$ValidNodes(V, use\_cont) = \{v \in V \mid \exists\ (v_1, v_2, k) \in ValidArcs(I, use\_cont)$ *such as* $v_1 = v \lor v_2 = v\}$.

Once valid influences and valid notions are determined using the use context, they form a new map, simpler and that fits more for the user.

Definition 9 (Restricted map). *Let N be a set of notions. Let S be a set of symbols. Let K be a set of contexts defined on a support T_K. Let $M = (V, label_V, I, label_I)$ be a contextual cognitive map defined on N, S, T_K, K. Let use_cont be a use context of M.*
The restricted map *of M for use_cont is the contextual cognitive map*
$(ValidNodes(V, use\_cont), label_V, ValidArcs(I, use\_cont), label_I)$.

Example 4. The purpose of the cognitive map of figure 5 is to increase pedestrians's awareness of road problems. This restricted map is obtained by masking the influences for which there is no projection from their associated conceptual graph to the use context: *"A pedestrian in a town during the night"*. Notice that there is only a projection from the contexts *A*, *C*, *E*, *G* to *"A pedestrian in a town during the night"*. In a context of sensitizing pedestrians to the road problems, notions and influences that are related to the use of vehicles as for example the influence of the *Speed* on the *Fatal accident* are masked. The map is then simpler, and more adjusted to this use. The figure 6 shows the restricted map for *"A motorist on a motorway during the day"*. There is a projection from each context *A*, *B*, *F*, *H* to the use context *"A motorist on a motorway during the day"*.

5 Influence Propagation in a Contextual Cognitive Map

An inference mechanism can be defined to determine the propagated influence of any notion to another in a contextual cognitive map. Notice that these map can be a restricted map of another for a use context. The propagated influence from a notion to another is computed according to the paths existing between the nodes labeled by these notions. We call them influence paths.

Definition 10 (Influence path). *Let N be a set of notions. Let S be a set of symbols. Let K be a set of contexts defined on a support T_K. Let $M = (V, label_V, I, label_I)$ be a contextual cognitive map defined on N, S, T_K, K. Let n_1, n_2 be two notions of N.*

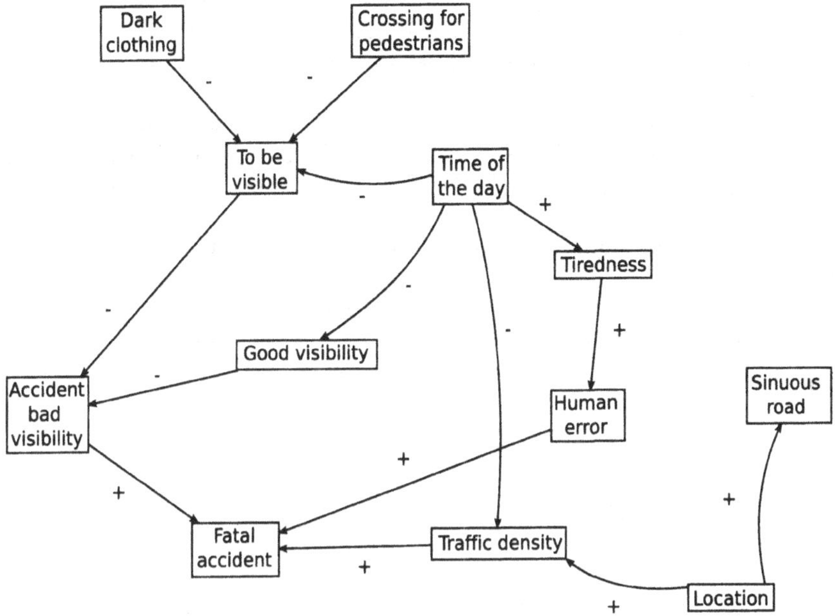

Fig. 5. Use of the map for a pedestrian in a town during the night

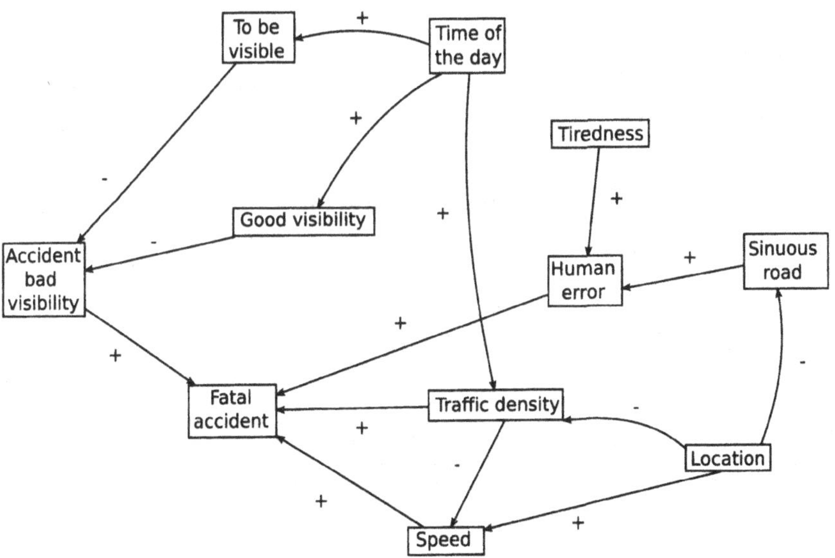

Fig. 6. Use of the map for a motorist during the day

– *An* influence path *from* n_1 *to* n_2 *is a sequence (of length* l*) of arc* $(u_i, v_i, k) \in I$ *such that* $u_1 = label_V^{-1}(n_1)$, $v_l = label_V^{-1}(n_2)$ *and* $\forall i \in [1..l-1], v_i = u_{i+1}$.

- *An influence path P from n_1 to n_2 is* minimal *iff there is no influence path P' from n_1 to n_2 such that P' is a subsequence of P*
- *We note $\mathcal{P}_{n_1,n_2}$ the set of minimal influence paths from n_1 to n_2.*

So as to determine the effect of a notion on another, the propagated influence in an influence path must be evaluated. To do that, symbols of each influence of an influence path are aggregated. The set of symbols used in this paper is $\{+,-\}$.

Definition 11 (propagated influence for an influence path). *Let N be a set of notions. Let K be a set of contexts defined on a support T_K. Let $M = (V, label_V, I, label_I)$ be a contextual cognitive map defined on N, $\{+,-\}$, T_K, K.*
The propagated influence for an influence path P *is:*

$$\mathcal{I}_P(P) = \bigwedge_{i \ of \ P} label_I(i)$$

with $\bigwedge$ a function defined on $\{+,-\} \times \{+,-\} \mapsto \{+,-\}$ represented by the matrix:

$\bigwedge$	+	−
+	+	−
−	−	+

The propagated influence mechanism between two notions aggregates propagated influence of each minimal influence paths existing between these notions. The value returned by this mechanism can be positive (noted +), negative (−), null (0) or ambiguous (?).

Definition 12 (Propagated influence between two notions of a contextual cognitive map). *Let N be a set of notions. Let K be a set of contexts defined on a support T_K. Let $M = (V, label_V, I, label_I)$ be a contextual cognitive map defined on N, $\{+,-\}$, T_K, K.*
The propagated influence *between two notions is a function $\mathcal{I}$ defined on $N \times N \mapsto \{0,+,-,?\}$ such that:*

$$\mathcal{I}(n_1,n_2) = 0 \ if \ \ \mathcal{P}_{n_1,n_2} = \emptyset$$

$$\mathcal{I}(n_1,n_2) = \bigvee_{P \in \mathcal{P}_{n_1,n_2}} \mathcal{I}_P(P) \ if \ \ \mathcal{P}_{n_1,n_2} \neq \emptyset$$

where $\bigvee$ is a function defined on $\{+,-,?\} \times \{+,-,?\} \mapsto \{+,-,?\}$ represented by the matrix:

$\bigvee$	+	−	?
+	+	?	?
−	?	−	?
?	?	?	?

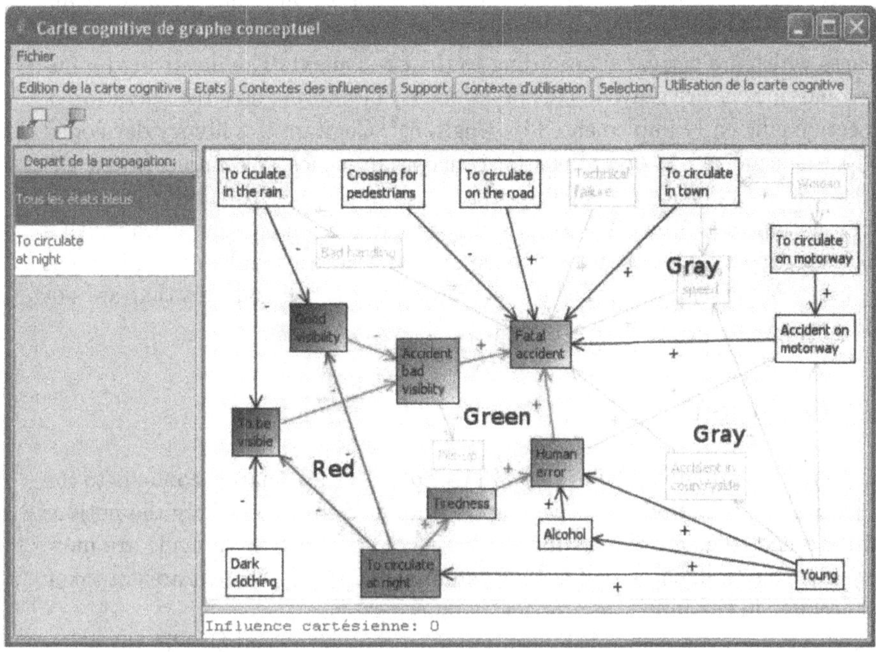

Fig. 7. Prototype : Use of a cognitive map

Example 5. In the map of figure 2 there is only one influence path between *Good visibility* and *Fatal accident*. This influence path is composed of a negative influence of the notion *Good visibility* on the notion *Accident bad visibility* and a positive influence of the notion *Accident bad visibility* on the notion *Fatal accident*. The propagated influence from the notion *Good visibility* to the notion *Fatal accident* is then negative (-). It can be interpreted as: *"A good visibility decreases the risk of having a fatal accident"*.

The filtering mechanism enables the propagated influence mechanism to be more precise. For example, the propagated influence from *Time of the day* to the *Good visibility* is ambiguous because there are two influences with different symbols between these notions. By using the restriction of the map for the use context *"A pedestrian in a town during the night"* (figure 5), only the negative influence associated to the context E (*Night*) is taken account. The other influence is not used because there is no projection from the context F (*Day*) to the use context. For this use context, the propagated influence from the *Time of the day* to the *Good visibility* is then negative.

6 Prototype

We have developed a prototype[1] (figure 7) in Java that allows to build and handle contextual cognitive maps. Various graphical components used to represent

[1] This prototype is downloadable at:
`http://forge.info.univ-angers.fr/~lionelc/CCdeGCjava/`

cognitive maps, conceptual graphs and the support are implemented using JGraph[2], a graph visualization library. The user sees the notions and the influences which are activated according to the use context. The inactivated influences and notions are grayed. This functionality of filtering uses the operation of projection efficiently implemented by Cogitant[3]. Cogitant is a library developed in C++ specialized in operations on conceptual graphs. We implemented the mechanism of influence propagation that enables the user to ask for the effect from a notion to another. The results are presented in an ergonomic way with a color code: green for a positive influence, red for a negative influence and orange for an ambiguous influence. For a use context, the nodes and arcs that are not valid are grayed.

7 Conclusion

The cognitive map model is improved in this paper by associating contexts to the influences. A contextual cognitive map is simplified to present only the notions and influences that are interesting for a use context. Inferences made are more precise because influences that are not interesting for a use context are not taken into account in the influence propagation mechanism.

In this paper, the choice has been made to represent symbolically the effect of an influence by $\{+, -\}$. This choice enables a human to understand easily the influence propagation mechanism. Some works define other means to propagate influences. Their objective is often to solve automatically ambiguities. Some works use fuzzy representation [12]. Kosko [9][13] has proposed a fuzzy cognitive map model in which the influence effect is not represented with a symbol but with a numerical value. The inference mechanism for fuzzy cognitive maps uses a threshold fonction. Taber [14], Perushich [15] and Kosko [9] have experimented different threshold fonctions in their works. Carvalho and Tomé [16] presents an extension of fuzzy cognitive maps where the inference behaviour uses fuzzy logic rules defined in each concept of the map. Liu and Zhang [17] provides a comparaison of papers on fuzzy congitive maps. These mechanisms are interesting when a cognitive map is used as an autonomous system but the model and the results are difficult to understand by the user. Nevertheless our contextual approach can be adapted to fuzzy cognitive maps.

The extension of cognitive maps presented in this paper can be associated with the model of cognitive map of conceptual graphs [3][18]. In our model the semantics of a notion is defined in natural language. Several persons could have different interpretations of a same notion and consequently of the same map. Contrary to our model, the model of cognitive map of conceptual graphs do not focus on the influences but it specifies the semantic of the notions in using conceptual graphs. In this model the projection provides other decision mechanisms such as selection of notions semantically bound and an inference mechanism between two sets of notions.

[2] JGraph website: `http://www.jgraph.com`

[3] Cogitant website: `http://cogitant.sourceforge.net`

References

1. Tolman, E.C.: Cognitive maps in rats and men. The Psychological Review 55(4), 189–208 (1948)
2. Axelrod, R.: Structure of decision: the cognitive maps of political elites. Princeton University Press, New Jersey (1976)
3. Aissaoui, G., Loiseau, S., Genest, D.: Cognitive map of conceptual graphs. In: Ganter, B., de Moor, A., Lex, W. (eds.) ICCS 2003. LNCS (LNAI), vol. 2746, pp. 337–350. Springer, Heidelberg (2003)
4. Eden, C., Ackermann, F.: Cognitive mapping expert views for policy analysis in the public sector. European Journal of Operational Research 152, 615–630 (2004)
5. Touretzky, D., Redish, A.: Landmark arrays and the hippocampal cognitive map. Current Trends in Connectionism, 1–13 (1995)
6. Celik, F.D., Ozesmi, U., Akdogan, A.: Participatory ecosystem management planning at tuzla lake (turkey) using fuzzy cognitive mapping (October 2005) eprint arXiv:q-bio/0510015
7. Poignonec, D.: Apport de la combinaison cartographie cognitive/ontologie dans la compréhension de la perception du fonctionnement d'un écosystème récifo-lagonaire de Nouvelle-Calédonie par les acteurs locaux. PhD thesis, ENSA Rennes France (2006)
8. Cossette, P.: Introduction, Cartes cognitives et organisations. Cossette edn. Les presses de l'université de Laval (1994)
9. Kosko, B.: Neural networks and fuzzy systems: a dynamical systems approach to mahine intelligence. Prentice-Hall, Engelwood Cliffs (1992)
10. Mugnier, M.L.: Knowledge representation and reasonings based on graph homomorphism. In: International Conference on Conceptual Structures, pp. 172–192 (2000)
11. Mugnier, M., Chein, M.: Représenter des connaissances et raisonner avec des graphes. Revue d'intelligence artificielle 10, 7–56 (1996)
12. Zadeh, L.: Outline of a new approach to the analysis of of complex systems and decision processes (1973)
13. Kosko, B.: Fuzzy cognitive maps. International Journal of ManMachines Studies 25 (1992)
14. Taber, R.: Knowledge processing with fuzzy cognitive maps. In: Expert systems with application edn. vol. 2 (1991)
15. Perusich, K.: Fuzzy cognitive maps for policy analysis 10, 369–373 (1996)
16. Carvalho, J., Tomé, J.: Rule based fuzzy cognitive maps – qualitative systems dynamics (1999)
17. Liu, Z.Q., Zhang, J.Y.: Interrogating the structure of fuzzy cognitive maps. Soft Comput. 7(3), 148–153 (2003)
18. Genest, D., Loiseau, S.: Modélisation, classification et propagation dans des réseaux d'influence. Technique et Science Informatiques 26(3-4), 471–496 (2007)

Employing a Domain Specific Ontology to Perform Semantic Search

Maxime Morneau and Guy W. Mineau

Département d'informatique et de génie logiciel
Pavillon Adrien-Pouliot, 1065, av. de la Médecine, Université Laval
Quebec City (Québec), G1V 0A6, Canada
maxime.morneau.1@ulaval.ca, guy.mineau@ift.ulaval.ca

Abstract. Increasing the relevancy of Web search results has been a major concern in research over the last years. Boolean search, metadata, natural language based processing and various other techniques have been applied to improve the quality of search results sent to a user. Ontology-based methods were proposed to refine the information extraction process but they have not yet achieved wide adoption by search engines. This is mainly due to the fact that the ontology building process is time consuming. An all inclusive ontology for the entire World Wide Web might be difficult if not impossible to construct, but a specific domain ontology can be automatically built using statistical and machine learning techniques, as done with our tool: SeseiOnto. In this paper, we describe how we adapted the SeseiOnto software to perform Web search on the Wikipedia page on climate change. SeseiOnto, by using conceptual graphs to represent natural language and an ontology to extract links between concepts, manages to properly answer natural language queries about climate change. Our tests show that SeseiOnto has the potential to be used in domain specific Web search as well as in corporate intranets.

1 Introduction

Succeeding in the management of information is nowadays all about coping with the tremendous amount of available knowledge. Huge corporations, small organizations as well as individuals are all confronted to an overload of data. Information retrieval is a young science and methods to extract documents from the Web or from corpora are not flawless. Boolean search is still the preferred way to retrieve data. This approach, although efficient, has the disadvantage of not being easy to use for specific queries since the choice of logical operators most relevant to the query is not straightforward [8].

To sort through the enormous amount of information available on the Web, researchers proposed a semantic approach to the problem. Data on the Web and in corporate intranets structured using HTML could be stored together with semantic description of its content. That way, information retrieval would be greatly facilitated [16]. Hence, a proposed solution is to rely on an ontology to extract

P. Eklund and O. Haemmerlé (Eds.): ICCS 2008, LNAI 5113, pp. 242–254, 2008.

concepts and their relations from these pages. The construction of an ontology is however time consuming, particularly with large document databases [5]. However for a restricted field of knowledge, it is possible to consider employing an ontology since the core concepts and relations of a small domain are usually more constrained [4]. The use of an ontology allows the query language to accept natural language-based sentences. Moreover, the automatic creation and update of the ontology could provide a way to manage the information in an evolving corpus as well as improving domain restricted search done on the World Wide Web.

In this article, we present the SeseiOnto software, an information retrieval tool that uses natural language processing (NLP) as its search interface and an automatically generated ontology to obtain semantics about a domain. SeseiOnto uses conceptual graphs to process natural language and to evaluate the relation between a query and a document.

We applied this method in the context of the 2008 ICCS Challenge. The goal of this challenge was to see how a tool that uses conceptual graphs could be used to support research on climate change. Consequently, the challenge required the tool to be evaluated using data taken from the Wikipedia page on climate change. We found out that SeseiOnto can correctly pinpoint significant answers to natural language queries about climate change.

Thus, Section 2, presents similar approaches to SeseiOnto. In Section 3, we detail how the software works. In Section 4, we analyze the different results obtained by SeseiOnto in the context of the ICCS Challenge. Section 5 is a review of SeseiOnto main strengths and weaknesses, and provides an introduction on future work.

2 Similar Approaches

There already exist different semantic information retrieval methods and systems, each one having its own advantages and limitations. In this section, we briefly present similar work to our own.

In [17], the authors presents a system that sorts documents returned by Google using a dynamically created taxonomy. This taxonomy is built using the same documents that are returned by Google for a specific user query. This relates considerably to the method that was used by the Sesei software [15], the predecessor of SeseiOnto. Therefore, this taxonomy is employed to improve the user's search experience by returning documents that are the most significant with regard to his query. One of the drawbacks of this approach is that it may not be necessary to build an ontology dynamically for every query. A domain ontology, although possibly less oriented towards the user's query, could correctly provide an appropriate answer. Furthermore, documents returned from the Web will probably contain a lot of noise and information that is by no means related to the topic. In our opinion, a domain ontology contains enough semantics to cover a wide range of queries. Moreover, it needs to be updated only when the document base evolves, which is much less frequently than with each query.

In [4], the author presents a method to build an ontology using "expert-created" sources containing similar information. The source for this hierarchy construction process is made of tables coming from Web pages. The ontology construction process presented in this work begins with a small human-built ontology requiring a thorough knowledge of the domain. Nevertheless, an interesting point made by this research is that a lot of emphasis seems to have been put on evolutionary data, which is particularly important in the context of corporate Intranets where the content constantly evolves.

Another method presented in [3] is focusing on building hierarchical representation of natural language sentences using a set of rewrite rules. These rules describe subsumption relations between various text representations. The hierarchy is then employed to determine if the meaning of a given sentence entails that of another. The main similarity of this work with ours is that their analysis of text is based on a type of "transformation rules". However, this approach could be time consuming if large corpora of texts were used.

In [2], a sophisticated question answering system used for passage retrieval is described. This approach employs a fuzzy relation matching technique to answer queries. Similar grammatical relations are identified between queries and passages to evaluate their degree of relevancy. This system was tested in the context of the Text REtrieval Conference (TREC). Their results indicate that sophisticated relation matching techniques seems to have a strong potential for natural language question answering. The inclusion of fuzzy CGs in our algorithm is to be explored

3 SeseiOnto

SeseiOnto [12][11] is a standalone application used to perform semantic search on a corpus of textual documents. It aims at being an alternative to traditional Boolean search engines by providing a mean of integrating NLP-based querying and ontology extraction. Natural language is processed using the representation power of conceptual graphs; and ontologies are automatically built using the Text-To-Onto software.

Figure 1 shows SeseiOnto global process. Natural language-based queries as well as the presumably relevant ones from the corpus are processed by the Connexor syntactic analyzer [7]. Connexor's output is converted to CGs using a set of 76 ad hoc transformation rules [13]. The ontology is generated by the Text-To-Onto software and applied on a subset of the documents from the corpus. Using CGs that represent both the query and parts of the documents and employing a domain ontology, SeseiOnto tries to identify potential matches[1].

3.1 SeseiOnto's Search Process

SeseiOnto is mainly based on the Sesei software [15]. Sesei was built to answer natural language queries on the World Wide Web using an ontology specific to the user's query. Using definitions from WordNet [10], the user has to disambiguate words composing his query. A type hierarchy is created using the definitions provided by the user and the concept hierarchy of WordNet.

[1] This process will be explained in more details in the current section.

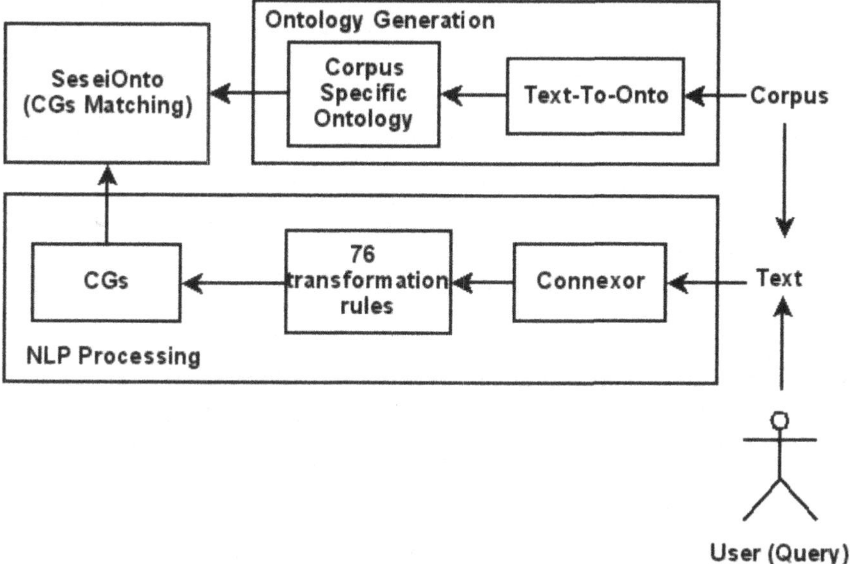

Fig. 1. SeseiOnto's global process

As for SeseiOnto, it initially takes a user query as input. This query is then sent to the Connexor [7] syntactic analyzer. Prepositions and articles are removed and words are lemmatized. Words are matched to the ontology, which is viewed as a type hierarchy by SeseiOnto. If a word from the query is not identified in the ontology, it is added to it. The query is then converted to CGs using the set of 76 transformation rules. According to previous tests [14], this set of rules is broad enough to represent a sufficient number of semantic phenomena. An example of a rule is: *when a noun(A) is the subject of a verb(B) at active voice in a sentence, it should then be converted to a CG stating that a concept of type A is the agent of concept of type B.* Afterwards, sentences from the documents in the corpus, that is our *resource documents*, are converted to CGs using the same process as with the query.

The next step is identifying the quantity of information shared by the query and the resource documents to know which documents are the most relevant. This goal is achieved by calculating a *semantic score* between the query sentence and sentences from the resource documents. To obtain this semantic score, a set of generalizations of each concept and relation are created using the ontology. To compare two concepts, or two relations, SeseiOnto will try to find the most common generalization between them. The more specific the generalization, the higher the semantic score will be. An example of this process can be seen in Figure 2, extracted from [15]. The query is "Who offers a cure for cancer?" and the resource sentence is "a big company will market a sedative".

A generalization will only be evaluated by SeseiOnto if the concepts from the query graph and the resource graph are linked by relations of the same type.

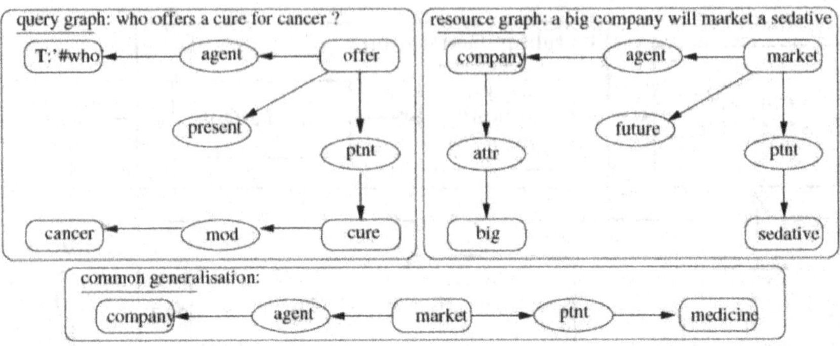

Fig. 2. A query graph, a resource graph and their common generalization. The generalization of *market* and *offer* is *market* in the type hierarchy: $\top \rightarrow trade, merchandise \rightarrow market \rightarrow offer \rightarrow \ldots \rightarrow \bot$. The generalization of *company* and $\top$ is *company*, per definition.

Of course, the ontology needs to be extensive enough to make this common generalization search process possible. For more details about this step, see [15].

Hence, every document from the corpus will be assigned a semantic score. Documents and sentences deemed the most relevant, that is, sentences that seem related to the user's query, will then be returned by the system. Most relevant sentences within documents are pinpointed, sorted by their semantic score and returned to the user.

Additionally, SeseiOnto needs some sort of threshold to discriminate relevant and irrelevant sentences from the resource documents. In SeseiOnto, this threshold is influenced by the domain ontology and SeseiOnto must first set it. Therefore, to define it, a set of queries already matched to relevant documents in the corpus needs to be available. Such a matching can be obtained through manual evaluation of queries and documents by domain experts. Queries are sent to SeseiOnto which will assign a semantic score to each resource documents. Knowing the relevancy of each document, the precision and recall of the output of SeseiOnto can be calculated for every query. This way, SeseiOnto can determine the threshold in terms of the semantic score that maximizes, in average over all queries, the precision and recall. Using a training set and a test set, together with K-Fold Cross Validation, a threshold is established and is used for subsequent user queries. More details about SeseiOnto's search method can be found in and [11] and [12].

Such a process proved to be quite effective to compare similarity between two sentences. The next step in our research was to replace the type hierarchy dynamically created using words from the query by an ontology. If queries were related to the same field of knowledge, the same ontology could probably be reused to perform the query-document matching. Instead of creating our own ontologies, we determined we could learn them automatically.

3.2 Ontology Learning

Building an ontology from scratch can easily become a time consuming task. It can require experts from a specific domain and thorough and extensive reasoning to create concepts and relations describing the field. The broader the domain, the harder it can get to assemble the initial ontology. Maintaining an ontology is also a difficult task. A possible solution to ontology creation might be to construct it automatically [8].

Researchers have employed different types of procedures to learn an ontology automatically. Most of them rely on learning an ontology from structured information such as databases, knowledge bases and dictionaries [17]. Others think that unstructured information such as Web pages can provide a powerful mean of creating ontologies from scratch. To be able to perform such a task, a system needs to have strong natural language processing capabilities to create adequate ontologies. A valuable starting point for an ontology containing different types of relations between concepts is a taxonomy. A taxonomy is frequently defined as a hierarchical structure comprised of "is a" links between concepts that describes a specific environment.

The ontology is a crucial component of our search process. To build this ontology automatically, we employ the Text-To-Onto software [9]. This ontology is constructed by using a text corpus containing documents pertaining to the same field of knowledge. Text-To-Onto allows the knowledge engineer to use a general ontology, like WordNet, and adapt it to the domain. Domain specific concepts will be added to the ontology while superfluous ones will be pruned from it. To build this ontology automatically, Text-To-Onto uses linguistic patterns and machine learning methods. Text-To-Onto also allows the building of an ontology from scratch by using its TaxoBuilder module. TaxoBuilder also uses machine learning techniques in conjunction with linguistic patterns to create the ontology.

The ontology building method depends mostly of what is favored by the user between recall and precision. Deeper ontologies, i.e., with more generalizations/specializations, usually require far more computing time and produce a better recall. Shallow ontologies with many leaf nodes starting from the root tends to generally produce better precision with a shorter processing time.

Using this ontology learning approach, we assumed that an ontology built using documents from the corpus provided enough semantics to represent information contained in unseen queries about the domain.

SeseiOnto typically uses seven types of ontologies:

1. *FCA ontology with lexicographer classes*: Ontology built using Formal Concept Analysis and with the 15 WordNet verbal lexicographer classes [10] as root nodes.
2. *FCA ontology without lexicographer classes*: Ontology built using Formal Concept Analysis without any particular root nodes specified.
3. *Vertical Relation Heuristics ontology*: Ontology built using compound words found in documents from the corpus.
4. *Hearst patterns ontology*: Ontology built using Hearst linguistic patterns [6] to build "is a" relationships.

5. *Combination of Vertical Relation Heuristics and Hearst patterns ontology*: Ontology built using both previous methods.
6. *Combination of Vertical Relation Heuristics, Hearst patterns and WordNet ontology*: Ontology built using Vertical Relation Heuristics, Hearst patterns and WordNet to build the ontology.
7. *Domain adapted WordNet*: A modified version of WordNet where specific concepts from the corpus have been added and concepts too general from WordNet have been removed.

For more details about Text-To-Onto and TaxoBuilder ontology construction methods, see [1].

After that our method was clearly defined, we then needed a proof-of-concept in a real environment. We therefore had to develop a strategy to analyze the potential of SeseiOnto.

3.3 Past Results

We had the oportunity to test SeseiOnto on one corpus in the past, the Cystic Fibrosis Database (*CF Database*) [18]. The CF Database contains a set of 1,239 documents together with a set of 100 queries, each one individually matched to corresponding relevant documents. These documents are all abstract of scientific papers about research made on cystic fibrosis during the 1970's. Domain experts have performed the matching between queries and documents.

We managed to achieved interesting results, compared to the ones obtained by a "classical" Boolean search engine, Coveo[2]. To evaluate SeseiOnto's performance, we used recall, precision and the F-Measure [19] as a combined metric for that particular purpose. The following formula defines it:

$$F = \frac{(\beta^2 + 1) \times Precision \times Recall}{\beta^2 Precision + Recall}$$

The F-Measure can therefore be considered as the weighted harmonic mean of precision and recall. The weight given to either recall or precision in the formula is expressed with the β symbol. In the measure, a β lower than 1 gives more importance to precision while a β higher than 1 gives more importance to recall. In our tests, we used a β of 0.5 to emphasize the importance of precision over recall in our type of application domain.

SeseiOnto managed to achieve a recall of 44% and a precision of 41%, which gives an F-Measure of 42%. As for Coveo, the search engines manage to reach a recall of 11% and a precision of 35%, yielding an F-Measure of 25%.

However, in terms of processing time, Coveo performs better than SeseiOnto. Coveo can answer a query in milliseconds while SeseiOnto can take up to five minutes. Nevertheless, SeseiOnto has the major advantage of being able to identify precise sentences within a document that indicate to the user where exactly is the information he is looking for. SeseiOnto also remains a research prototype

[2] www.coveo.com

and we are convinced search time could easily be improved by using parallel computing, search indexing and preprocessing of documents in the corpus.

Furthermore, SeseiOnto provides a natural language search interface which is much more intuitive for a user than using keywords and Boolean operators. SeseiOnto is able to link two different concepts without them necessarily being homographs, thus improving recall. A regular search will usually eliminate a document if it does not contain one of the keyword contained in the initial query. By taking into account the semantic structure of the sentences (through Connexor and transformation rules), SeseiOnto manages to improve precision. Having seen that that SeseiOnto had potential, we thereafter started experimentations on other databases.

4 Tests and Results

To evaluate SeseiOnto in a new environment, we selected the 2008 ICCS Challenge as our test bed. We wanted to apply the SeseiOnto techniques on the Wikipedia page on climate change[3]. All our tests were performed with the page that was available on Wikipedia on November 28th, 2007.

The reader can see in Figure 3 the SeseiOnto's workflow, in the context of using the software in a restricted domain Web environment. Hence, the user must initially start by sending a natural language query to SeseiOnto and selecting an ontology to perform his search. Afterwards, the query sentence is parsed using Connexor, query words are matched to the selected ontology and a CG is obtained using the set of transformation rules of SeseiOnto. As for resource CGs, documents are obtained from the appropriate Web page (in our case, the Wikipedia page on climate change) and sentences are extracted from the resource document. Resource CGs are built using the same process as with the query CGs. Resource and query CGs are then compared using the method presented in Section 3. Sentences with the highest semantic score are returned to the user.

To perform our tests, we applied an empirical approach. We used a set of finite queries on climate change taken from the World Wide Web[4]and we manually evaluated how SeseiOnto could answer them. We compared different ontologies to see how they each individually performed at answering these specific queries. We also wanted to compare the approach of manually disambiguating the words from query using the definitions from WordNet, i.e., using the search approach of Sesei, SeseiOnto's "ancestor".

We are showing below examples of how SeseiOnto performed at answering some of these queries. For the question "How could climate change affect us in the future?", SeseiOnto gave the three following sentences as the first three answers:

- An October 29, 2006 report by former Chief Economist and Senior Vice-President of the World Bank Nicholas Stern states that climate change could affect growth, which could be cut by one-fifth unless drastic action is taken

[3] `http://en.wikipedia.org/wiki/Climate_change`

[4] `http://www.greenfacts.org/studies/climate_change/index.htm`
`http://www.gcrio.org/ipcc/qa/index.htm`

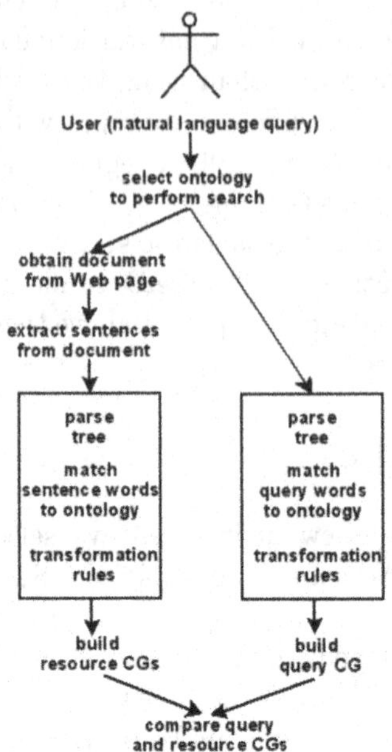

Fig. 3. Overview of SeseiOnto's workflow

- A single eruption of the kind that occurs several times per century can affect climate causing cooling for a period of a few years
- In short climate change can be a self-perpetuating process because different aspects of the environment respond at different rates and in different ways to the fluctuations that inevitably occur

It is possible to see that SeseiOnto correctly identified that climate change could affect economical growth. To the question "How could greenhouse gas emissions be reduced?", SeseiOnto answered:

- According to a 2006 United Nations report, Livestock's Long Shadow, livestock is responsible for 18% of the world's greenhouse gas emissions as measured in CO2 equivalents
- Similarly rising temperatures caused for example by anthropogenic emissions of greenhouse gases could lead to retreating snow lines revealing darker ground underneath and consequently result in more absorption of sunlight
- These principals can be observed as bubbles which rise in a pot of water heated on a stove or in a glass of cold beer allowed to sit at room temperature gases dissolved in liquids are released under certain circumstances

One can deduce that livestock and human activity ("anthropogenic emissions") influenced greenhouse gas emissions. To the question "Why should a few degrees of warming be a cause for concern?", SeseiOnto answered:

- According to these studies, the greenhouse effect, which is the warming produced as greenhouse gases trap heat, plays a key role in regulating Earth's temperature
- There are several examples of rapid changes in the concentrations of greenhouse gases in the Earth's atmosphere that do appear to correlate to strong warming, including the PaleoceneEocene thermal maximum, the Permian-Triassic extinction event, and the end of the Varangian snowball earth event
- The biggest factor of present concern is the increase in CO2 levels due to emissions from fossil fuel combustion followed by aerosols matter in the which exerts a cooling effect and cement manufacture

With the first answers, we understand that greenhouse gases play a key role in regulating Earth's temperature. To the question "How do we know that the atmospheric build-up of greenhouse gases is due to human activity?", SeseiOnto answered:

- As far as is known the climate system is generally stable with respect to these feedbacks positive feedbacks do not
- Similarly rising temperatures caused for example by anthropogenic emissions of greenhouse gases could lead to retreating snow lines revealing darker ground underneath and consequently result in more absorption of sunlight
- According to a 2006 United Nations report, Livestock's Long Shadow, livestock is responsible for 18% of the world's greenhouse gas emissions as measured in CO2 equivalents.

Once again, livestock and human activity is identified as a source of greenhouse gases.

Thus we can see that SeseiOnto has the potential to answer natural language queries that contain many different linguistic phenomena. The processing time for each query varies between two to three minutes.

The answers presented here were produced using WordNet concept hierarchy. Therefore, we had to disambiguate the words composing the query using WordNet's definitions. We also did some tests using other ontologies generated by Text-To-Onto and which were based on the Wikipedia page on climate change. We obtained small differences by using these ontologies. Answers were similar to the one presented here. The main difference was the order they were presented to the user.

The primary advantage of using a Text-To-Onto ontology with SeseiOnto is that the user does not need to disambiguate words composing his query. This can be important because it eases the search process while at the same time preventing the hassle of selecting the correct definitions from WordNet. The difference between definitions is often subtle and different users could choose separate definitions with the same concept meaning in mind. For that reason, we would recommend using a Text-To-Onto generated ontology to perform this type of Web search but thorough testing is necessary to prove this theory.

The ontology used by SeseiOnto, whether it be WordNet or Text-To-Onto, should not be considered as a thorough representation of the knowledge contained in the domain. It should be viewed as a sufficient semantic representation that allows the system to answer natural language queries. Certainly, an accurate and extensive human-built ontology would provide more information about the domain. However, we assumed that by using a WordNet or a Text-To-Onto ontology, we could more rapidly test our system with different corpora without needing to create an ontology by hand for each domain.

Consequently, we can suppose that SeseiOnto has the potential to be employed in the context of Web searches done on a specific field of knowledge. SeseiOnto is able to answer the user's query by providing pointers to relevant sentences within a document. Past results have shown that SeseiOnto could also be used in the context of larger documents collections [11].

5 Conclusion

With this research, we provide a unique analysis on how SeseiOnto can answer a query by selecting sentences within a relatively long document (more than 5,000 words), the Wikipedia page on climate change. This type of tests was never done with SeseiOnto in the past. We found out that within the first three answers given by our system, at least one is relevant to the query in most cases.

The main advantages of SeseiOnto are:

- It provides a natural language interface;
- it can pinpoint exact relevant sentences within a document to help the user answer his query;
- it provides an automatic ontology construction mechanism that can adapt to corpus updates;
- it seems to improve precision and recall for domain specific corpora or restricted domain Web search;
- it does not require experts' interventions;

Its main weaknesses are:

- Its processing time is relatively long (between two or three minutes);
- it needs to have a simple mechanism to select the correct ontology for a particular corpus;
- when WordNet is used, the disambiguation process can be confusing for the user.

Despite these drawbacks, we think that with additional testing and minor improvements, we could easily achieve even better results and apply the search methods of SeseiOnto in a professional environment.

5.1 Future Work

For the time being, SeseiOnto only functions as a standalone application. In the coming months, we intend to make it publicly available on the Web. We will start by making a system that can answer any queries relating to the current

Wikipedia page on climate change. The user will be able to select the method he wants to use to perform his query, i.e., selecting a particular ontology or using WordNet to disambiguate words from his question.

Although we think that SeseiOnto was up to task of the ICCS Challenge, extensive testing is still necessary to assess its full potential. In the case of Wikipedia, including sub-pages that are referenced by a link on the climate change page could provide an interesting way of seeing if SeseiOnto still performs well with additional documents. Testing SeseiOnto on new corpora would also provide an interesting feedback about its possibilities.

Moreover, the development of Text-To-Onto is now stopped and has been replaced by the new ontology creation framework, Text2Onto. It could be very interesting to use SeseiOnto with this new environment.

Incorporating formal ontologies to our application could also be pertinent since it would more easily permit the evaluation of the coherence of ontologies generated with Text-To-Onto.

Improving SeseiOnto processing time is very important if we ever want it to make it publicly available. To do so, indexing documents from the corpus and pre-converting documents' sentences to CGs would assuredly reduce its search time.

In conclusion, SeseiOnto shows that conceptual graphs have an immense potential to represent natural language. Although completely hidden to the user in our software, they are a key element to converting the syntactic representation given by Connexor to a semantic one. The ontology used by SeseiOnto is either automatically constructed with Text-To-Onto or taken from WordNet following the user's query words disambiguation. With this ontology, we obtain a simple yet effective way of comparing a query with many sentences coming from the corpus the search is being made on.

SeseiOnto was assembled using tools coming from the industry as well as the open-source, research and CG communities. This software is a concrete example on how conceptual structures can be used in an application and how their representation power can be employed to process information. With additional thorough testing, a tool such as SeseiOnto could probably be coupled to other information retrieval and information extraction applications. Such a coupling could provide a whole new range of possibilities in the Semantic Web context.

References

1. Bloehdorn, S., Cimiano, P., Hotho, A., Staab, S.: An ontology-based framework for text mining. GLDV-Journal for Computational Linguistics and Language Technology 20, 87–112 (2005)
2. Cui, H., Sun, R., Li, K., Kan, M.-Y., Chua, T.-S.: Question answering passage retrieval using dependency relations. In: International Conference on Research and Development in Information Retrieval (SIGIR), Salvador, Brazil, pp. 400–407. ACM Press, New York (2005)
3. de Salvo Braz, R., Girju, R., Punyakanok, V., Roth, D., Sammons, M.: An inference model for semantic entailment in natural language. In: Machine Learning Challenges Workshop, Pittsburgh, USA, pp. 261–286 (2005)

4. Embley, D.W.: Towards semantic understanding – an approach based on information extraction ontologies. In: Database Technologies 2004, Proceedings of the fifteenth Australasian database conference, Dunedin, New Zealand (2004)
5. Gómez-Pérez, A., Fernández-López, M., Corcho, O.: Ontological Engineering with examples from the areas of Knowledge Management, e-Commerce and the Semantic Web. Springer, Heidelberg (2004)
6. Hearst, M.: Automated discovery of wordnet relations. In: Fellbaum, C. (ed.) In WordNet: An Electronic Lexical Database and Some of its Applications. MIT Press, Cambridge (1998)
7. Järvinen, T., Tapanainen, P.: Towards an implementable dependency grammar. CoRR cmp-lg/9809001 (1998)
8. Maedche, A.: Ontology Learning for the Semantic Web. Kluwer Academic Publishers, Norwell, USA (2002)
9. Maedche, A., Staab, S.: Mining ontologies from text. In: Dieng, R., Corby, O. (eds.) EKAW 2000. LNCS (LNAI), vol. 1937, pp. 189–202. Springer, Heidelberg (2000)
10. Miller, G., Beckwith, R., Fellbaum, C., Gross, D., Miller, K.: Introduction to WordNet: An on-line lexical database. Journal of Lexicography 3(4), 234–244 (1990), `ftp://ftp.cogsci.princeton.edu/pub/wordnet/5papers.ps`
11. Morneau, M., Mineau, G.W., Corbett, D.: SeseiOnto: Interfacing NLP and ontology extraction. In: WI 2006: Proceedings of the 2006 IEEE/WIC/ACM International Conference on Web Intelligence, Hong Kong, China, pp. 449–455. IEEE Computer Society Press, Los Alamitos (2006)
12. Morneau, M., Mineau, G.W., Corbett, D.: Using an automatically generated ontology to improve information retrieval. In: First Conceptual Structures Tool Interoperability Workshop (CS-TIW 2006), Aalborg, Denmark, pp. 119–134. Aalborg University Press (2006)
13. Nicolas, S.: Sesei: un filtre semantique pour les moteurs de recherche conventionnels par comparaison de structures de connaissance extraites depuis des textes en langage naturel. Master's thesis, Département d'informatique et de génie logiciel, Université Laval (2003)
14. Nicolas, S., Mineau, G., Moulin, B.: Extracting conceptual structures from english texts using a lexical ontology and a grammatical parser. In: Sup.Proc. of 10th International Conference on Conceptual Structures, ICCS 2002, Borovets, Bulgaria (2002)
15. Nicolas, S., Moulin, B., Mineau, G.W.: Sesei: A CG-based filter for internet search engines. In: Conceptual Structures for Knowledge Creation and Communication. 11th International Conference on Conceptual Structures, Dresden, Germany, pp. 362–377. Springer, Heidelberg (2003)
16. Paliouras, G.: On the need to bootstrap ontology learning with extraction grammar learning. In: Dau, F., Mugnier, M.-L., Stumme, G. (eds.) ICCS 2005. LNCS (LNAI), vol. 3596, pp. 119–135. Springer, Heidelberg (2005)
17. Sánchez, D., Moreno, A.: Automatic generation of taxonomies from the WWW. In: Karagiannis, D., Reimer, U. (eds.) PAKM 2004. LNCS (LNAI), vol. 3336, pp. 208–219. Springer, Heidelberg (2004)
18. Shaw, W., Wood, J., Wood, R., Tibbo, H.: The cystic fibrosis database: Content and research opportunities. Library and Information Science Research 13, 347–366 (1991)
19. Yang, Y., Liu, X.: A re-examination of text categorization methods. In: SIGIR 1999: Proceedings of the 22nd annual international ACM SIGIR conference on research and development in information retrieval, pp. 42–49. ACM Press, New York (1999)

Concept Similarity and Related Categories in SearchSleuth

Frithjof Dau, Jon Ducrou, and Peter Eklund

School of Information Systems and Technology
University of Wollongong
Northfields Ave, Wollongong, NSW 2522, Australia
dau@dr-dau.net, jonducrou@gmail.com, peklund@uow.edu.au

Abstract. SearchSleuth is a program developed to experiment with the automated local analysis of Web search using formal concept analysis. SearchSleuth extends a standard search interface to include a conceptual neighborhood centered on a formal concept derived from the initial query. This neighborhood of the concept derived from the search terms is decorated with its upper and lower neighbors representing more general and specialized concepts respectively. In SearchSleuth, the notion of related categories – which are themselves formal concepts – is also introduced. This allows the retrieval focus to shift to a new formal concept called a sibling. This movement across the concept lattice needs to relate one formal concept to another in a principled way. This paper presents the issues concerning exploring and ordering the space of related categories.

1 Introduction

There are several Formal Concept Analysis-based Web search applications which provide automatic local analysis of search results for query refinement and labeled clustering [1,2,3]. These systems work via the creation of a conceptual space from polled search results which are displayed in various ways. The method is limited in that the systems fail to create a concept representing the query itself within the information space – meaning the space is representative of the results returned from the query terms, but not to the query terms themselves. SearchSleuth [4] overcomes this problem by creating a conceptual space as a neighborhood of the *search concept*: the formal concept derived from the search terms. The resulting neighborhood is comprised of generalisations (upper neighbors), specializations (lower neighbors) and related categories (called siblings). Fig. 1 shows the interface and these components.

By centering the conceptual space around the search concept, the resulting query refinement operations are more closely coupled to the search terms used in the creation of the space. SearchSleuth was first presented at the concept lattice applications conference in October 2007 [4], in that paper we discussed some of the preliminaries of search in the conceptual neighborhood of a query and go on to differentiate SearchSleuth work from other FCA-based web search tool such as *CREDO* [1] and FooCA[2,3]. In this paper, we re-iterate some of

P. Eklund and O. Haemmerlé (Eds.): ICCS 2008, LNAI 5113, pp. 255–268, 2008.

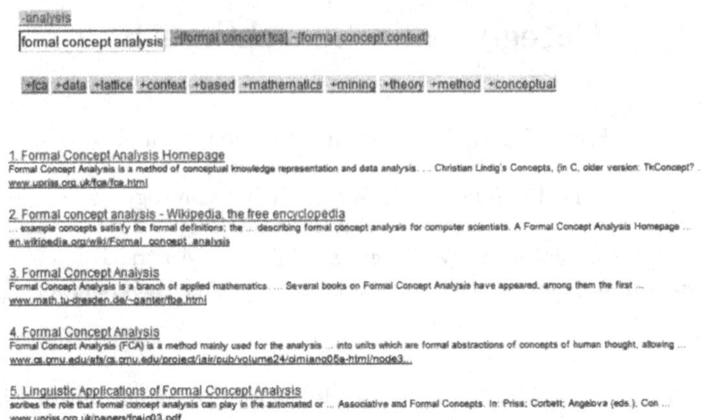

Fig. 1. SearchSleuth display, including top results, after a search for '`formal concept analysis`'. Generalization/specialization formal concepts shown above/below the search box resp. The related categories or siblings are to the right of search box.

the fundamentals of SearchSleuth, so that the paper is self-contained, however our contribution is in terms of explorations of the category space: namely how alternative formal concepts in the neighborhood of the current query concept are derived. Our presentation includes the analysis of lattice-theoretic and set-theoretic notions of proximity and we conclude that these two ideas are orthogonal but complementary. The outcome is of the analysis is reflected in the design of SearchSleuth.

2 Navigation and Conceptual Neighborhoods

Kim and Compton [5,6] presented a document navigation paradigm using FCA and a neighborhood display. Their program, *KANavigator* uses annotated documents that can be browsed by keyword and displays the direct neighborhood (in particular the lower neighbors) as its interface. Kim and Compton's system emphasised the use of textual labels as representations of single formal concepts as opposed to a line diagram of the concept lattice.

ImageSleuth [7] used a similar interface design to allow exploration of image collections. By showing upper and lower neighbors of the current concept and allowing navigations to these concepts, users could refine or generalise their position in the information space. This is aided by the use of pre-defined conceptual scales that could be combined to define the attribute set of the lattice which forms the information space (see Fig. 2 (left)).

ImageSleuth uses most of its interface (shown in Fig. 2) to show thumbnails of images in the extent of the chosen concept. As a result the user never sees the line diagram of a concept lattice. Instead, the lattice structure around the current concept is represented through the list of upper and lower neighbors that allow the user to move to super- or sub-concepts. For every upper neighbor

(C, D) of the current concept (A, B) the user is offered to remove the set $B \setminus D$ of attributes from the current intent. Dually, for every lower neighbor (E, F) the user may include the set $F \setminus B$ of attributes which takes her to this lower neighbor. By offering the sets $B \setminus D$ and $F \setminus B$ dependencies between these attributes are shown. Moving to the next concept not having a chosen attribute in its intent may imply the removal of a whole set of attributes. ImageSleuth was usability tested and results indicated that the approach aided navigation in image collections [8,9].

SearchSleuth follows from ImageSleuth and employs the same conceptual neighborhood paradigm for display purposes. Unlike ImageSleuth, SearchSleuth's context is not static, so the space is rebuilt with each navigation step. This is because computing the entire domain, the Internet, as a conceptual neighborhood would be computationally prohibitive.

3 Design Approach of SearchSleuth: Context Building

For the Web, result sets from search engines usually take the form of the lists of URLs, each with the document title, a short summary of the document (or snippet) and various details such as date last accessed. Formal Concept Analysis-based Web search tools use the text-based components of the result set to create a formal context of the results. This context is then the basis for the conceptual space to be navigated. One problem with the transformation from Web search results to formal context is that ranking information on the result set is lost. All results are treated equally, this issue is usually addressed by re-introducing the rank ordering from the search engine on any result set that is realized from the concept lattice.

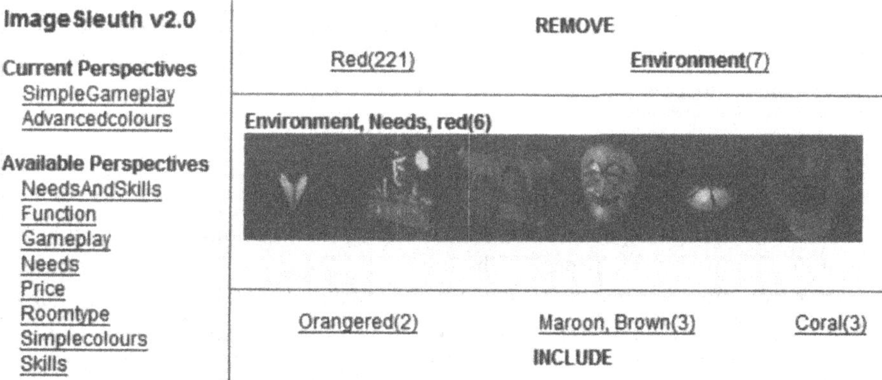

Fig. 2. ImageSleuth: the interface presents only the extent of the current concept as thumbnails and generalizations/specializations by removal/addition of attributes to reach the upper and lower neighbors (shown to the top/bottom of the thumbnails). Pre-defined scales (perspectives) are displayed on the left.

Another difficulty experienced with Web search using FCA is that ranking methods use techniques such as link structure, page popularity and analysis of referring pages. As such, we cannot assume that all results of a multiple term query will contain all the queried terms used. Even a single query term may yield a page that does not contain the search term entered. This seems counter intuitive, but if there are enough Web pages linked to the result page that *do* contain the search term, that page's rank may be inflated enough to feature in the result set.

SearchSleuth uses the 'result has term' representation to build a formal context. The formal context for SearchSleuth is created on demand for each query; this suits the dynamic nature of the Internet. The formal objects are the individual results, and the formal attributes are the terms contained in the title and summary of each result. Terms are extracted from the title and summary after stemming and stop-word filtering has been performed. Stemming reduces words to their lexical root (e.g. `jump`, `jumping` and `jumps` are all reduced to `jump`). Stop-word filtering removes words without individual semantic value, for example `a`, `the` and `another`. Removing these words reduces the complexity of the context without noticeable reduction in semantic quality.

The context is then reduced by removing attributes with low support. Every attribute that has less than 5% of the objects in the incidence relation is removed. This decreases the computational overhead of involved in computing the concept lattice. Experience shows that this reduction rarely effects the computed conceptual neighborhood as the terms removed are scarce within the information.

Once the formal context is constructed, the search concept is created. This is done by taking the provided query terms as attributes and deriving the formal concept. The upper neighbors of this formal concept are then derived and used to expand the context. This is done by querying the search engine with the attributes of each upper neighbor and inserting the results into the context. Results for these ancillary searches are limited to fewer results.

This process of building the context increases the number of terms in the information space based on a single level of generalisation. It makes the information space larger and richer.

4 Building the Information Space

Once the context is expanded, the search concept is recomputed as it may have been invalidated by this process. The upper and lower neighbors are computed next. A concept A is said to be the upper neighbor (or cover) of a iff we have $A > B$, and/but there is no concept C with $A > C > B$. A concept A is said to be the lower neighbor of (or covered by) a concept B iff we have $A < B$, and/but there is no concept C with $A < C < B$. The DownSet (DS) and UpSet (US) are defined as follows;

$$DS(X) := \{y \mid y \leq x \text{ for an } x \in X\} \qquad US(X) := \{y \mid y \geq x \text{ for an } x \in X\}$$

Upper and lower neighbors of a concept C are written as $UN(C)$ and $LN(C)$ respectively. Consider now the set of concepts X, $UN(X)$ is defined as the union of all upper neighbors of the concepts in X. Dually, consider the set of concepts X, $LN(X)$ is defined as the union of all lower neighbors of the concepts in X.

$$UN(X) := \bigcup\{UN(C) \mid C \in X\} \qquad LN(X) := \bigcup\{LN(C) \mid C \in X\}$$

The next step is to compute the related categories or sibling concepts. Sibling concepts are then calculated by finding all of the lower neighbors of upper neighbors which are upper neighbors of lower neighbors. Put another way, *siblings* constitute formal concepts created by the removal of an attribute (or attributes) that define an upper neighbor (UN), and the inclusion of an attribute (or attributes) that defines a lower neighbor (LN). Child Siblings (CS) and Parent siblings (PS) defined as: are defined:

$$CS(C) := UN(LN(C))\backslash\{C\} \qquad PS(C) := LN(UN(C))\backslash\{C\}$$

Exact Siblings (ES or Type I siblings) are those which are both Parent and Child siblings, Since they represent a stricter version of the notion of siblings they are referred to as Type I siblings and PS and CS are termed Type II siblings:

$$ES(C) := [LN(UN(C)) \cap UN(LN(C))]\backslash\{C\}$$

General Siblings (GS – Type III) define an even broader set of sibling concepts and are defined:

$$GS(C) := [DS(UN(C)) \cap US(LN(C))]\backslash(\{C\} \cup UN(C) \cup LN(C))$$

namely, anything strictly between some lower and some upper neighbor.

Child Siblings (CS), Parent Siblings (PS), and Exact Siblings (ES) form anti-chains, but General Siblings (GS) do not.

An example is shown in Fig. 3; concepts with a grey backing are Exact Siblings (ES) of the concept marked C.

Using the same labeling scheme as ImageSleuth for upper and lower neighbors and using the full intent as labels of sibling concepts, a display is rendered for the user (shown in Fig. 1).

Upper neighbors are shown above this text entry box, displayed as text labels (shown in Fig. 1). The labels are the attributes which would be removed to navigate to that upper neighbor. These labels are preceeded by a minus symbol (-) to reinforce the notion of *removal.*

Lower neighbors are similarly displayed (also indicated with arrows in Fig. 1), but placed below the text entry box. These labels are the attributes which would be added to navigate to that lower neighbor. Like upper neighbor labels, these labels are preceeded by a symbol to reinforce the labels meaning, namely the plus symbol (+) and the notion of *include.*

The display order of the upper and lower neighbors is defined by extent size, larger extents displayed first (left-most). Extent is representative of the importance or prominence within the current information space. Extent is also used to aid in the coloring of the labels background. The higher the extent on a lower

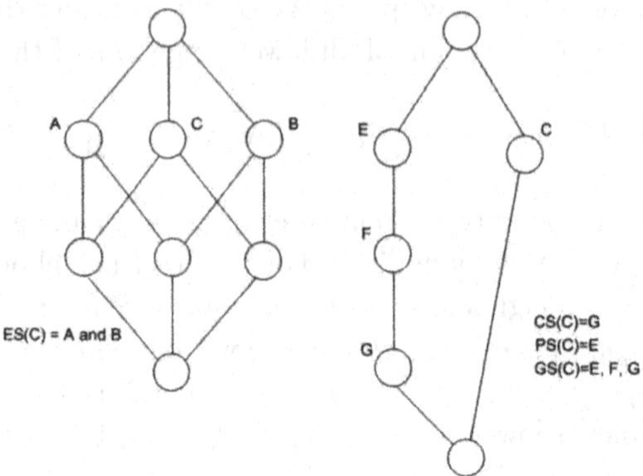

Fig. 3. Diagram demonstrating the *Parent Sibling (PS), Child Siblings (CS), Exact Sibling (ES)* and *General Siblings (GS)* concepts of the concept labeled with a C in two lattices

neighbor, the deeper the blue block shade behind that concepts label. Upper neighbors are displayed with the same principle but with red block shade.

One method for dealing with the return of empty-extents from term-based searching is to provide users with a list of the terms entered so that they can incrementally remove terms to unconstrain the search. SearchSleuth explores an approach based on variations on defined distance [10] and similarity [11] metrics in the FCA literature in order to find similar relevant concepts.

Exact Siblings (ES) are shown to the right of the text entry box (indicated with arrows in Fig. 1) and are indicative of related concepts. The complete intent of these concepts is displayed within square brackets preceded by a tilde (`~[...]`). This helps group the concept intents and aids distinguishing between related concepts. Unlike upper and lower neighbors, Exact Siblings are ordered by *similarity.* The similarity metric is based on work by Lengnink [10] and was initially adapted for ImageSleuth. It uses the size of the common objects and attributes of the concepts. For two concepts (A, B) and (C, D), we set:

$$s((A,B),(C,D)) := \frac{1}{2}\left(\frac{|A \cap C|}{|A \cup C|} + \frac{|B \cap D|}{|B \cup D|}\right). \tag{1}$$

The similarity metric is used to order the exact sibling concepts, while highlighting remains based on extent size. Coloring on sibling labels is based on grey block shades.

By clicking any of the possible concept labels, the query is set to the intent of the selected concept and the query process is restarted. This is an important restructuring step as a change in the query will change the result set, and in order for the information to be valid it needs to be recomputed.

Looking back to Fig. 1, we see the search concept shown is based on the query **`formal concept analysis`**. It shows a single upper neighbor **`analysis`** which interestingly shows an implication that **`formal`** and **`concept`** are implied by **analysis**. The first of the lower neighbors is the acronym `fca`. This is followed by terms such as `lattice`, `mathematics` and `theory`. These terms are good examples of specialisation from the concept of Formal Concept Analysis. This neighborhood is based on 115 formal objects. The initial number of formal attributes for this example was 623, after reducing the context this was lowered to 40. This offers a tremendous reduction in context complexity, and therefore computation time but these numbers also reflect the need to search a subset of conceptual neighborhood.

A main question in the design of SearchSleuth is whether the definition of Exact Siblings provides sufficient space for proximity search of neighboring categories. The remainder of the paper addresses this issue in detail.

5 Distance, Similarity and Siblings

We have two measures to consider the proximity of formal concepts. In addition to similarity (s) defined in Eqn. (1) we also have for two formal concepts (A, B), (C, D),

$$d((A,B),(C,D)) := \frac{1}{2}\left(\frac{|A \backslash C| + |C \backslash A|}{|G|} + \frac{|B \backslash D| + |D \backslash B|}{|M|}\right)$$

where d the *distance* of the concepts (A, B), (C, D) [10]. To ease comparison between the two measures, let

$$s'((A,B),(C,D)) := 1 - s((A,B),(C,D))$$

Let us first note that s' and d are metrics in the mathematical understanding. That is, d satisfies for arbitrary concepts x, y, z: $d(x,y) \geq 0$ and $d(x,y) = 0 \Leftrightarrow x = y$ (non-negativity and identity of indiscernibles), $d(x,y) = d(y,x)$ (symmetry), and $d(x,z) \leq d(x,y) + d(y,z)$ (triangle inequality). The triangle inequalities can easily be shown by straight-forward computations, and the remaining properties are easily to be seen.

Next, note that we have

$$s'((A,B),(C,D)) = \frac{1}{2}\left(\frac{|A \cup C| - |A \cap C|}{|A \cup C|} + \frac{|B \cup D| - |B \cap D|}{|B \cup D|}\right) \quad (2)$$

$$d((A,B),(C,D)) = \frac{1}{2}\left(\frac{|A \cup C| - |A \cap C|}{|G|} + \frac{|B \cup D| - |B \cap D|}{|M|}\right) \quad (3)$$

Comparing Eqns. (2) and (3), we see that they differ in that in (2), we divide through $|A \cup C|$ and $|B \cup D|$, whereas in (3), we divide through $|G|$ and $|M|$. Therefore s' is a local distance, focusing on the shared attributes and objects of the two formal concepts being compared, and d is a global distance, using all the attributes and objects in the context. The choice of measurement to use therefore depends on the sensitivity of the proximity measure required. The preferred approach for SearchSleuth is proximity in the conceptual neighborhood to the current formal concept. Therefore the local measure is considered most suitable.

One can however easily combine the two measures. Let $l \in [0,1]$, measuring the desire of a local point of view ($l = 0$ means the user wants a purely global point of view, and $l = 1$ means the user wants a purely local point of view). Then the corresponding distance (*dist*) measure is,

$$dist((A,B),(C,D)) := l \cdot s'((A,B),(C,D)) + (1-l) \cdot d((A,B),(C,D)).$$

6 Relationship between Metric and Sibling Explored

The basic approach of SearchSleuth is to explore the 'conceptual neighborhood' of a given concept. To grasp this 'conceptual neighborhood', SearchSleuth takes advantage of two fundamentally different notions of neighboorhood. On the one hand, we use the lattice-theoretic notions of siblings, which do do not take the sizes of the concept-extents or intents into account. On the other hand, we use the notions of similarity and distance metrics are set-theoretic notions (they do not take the lattice-order into account). In the next two pararaphs, we first investigate the different types of siblings, and then the two kinds of similarity metrics.

The notions of siblings is more fine-grained divided into exact siblings (Type I), parent- and child siblings (Type II), and general siblings (Type III). Obviously, this is a hierarchy of types: Each Type I sibling is a Type II sibling, and each Type II sibling is a Type III sibling. Besides this inclusions, we cannot provide any general estimations on the number of the different types of siblings. To be more precise: If $n_{\text{I}}, n_{\text{II}}, n_{\text{III}} \in \mathbb{N}_0$ are three numbers with $n_{\text{I}} \leq n_{\text{II}} \leq n_{\text{III}}$ and $n_{\text{II}} \neq 1$, then there exists a lattice with an element c which has n_{I} Type I siblings, n_{II} Type I siblings, and n_{III} Type I siblings. An example for such a lattice is given below. In the diagram, for each sibling of c, the most special type the sibling belongs to is inscribed into its node. That is in the diagram, n_{I} nodes are labelled with 'I', $n_{\text{II}} - n_{\text{I}}$ nodes are labelled with 'II', and $n_{\text{III}} - n_{\text{II}} - n_{\text{I}}$ nodes are labelled with 'III'.

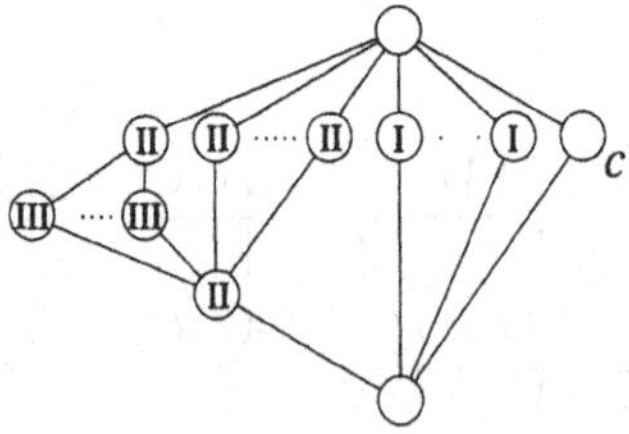

For the notions of local (s') and global (d) distance, a somewhat similar consideration applies. Due to Eqns. (2) and (3), in each lattice, for any concepts x, y, we have

$$s'(x,y) \geq d(x,y) \quad .$$

On the other hand, there are examples (one is given in the following subsection) of lattices where there are two concepts c, n which are arbitrary close with respect to the local, but arbitrary distant with respect to the global distance.

The question remains whether there are dependencies between the lattice-theoretic (i.e., siblings) and the set-theoretic (i.e., metrics) notions of conceptual neighborhood. We will investigate some examples in the following sections. As these examples will show, the two notions are somewhat orthogonal but complementary in determining the most appropriate related categories in SearchSleuth.

6.1 Exact Siblings (ES) and Proximity Metrics

We first consider an example where we have an exact sibling n of a concept c, and we investigate whether we can draw some conclusions about the local or global distance between c and n. The example we consider is the following concept-lattice:

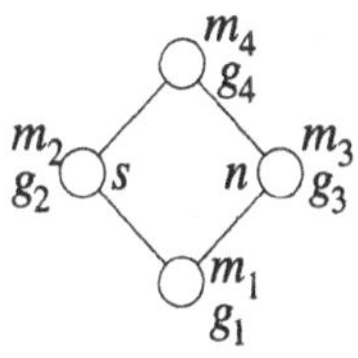

In this diagram, g_1, g_2, g_3, g_4 resp. m_1, m_2, m_3, m_4 do *not* denote objects or attributes, but the *numbers* of objects resp. attributes which generate the concept. For example, for $c = (G_2, M_2)$, we have $g_2 = |G_2| - |G_1|$. That is, the g_i and m_i are the numbers of objects and attributes in the common diagrams of concept lattices. Two concepts are given names, namely c and n. We have:

$$s(c,n) = \frac{1}{2}\left(\frac{g_1}{g_1+g_2+g_3} + \frac{m_4}{m_2+m_3+m_4}\right)$$
$$d(c,n) = \frac{1}{2}\left(\frac{g_2+g_3}{g_1+g_2+g_3+g_4} + \frac{m_2+m_3}{m_1+m_2+m_3+m_4}\right)$$

For fixed $g_2, g_3, g_4, m_1, m_2, m_3$ (e.g., $g_2 = g_3 = g_4 = m_1 = m_2 = m_3 = 1$), we have

$$\lim_{\substack{g_1\to\infty\\ m_4\to\infty}} s'(c,n) = 1 - \frac{1}{2}(1+1) = 0 \qquad \text{and} \qquad \lim_{\substack{g_1\to\infty\\ m_4\to\infty}} d(c,n) = \frac{1}{2}(0+0) = 0$$

i.e., c and n can be arbitrarily similar with respect to both s' and d.

On the other hand, for fixed g_1, g_3, m_3, m_4, we have

$$\lim_{\substack{g_2\to\infty\\ m_2\to\infty}} s'(c,n) = 1 - \frac{1}{2}(0+0) = 1 \qquad \text{and} \qquad \lim_{\substack{g_2\to\infty\\ m_2\to\infty}} d(c,n) = \frac{1}{2}(1+1) = 1$$

(similar for g_2, m_3, and g_3, m_2, and g_3, m_3). That is, c and n can be arbitrarily different (again with respect to s and to d).

Now let $\varepsilon_1, \varepsilon_2 > 0$. Let g_3, g_4, m_1, m_3 be fixed. By *first* choosing g_2 and m_2 sufficiently large, we can achieve $s(x,n) < \varepsilon_1$, and by *then* choosing g_1, m_4 sufficiently large (which does not affect $s(c,n)$), we can achieve $d(c,n) < \varepsilon_2$.

That is, we can achieve that in a local understanding (i.e., w.r.t. s'), the concepts c and n are very similar, whereas in a global understanding ((i.e., w.r.t. d), the concepts c and n are very distant.

To summarize this example: even for the most special case of being an exact sibling n of a given concept c, we cannot draw any conclusion about the local or global distance between c and n.

6.2 The Proximity of Type I Siblings Versus Non Siblings

A concept n which is a sibling for a given concept c belongs, from a lattice-theoretic point of view, to the conceptual neighborhood of c; a concept x which is not a sibling of c does not belong to the conceptual neighborhood. Is this property reflected by the distances s' and d? We consider again an example where n even is an exact sibling of c.

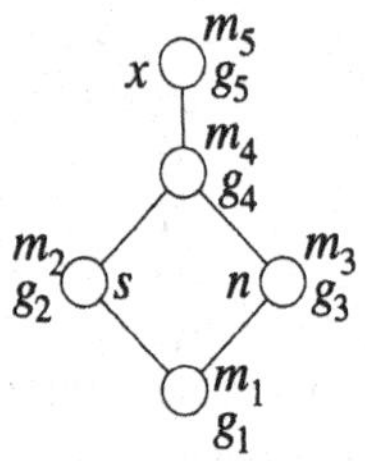

In terms of similarity we have:

$$s_n := s(c,n) = \frac{1}{2}\left(\frac{g_1}{g_1+g_2+g_3} + \frac{m_4+m_5}{m_2+m_3+m_4+m_5}\right)$$

$$s_x := s(c,x) = \frac{1}{2}\left(\frac{g_1+g_2}{g_1+g_2+g_4+g_5} + \frac{m_5}{m_2+m_4+m_5}\right)$$

Note that we can have $g_1 = 0$, $m_1 = 0$, $g_4 = 0$, and $m_5 = 0$, but all other numbers must be ≥ 1. Now, n could be more similar to c than x, equally similar, or less similar, as the following examples show.

g_1	g_2	g_3	g_4	g_5	m_1	m_2	m_3	m_4	m_5	$2 \cdot s_n = 2 \cdot s(c,n)$	$2 \cdot s_x = 2 \cdot s(c,x)$
1	1	1	0	1	1	1	1	1	1	$1/3 + 2/4 = 5/6$	$2/3 + 1/3 = 1$
1	1	1	1	1	0	1	1	1	1	$1/3 + 2/4 = 5/6$	$2/4 + 1/3 = 5/6$
1	1	1	1	1	0	1	1	1	0	$1/3 + 1/3 = 2/3$	$2/4 + 0/2 = 1/2$

Running a computer-program checking all values for g_i and m_i with a threshold of 8 yields:

$s(c,n) > s(c,x)$	$s(c,n) < s(c,x)$	$s(c,n) = s(c,x)$
804.068.208	913.112.127	2.746.449

Therefore the cases in which $s(c,n) > s(c,x)$ and $s(c,n) < s(c,x)$ do not significantly differ and we cannot conclude (at least for this toy-example) that siblings are *generally* more similar than non-siblings.

Similarly, repeating the analysis in terms of the distance metric d, we have:

$$d_n := s(c,n) = \frac{1}{2}\left(\frac{g_2+g_3}{g_1+\cdots+g_5} + \frac{m_3+m_4}{m_1+\cdots+m_5}\right)$$
$$d_x := s(c,x) == \frac{1}{2}\left(\frac{g_4+g_5}{g_1+\cdots+g_5} + \frac{m_2+m_4}{m_1+\cdots+m_5}\right)$$

$g_1g_2g_3g_4g_5$	$m_1m_2m_3m_4m_5$	$2 \cdot d_n = 2 \cdot d(c,n)$	$2 \cdot d_x = 2 \cdot d(c,x)$	result
0 1 1 0 2	0 1 1 2 0	$2/4+2/4=1$	$2/4+3/4=5/4$	$d_1 < d_2$
0 1 1 1 1	0 1 1 1 0	$2/4+2/3=7/6$	$2/4+2/3=7/6$	$d_1 = d_2$
0 1 1 0 1	0 1 1 1 0	$2/3+2/3=4/3$	$1/3+2/3=1$	$d_1 > d_2$

$d(c,n) > d(c,x)$	$d(c,n) < d(c,x)$	$d(c,n) = d(c,x)$
908.328.121	788.136.280	23.462.383

Again the cases $d(c,n) > d(c,x)$ and $d(c,n) < d(c,x)$ do not significantly differ.

To summarize this example: even for the most special case of being an exact sibling n of a given concept c, we cannot draw any conclusion that n is closer to c compared to a non-sibling.

6.3 The Proximity of Type II Versus Type III Siblings

We have different strengths of being a sibling. We still could hope that this is reflected by the metrics. In the following example, we consider Type II siblings of a concept c with the more general Type III siblings and check whether the Type II siblings are closer to c than the Type III siblings.

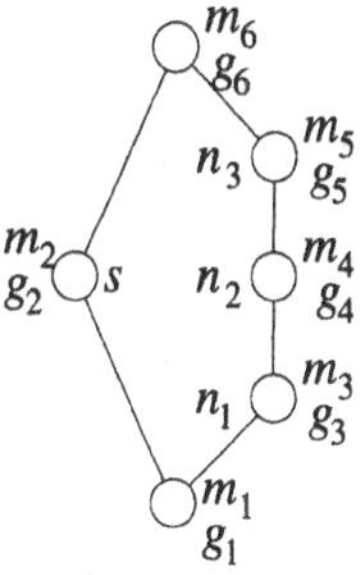

In terms of similarity we have:

$$s_1 := s(c,n_1) = \frac{1}{2}\left(\frac{g_1}{g_1+g_2+g_3} + \frac{m_6}{m_2+m_3+m_4+m_5+m_6}\right)$$
$$s_2 := s(c,n_2) = \frac{1}{2}\left(\frac{g_1}{g_1+g_2+g_3+g_4} + \frac{m_6}{m_2+m_4+m_5+m_6}\right)$$
$$s_3 := s(c,n_3) = \frac{1}{2}\left(\frac{g_1}{g_1+g_2+g_3+g_4+g_5} + \frac{m_6}{m_2+m_5+m_6}\right)$$

In this example, there is no order relationship between s_1, s_2, and s_3. We can have $s_1 < s_2 < s_3$ or $s_3 < s_1 < s_2$ or $s_1 = s_2 < s_3$ etc. Any combination is possible. The following table shows examples for all possible strict orders of s_1, s_2, s_3 (examples for cases like $s_1 = s_2 < s_3$ are left out due to space limitations).

$g_1g_2g_3g_4g_5g_6$	$m_1m_2m_3m_4m_5m_6$	$2 \cdot s_1$	$2 \cdot s_2$	$2 \cdot s_3$	result
1 1 1 1 1 1	1 1 2 1 1 2	$1/3+2/7$	$1/4+2/5$	$1/5+2/4$	$s_1 < s_2 < s_3$
1 1 2 1 2 1	1 2 2 1 2 2	$1/4+2/9$	$1/5+2/7$	$1/7+2/6$	$s_1 < s_3 < s_2$
1 1 1 1 1 1	1 1 1 1 1 2	$1/3+2/6$	$1/4+2/5$	$1/5+2/4$	$s_2 < s_1 < s_3$
1 1 1 1 1 1	1 1 1 1 2 2	$1/3+2/7$	$1/4+2/6$	$1/5+2/5$	$s_2 < s_3 < s_1$
2 2 2 1 2 1	1 1 2 1 2 1	$2/6+1/7$	$2/7+1/5$	$2/9+1/4$	$s_3 < s_1 < s_2$
1 1 1 1 1 1	1 2 1 1 2 1	$1/3+1/7$	$1/4+1/6$	$1/5+1/5$	$s_3 < s_2 < s_1$

Similarly, repeating the analysis in terms of the distance metric, we have:

$$d_1 := d(c, n_2) = \frac{1}{2}\left(\frac{g_2 + g_3}{g1 + \cdots + g_6} + \frac{m_2 + m_3 + m_4 + m_5}{m_1 + \cdots + m_6}\right)$$

$$d_2 := d(c, n_2) = \frac{1}{2}\left(\frac{g_2 + g_3 + g_4}{g1 + \cdots + g_6} + \frac{m_2 + m_4 + m_5}{m_1 + \cdots + m_6}\right)$$

$$d_3 := d(c, n_3) = \frac{1}{2}\left(\frac{g_2 + g_3 + g_4 + g_5}{g1 + \cdots + g_6} + \frac{m_2 + m_5}{m_1 + \cdots + m_6}\right)$$

Again here is no relationship between d_1, d_2, and d_3, and any combination is possible, as the following table shows:

$g_1g_2g_3g_4g_5g_6$	$m_1m_2m_3m_4m_5m_6$	$2 \cdot d_1$	$2 \cdot d_2$	$2 \cdot d_3$	result
1 1 1 1 1 1	1 1 1 1 2 1	$2/6+5/7$	$3/6+4/7$	$4/6+3/7$	$d_1 < d_2 < d_3$
1 1 1 1 1 1	1 2 1 2 2 2	$2/6+7/10$	$3/6+6/10$	$4/6+4/10$	$d_1 < d_3 < d_2$
1 1 1 1 1 1	1 2 2 1 2 2	$2/6+7/10$	$3/6+5/10$	$4/6+4/10$	$d_2 < d_1 < d_3$
1 1 1 1 1 1	1 1 2 1 1 1	$2/6+5/7$	$3/6+3/7$	$4/6+2/7$	$d_2 < d_3 < d_1$
1 1 1 1 1 1	1 1 1 2 1 1	$2/6+5/7$	$3/6+4/7$	$4/6+2/7$	$d_3 < d_1 < d_2$
1 1 1 1 1 1	1 1 2 2 1 1	$2/6+6/8$	$3/6+4/8$	$4/6+2/8$	$d_3 < d_2 < d_1$

To summarize the analysis: we cannot draw any conclusion that Type II siblings of a concept c are closer to c, using s' or d, than the weaker Type III siblings. That is, we cannot say that Type II siblings better represent related categories than Type III siblings.

6.4 Type I Versus Type II Versus Type III Siblings

This example compares now all three types of siblings are now

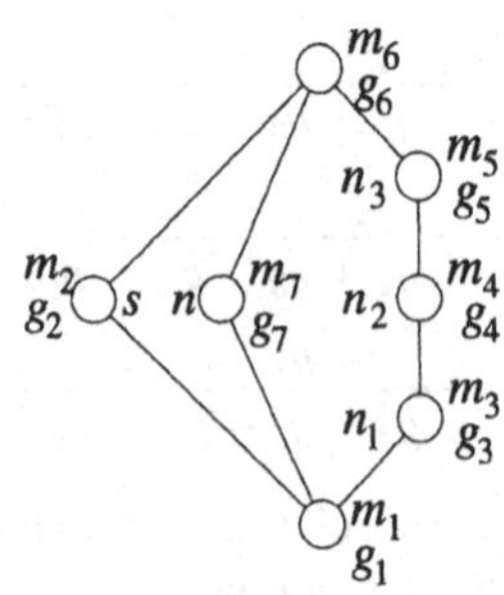

In terms of similarity we have:

$$s_n := s(c,n) = \frac{1}{2}\left(\frac{g_1}{g_1+g_2+g_7} + \frac{m_6}{m_2+m_6+m_7}\right)$$

$$s_1 := s(c,n_1) = \frac{1}{2}\left(\frac{g_1}{g_1+g_2+g_3} + \frac{m_6}{m_2+m_3+m_4+m_5+m_6}\right)$$

$$s_2 := s(c,n_2) = \frac{1}{2}\left(\frac{g_1}{g_1+g_2+g_3+g_4} + \frac{m_6}{m_2+m_4+m_5+m_6}\right)$$

$$s_3 := s(c,n_3) = \frac{1}{2}\left(\frac{g_1}{g_1+g_2+g_3+g_4+g_5} + \frac{m_6}{m_2+m_5+m_6}\right)$$

Note that changing g_7 and m_7 does not affect the similarity measures between c and n_1, n_2, n_3, resp. According to Section 6.1, for high values of g_7 and m_7, the similarity between c and n (i.e., d) decreases. So we easily can use the values for $g_1, \ldots, g_6, m_1, \ldots, m_6$ of the last example to get all possible orderings of s_1, s_2, s_3, and choose g_7 and m_7 such that $d < s_1, s_2, s_3$. That is, Type II siblings)are not necessarily more similar to c than Type III siblings.

In fact, we have again that for s, all 24 strict orders of s, n, s_1, s_2, s_3 can appear. And the same holds for d. (As we have both for s and d 24 such strict orders, thus 48 examples, these examples are not provided due to space limitations). In short, no general statements which render some preference for siblings used as Related Categories in terms of similarity and distance.

7 Conclusion

The notion of Type I, II and II siblings is a purely lattice-theoretic notion, whereas the notion of distance and similarity is a purely set-theoretic notion. As our examples show, these notions are somewhat complementary. In order to find similar concepts to a given concept (related categories), there is no hint that one should start with the immediate sibling neighbors of that concept. This might sound disappointing at a first glance, but in practice our observations lead to an important design feature in SearchSleuth. Computationally, the neighboring siblings to the current formal concept (whether of Type I, II or III) are the easiest concepts to compute and therefore represent natural candidates for related category search. In this case the search of the sibling space proceeds by considering related categories with the best distance and similarity stored in each of the neighboring siblings concepts for the current concept.

SearchSleuth, extends current FCA Internet search engines by positioning the user within the information space, rather than placing the user arbitrarily or presenting the entire space. This allows generalisation and categorisation operations to be performed against the current query concept. SearchSleuth overcomes a number of practical difficulties in the use of FCA for Internet Search, namely a practical approach to the construction of a sparse context and the categorisation operation, where the conceptual focus is moved to a sibling concept of the search concept. These paper explains how related categories are derived using a combination of order-theoretic notions of neighborhood in combination of set-theoretic definitions of concept similarity.

References

1. Carpineto, C., Romano, G.: Exploiting the potential of concept lattices for information retrieval with credo. Journal of Universal Computer Science 10(8), 985–1013 (2004)
2. Koester, B.: FooCA - Web Information Retrieval with Formal Concept Analysis. Diploma, Technische Universität Dresden (2006)
3. Koester, B.: Conceptual knowledge processing with google. In: Contributions to ICFCA 2006, pp. 178–183 (2005)
4. Ducrou, J., Eklund, P.: Searchsleuth: the conceptual neighbourhood of an internet query. In: 5th International Conference on Concept Lattices and Their Applications (CLA 2007), pp. 249–260 (2007), `http://www.CEUR-WS.Vol-331.Ducrou.pdf`
5. Kim, M., Compton, P.: Formal Concept Analysis for Domain-Specific Document Retrieval Systems. In: Australian Joint Conference on Artificial Intelligence, pp. 237–248 (2001)
6. Kim, M., Compton, P.: The perceived utility of standard ontologies in document management for specialized domains. International Journal of Human-Computer Studies 64(1), 15–26 (2006)
7. Ducrou, J., Vormbrock, B., Eklund, P.: FCA-based Browsing and Searching of a Collection of Images. In: Schärfe, H., Hitzler, P., Øhrstrøm, P. (eds.) ICCS 2006. LNCS (LNAI), vol. 4068, pp. 203–214. Springer, Heidelberg (2006)
8. Ducrou, J., Eklund, P.: Faceted document navigation using conceptual structures. In: Hitzler, P., Schärf, H. (eds.) Conceptual Structures in Practice, pp. 251–278. CRC Press, Boca Raton (2008)
9. Ducrou, J., Eklund, P.: An intelligent user interface for browsing and search mpeg-7 images using concept lattices. International Journal of Foundations of Computer Science 19(2), 359–381 (2008)
10. Lengnink, K.: Ahnlichkeit als Distanz in Begriffsverbänden. In: Stumme, G., Wille, R. (eds.) Begriffliche Wissensverarbeitung: Methoden und Anwendungen, pp. 57–71. Springer, Heidelberg (2001)
11. Saquer, J., Deogun, J.S.: Concept aproximations based on rough sets and similarity measures. Int. J. Appl. Math. Comput. Sci. 11, 655–674 (2001)

Grounded Conceptual Graph Models

Harry S. Delugach and Daniel M. Rochowiak

Computer Science Department
Univ. of Alabama in Huntsville
Huntsville, AL 35899
{delugach,drochowi}@cs.uah.edu

Abstract. The ability to represent real-world objects is an important feature of a practical knowledge system. Most knowledge systems involve informal or ad-hoc mappings from their internal symbols to objects and concepts in their environment. This work introduces a framework for formally associating symbols to their meanings, a process we call *grounding*. Two kinds of grounding are discussed with respect to conceptual graphs – active grounding, which involves actors to provide mappings to the environment, and terminological grounding, which involves actors that establish the basic elements of meaning with respect to a subject field's agreed-upon terminology. The work incorporates active knowledge systems and international terminological standards.

Keywords: active knowledge systems, grounding, logical interpretation, actors, terminology.

1 Introduction

In the broad field of knowledge representation, there have been many successes, one of which is the representation of conceptual graphs (CGs). Based on first-order logic, and now supported by a freely-available international standard [1], the strengths of CGs are well-known and familiar to most readers of these proceedings. (For additional details, see [2] [3].

This paper focuses on the use of knowledge representation for practical system development aimed at building solutions to real-world knowledge-intensive problems. The authors have pursued this effort for some years already, and we are beginning to understand some of the features needed for such development.

The limitations of purely first-order logic systems for practical development are well-documented; for a summary, see [4]. Two of the limitations mentioned there will be specifically addressed by this paper: first, reasoners often operate with deductions that are context-dependent, and second, reasoners often need to seek additional information beyond what has already been represented (from [5] quoted in [4]), as opposed to merely drawing conclusions based on what is at hand.

We propose using an Active Knowledge System (AKS) [6] to address these problems. It is our contention that an AKS can provide the foundation for the construction and communication of changeable knowledge driven by a user community. (A preliminary version of this work appears in [7].)

P. Eklund and O. Haemmerlé (Eds.): ICCS 2008, LNAI 5113, pp. 269–281, 2008.

Our focus is on developing and maintaining domain knowledge models after the developer is already constrained by those items that are already represented in some form in a domain. The paper is neither concerned with restructuring knowledge by producing a global *a priori* ontology, nor with generating a collection of partial structures through soft techniques. Rather we are concerned about how to build knowledge systems that can support development in subject fields of interest.

Knowledge is not a static collection of information, but a constantly changing collection that reflects new information and knowledge about the domain. An AKS provides "eyes" and "ears" for inter-acting (not just inter-face-ing) with external sources. Thus an AKS should provide two important capabilities:

- An AKS should represent the domain and provide answers to queries that reflect what is known at the time of the query.
- An AKS should provide grounded meaning where symbols are explicitly associated with the things they represent, the entities, relations and concepts in the world.

There are at least three major issues involved in such systems:

- How do different users of knowledge deal with their different terminologies, domains and purposes?
- How does the knowledge base ensure that the acquired knowledge is consistent (e.g., does not render previous conclusions invalid)?
- How does knowledge of the changing world become captured into the knowledge base's representation?

This paper explains one small step along the way to answering these questions, by describing how to ground conceptual graph models.

2 Related Work

This paper brings together three key ideas that support practical system development involving customers, users, and developers. System developers need *formal models* of software systems so that they can reason about the software's behavior and properties beyond what is possible by inspection, reviews and conventional testing. The concept of *grounding* a formal model is important because software systems ultimately must meet the needs of real-world people with respect to their real-world objects and activities. System developers are often interested in *standards* in order to both leverage a community's collective experience and to support interoperability among other systems.

The formal models in this framework are represented by *conceptual graphs* [2] [3] and are intended to be supported by freely-available tools such as CharGer [8], CoGITaNT [9] and Amine [10]. The effort to describe grounding in conceptual graphs actually began with the notion of a *demon* [11], and has been expanded more recently into *active knowledge systems* [6] [7], an extension to conceptual graphs which will be explored further in this paper. Two relevant *standards* are the Common Logic standard [1] and two basic terminology standards [12] [13]; these are valuable because conceptual graphs are a standardized dialect of Common Logic (Annex B) and the terminology standards are aimed at establishing a common vocabulary of concepts and terms in subject fields.

The remainder of the paper describes a framework with which grounding can be accomplished. We concentrate our efforts on two kinds of grounding:

Active grounding - Associating each symbol in a model with particular individuals or objects.

Terminological grounding - Associating a concept in a model with previously accepted terminology, including its intension (definitions) and extension (included set of individuals).

3 Active Grounding of a Knowledge Model

Any formal system (including a computer system) is composed of symbols and a set of rules for manipulating them. The nature of the rules depends on what kind of system is being developed. In a logic system, the rules are logical rules governing conclusions that can be derived from assumptions about the original symbols. So if we have the premises $(p \supset q)$ and p, we can conclude q by using a well-known rule called *modus ponens*.

Even in this simple example, there is a serious difficulty for practical developers: how do they come to know that the premises are true? The answer is: they don't! The premises are initial assumptions, taken on faith as it were; sometimes they are called axioms. If we assume that p = "Harry is a millionaire" and q = "Harry is happy", and we assume that Harry is a millionaire, we thereby infer that Harry is happy, but then there's a fundamental practical problem: Harry isn't a millionaire (at least the one writing this paper isn't!). That practical problem is not a concern of the logic itself (which only considers symbols and rules), but it's a major concern for the developers of a knowledge system. How ought the practical developer attach actual meaning to the premises, and hence to their conclusions?

We propose an answer to this issue: a process we call *active grounding*, which logicians call *interpretation*. Active grounding can be conceptualized as a procedure that maps individual symbols in a model to actual individuals in some domain of discourse. It also includes procedures that map the results of functions in the model to functions in the environment, etc. but in this paper we focus only on the mapping to individuals.

By *terminologically grounded model* we mean that there are concepts, intensions and extensions represented as in a terminology-based model. An *actively grounded model* is one in which actors (computer codes with pragmatic intent) can establish the correspondence between symbols or terms in the model and the objects or concepts to which they correspond. We illustrate this below in Figure 2 with a model of a sugar molecule in which actors establish the correspondence between the model's symbols for the concepts of atomic weights and the scientifically agreed upon values for those weights.

Both terminologically and actively grounded models are types of grounded models where, grounding is the process of establishing, for every symbol in the model, some individual (or relationship) in the "real world" to which it corresponds. In modeling the more formal term *universe of discourse* (or sometimes just *universe*) is often used. The grounded model is the software system's *environment.* We will also utilize the Common Logic standard [1], which recognizes that relations and functions, while not strictly mapped to individuals, will still map to relations and functions in the environment that are useful. These make up a potentially larger realm called the universe of reference. A logic system thereby provides predicates and functions, which allow the representation in a model of relationships between elements in a system. The rules of logical inference ensure that the truth of these predicates is preserved; e.g., if I assert that **dog(Spot)** is true and then perform any number of logical transformations, a (consistent) system should never allow me to derive **not dog(Spot)**. This is a necessary property of any system that is meant to preserve truth-values.

The practical system developer must also consider what the predicates mean to the users or stakeholders of the system. Logicians remind us that a predicate is assumed to be a primitive relation between symbols; a logical system's only obligation is to preserve those relationships throughout its logical operations. The symbols themselves, however, are arbitrary, insofar as the logic operations are concerned. For the practical system builder, however, these symbols are not arbitrary at all! Symbols used in a knowledge base stand for something, usually in the environment. We assume that the participants concur in at least partially establishing this sort of meaning and that these meanings act as bounds for the system to be developed. Figure 1(a) reminds us that symbols in a model are put there for a particular reason, which must also be known when symbols are "pulled out" after any transformations.

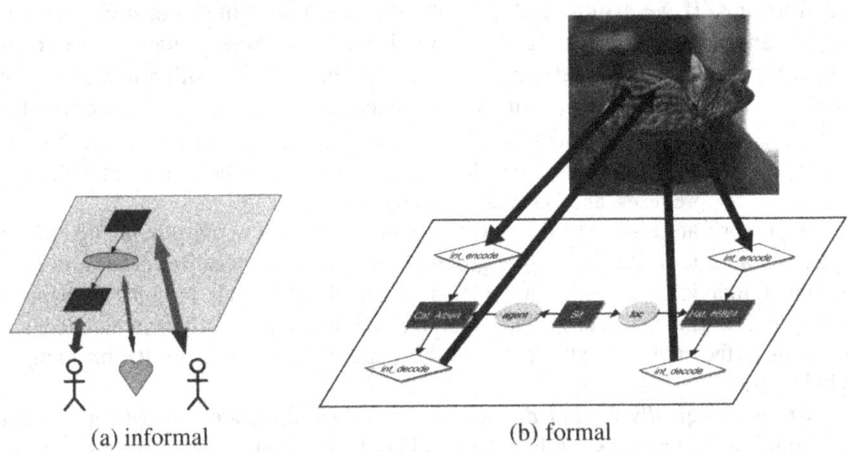

(a) informal (b) formal

Fig. 1. Interpretation of a conceptual graph

Figure 1(a) also indicates how a model relates to its environment. Things in the environment are represented in a knowledge-based system (shown as the "sheet of assertion" containing a conceptual graph) where the arrows show a two-way mapping between the model and the environment. Things in the environment are represented in the model ("encoded") using some procedure and are then available for representation, reasoning, or whatever operations are allowed in the conceptual graph model. These operations may cause new representations to arise within the system. These new representations' symbols can then be mapped back to the environment ("decoded") to capture their interpretation. The "encode" and "decode" procedures may be independent of each other.

There is a clear need to formalize models such as in Figure 1(a). We seek to create formal models in conceptual graphs, such as illustrated in Figure 1(b). Practical system builders require the system to preserve the symbols' meanings as they exist in the environment. Conceptual graphs provide these formal operations through the use of active relations called *actors*. Actors labeled "encode" and "decode" represent this formalization: namely, the ability to "encode" objects in the environment into formal symbols and then "decode" the (possibly transformed) symbols into environment objects they denote.

A more precise way of defining the relationship between a concept and its interpretation (i.e., its "counterpart" in the environment) is to use two actors for the interpretation procedure – one actor to provide the mapping from an individual in the environment to its concept in the graph; another actor to provide the reverse mapping from a concept in the graph to the individual(s) required by its interpretation. The formal structure is the "sheet" in Figure 1(b) where the actor **decode** provides the formal interpretation function, and the actor **encode** provides the inverse of the interpretation function. The actor's procedure may be any process needed to establish the counterpart; e.g., a database query, URI or even human interaction. The actor's presence pragmatically indicates that there is an anonymous procedure that will pick out "this item" where "this item" can be looked up. The lookup returns a "rigid designator" that is the same designator across all conceptualizations, contexts, microtheories, possible worlds etc. The actor tracks the handle through which the "real world" or "not a part of the formal system world" entity can be accessed. The practical system builder uses this handle to get the thing outside of the system's direct control.

An *actively grounded concept* is a concept that has fully specified both **decode** and **encode** actors. Once the actors have "fired" and established their groundings, the sheet of assertion comprises the basic knowledge model; i.e., the actors, the concepts and relations form a typical conceptual graph representing: *A cat Albert is sitting on a hat.* The identity of the referents is now known, since out of many cats named "Albert" the grounding actor will have established which one we mean. We can therefore perform the usual inferences, etc. within the formal model.

An interesting analogy to this situation can be found in the short novel *Flatland: A Romance of Many Dimensions* [14] which tells the story of a two-dimensional world, whose inhabitants are all two-dimensional geometric shapes; they cannot comprehend anything that might lie outside the (literal) plane of their existence. In a similar fashion, a conceptual graph exists on a "plane" of assertion that is necessarily "unaware" of anything outside of itself. Actors provide that "third dimension" in connecting the sheet of assertion with an outside environment.

The formalities of actor based interpretations are part of model theory and not within the scope of this paper (see [15] for the formal details). However, a key requirement for practical knowledge system developers and users is that the system has the capability to access its environment. Conceptual graphs [2] have this capability. A conceptual graph is considered an existential statement placed on a *sheet of assertion.* Within conceptual graphs, the capability to access the environment is realized through the use of an *actor*, which is a special kind of active relation; i.e., a function that provides a mapping from input concepts to output concepts.

The semantics of actors within a conceptual graph are well described in [2], with further elaborations in [11], [16] and [6]. In brief, an actor operates much as a node in a Petri Net [17]: when its input concepts are ready, the actor "fires", possibly changing its output concepts' referents (i.e., the individuals to which they refer). If the output concept is in fact a context (for an example, see Figure 4 and Figure 5 below), then the actor's function is to create or modify the entire sub-graph within that context. In this situation, the context itself is treated as a single concept whose referent is its enclosed graph.

A conceptual graph whose concepts denote values or identities obtained by actors in this way is called an *actively grounded conceptual graph*. Such graphs are a necessary ingredient in building true active knowledge systems. Figure 2 is an example of an actively grounded graph – in this case, a graph that represents the sugar molecule whose formula is $C_6H_{12}O_6$. There is a particular relationship between a molecule and its molecular weight, a relationship that is determined by specific experimental values for the atomic weights of its constituent atoms. The significance of the example is that a knowledge system cannot determine these values logically – the atomic weight values have been empirically determined. Furthermore, while it appears deterministic, there may in fact be more than one procedure (outside the graph itself of course) that is capable of determining atomic weights, and those values may vary based on which procedure is used.

Actors in a graph thereby provide "hooks" to things outside the sheet of assertion. Figure 2 shows several specific actors that may be used to provide meaning to the elements of a model. Some simple actors represent external procedures (i.e., in the environment) that provide behavior but require no access to specific individuals or other features from the environment. The **plus** actor represents a functional relationship between one or more input concepts' referents and an output referent that is required to represent the mathematical sum obtained by adding the input referents together. The **multiply** actor operates similarly in representing a mathematical product.

Other more interesting actors actually "hook" to more complicated external knowledge. Note the **lookup** actor, which performs the operations necessary to ensure that a particular concept referent will have a particular value obtained from a database. It contains aspects of both the **decode** operation ("find an individual record in a particular database") and **encode** ("set a referent in the graph to correspond to terms found in that record"). Figure 2 also shows the notion of a *context*, a grouping construct in conceptual graphs. The concepts representing the atoms in the molecule are grouped together into a context labeled **Molecule: Sugar**. This construct allows relationships and features between entire groups of related concepts, and enables future work in representing "possible worlds" (in the sense of Kripke [18]) and in microtheories [19].

An interpretation must provide a mapping from symbols in the model to a domain (universe) of individuals. This is shown in Figure 1(b) where the large arrows indicate the mapping of an interpretation. Conceptual graphs provide for an individual marker in a concept box; e.g., given **[Person: #3867]**, **#3867** is an individual marker denoting a particular individual of type **Person**. The definition for individual markers in conceptual graphs states that there must be a way to associate each distinct marker

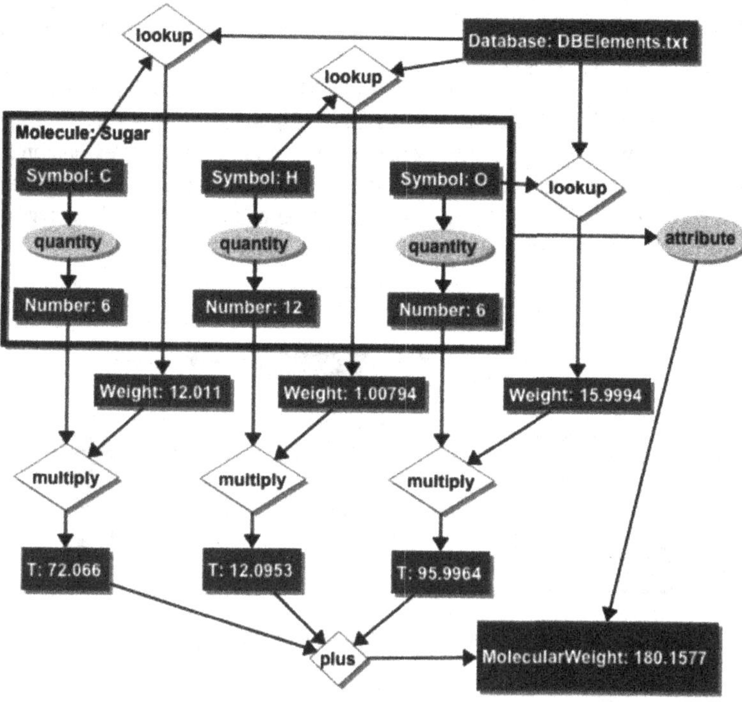

Fig. 2. Actively grounded conceptual graph for a sugar molecule

with a specific individual; this is precisely what the **encode/decode** operations capture. The notion of an individual marker can be extended to include any kind of identifier; useful ones are the Universal Resource Identifier (URI) [20] and the Digital Object identifier (DOI)[1].

The practical system developer can use this extended conceptual graph structure to model the domain and provide both the pragmatic and semantic content required by a system that relies on the domain knowledge. The use of agents for encoding and decoding provides formal "hooks" for the practical developer to attach both the terminology of the domain and the real world objects in the domain that are represented in the model.

4 Terminological Grounding of a Knowledge Model

Active knowledge systems address the grounding problem through the dual processes of encoding and decoding a model's interpretation. Since we are focusing on practical software development, we will assume that some subject field (domain) is being modeled for the purpose of communication and analysis and that there have been bounding and concurring processes conducted among developers and the stakeholders

[1] http://www.doi.org, accessed 30 Nov 2007.

to establish some boundaries for what belongs in the subject field. (Note: we are well aware that this second assumption is substantial; in future work, we intend to explore and formalize theses processes as well.)

We build upon existing terminology standards [12] [13] to build a conceptual model for subject fields. The conceptual graph in Figure 3 shows a framework for an example concept with three important relationships: its intension, its extension and its designator. These terms are taken from the definitions in [12]. Our purpose is to provide a framework whereby developers can formalize their model of the terms used in their software requirements, analysis, design and implementations

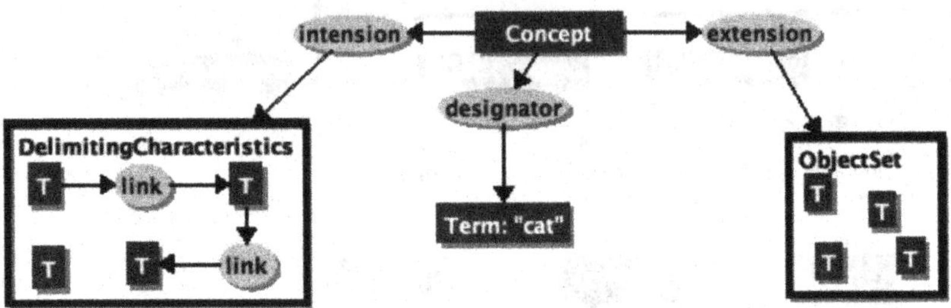

Fig. 3. Terminological model for a concept

The meaning of Figure 3 is that each concept is associated with the following:

- a designation (in this case, the term "cat") by which we refer to it. A concept may have more than one designation (e.g., an icon, terms/phrases in multiple languages)
- an extension consisting of the set of objects to which the concept corresponds,
- an intension consisting of the delimiting characteristics that define the concept.

This formal model can serve several purposes:

- An aid to communication and documentation
- Use of the intensional definition for data mining in order to identify new (and possibly unexpected) objects that belong in the extension of the concept
- Use of the extensional set to support graph matching and comparison processes that can identify delimiting characteristics.
- Use of both the intension and extension to identify a complete set of characteristics for the concept.
- When knowledge changes (designators, intension, or extension) use the others to validate (or invalidate) the changed knowledge.

To provide these capabilities, the practical system developer must have ways to establish the various parts of the conceptual model in Figure 3. We call this process *terminological grounding* and provide for it through the use of actors. There are several actors needed to terminologically ground these parts of the model, depending on which parts are present or absent in a developer's own initial model.

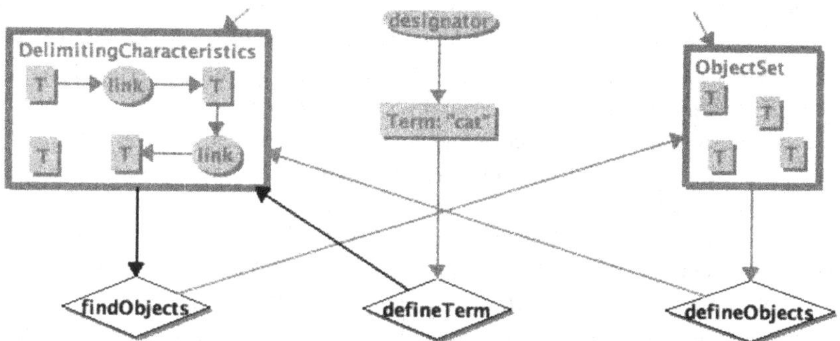

Fig. 4. Terminological grounding actors

Three actors for terminological grounding are shown in Figure 4. All of the actors are inherently procedural and able to access knowledge outside the formal model. That knowledge may be in the form of external databases, data mining procedures or sensors. The actors are:

defineTerm For a given term, provide its definition and determine its delimiting characteristics. This might be as simple as consulting a dictionary, but since we assume a specialized domain of interest, their terms may appear in a domain-specific dictionary or glossary.

defineObjects For a given set of objects, provide a set of delimiting characteristics that abstract them into a concept. An example procedure might be to perform formal concept analysis [21] on the perceived characteristics of the objects. To create the set of objects, this actor uses **decode** to find the actual objects, performs external procedures to determine their characteristics, and then uses **encode** to associate the resulting set of characteristics with intensions in the model, thus providing the active grounding described in the previous section.

findObjects For a given set of characteristics, seek out a set of objects that possess those characteristics, in effect establishing a extensional description of the concept. This actor uses **decode** for the characteristics, then searches through some external universe of individuals (its "environment") for objects that match them, so that it can use **encode** to associate the external individuals with particular concepts in the extension set.

The terminological grounding process thereby provides a way to blend a subject field's already defined terminology into the formal model.

5 Completely Grounding a Knowledge Model

Terminological grounding and active grounding actors together form the basis for grounding a complete knowledge model with respect to the system developer's domain. Combining the actors in Figure 4, it can be seen that from a single concept's term, actors can generate a set of delimiting characteristics from which a set of objects (the concept's extension) can be found. Our future work will further explore these actors and establish well-defined procedures for their operation.

Further elaboration of the terminologically based model can be achieved by considering concepts as types, which have superordinate concepts as supertypes. These superordinate concepts are represented as any other concept would be represented in the terminologically grounded model. Combining the actively grounded concept notion with the terminological notion of concepts related as super- and sub-ordinate, results in the richer knowledge model shown in Figure 5.

Figure 5 brings together all the actors into one framework. To eliminate clutter here, only a few **encode** and **decode** actors are shown, though they belong with every concept. The terminological grounding actors **defineTerm**, **findObjects**, and **defineObjects** are shown. Figure 5 also shows a **findSubset** actor, whose purpose is to discover a subset of the supertype's extension that conforms to the delimiting characteristics for the subtype – in effect to discover new subtypes.

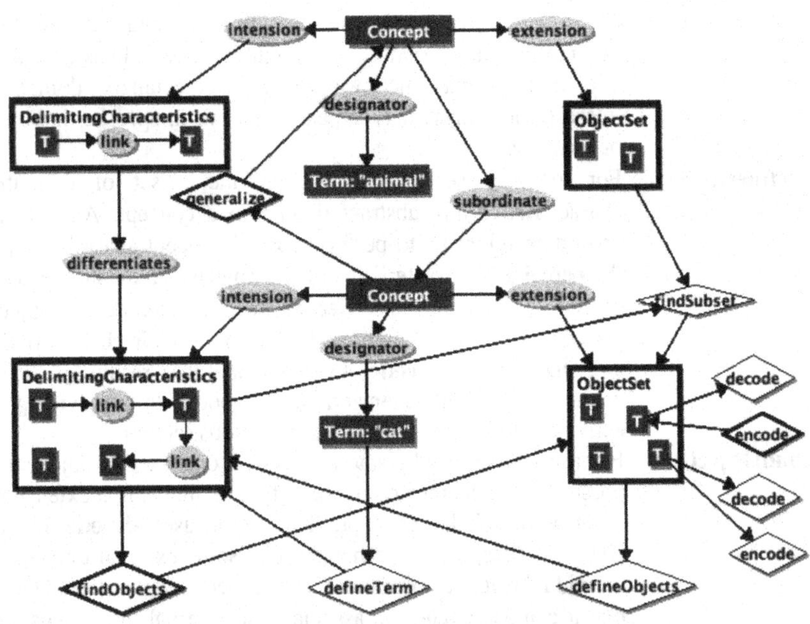

Fig. 5. A more complete grounding model

A complete terminologically and actively grounded knowledge model, constituted as we have outlined, could answer many interesting questions. Among these are:

- Given a concept, can its superordinate concepts be identified? This could be answered by actors comparing the given concept's intension and extension with other concepts' intensions and extensions.

- Given a change in the individuals in a concept's extension, is there a change in the concept's intensional definition?
- Given a change in a concept's intensional definition, how does that change the makeup of the concept's extension?

In addition to these specific questions, the terminologically based model can also be used to validate its own parts – for example, intensional definitions can be used to find its corresponding extensional set, which can be compared with the actual set determined by developers.

The next section gives several justifications for why developers are interested in the grounded model.

6 Why Does Grounding Matter to System Developers?

Given the generic view of a concept, the developer often finds himself in the position of lacking one or more of the pieces of the knowledge model puzzle. Our research is aimed at providing techniques for finding all the "pieces". In particular the grounding model links the declarative part of knowledge ("knowing that") to the pragmatic part of knowledge ("knowing how") [22].

Grounding concepts is a complex topic that touches empirical, cognitive science and formal, logical analysis. Following Harnad, the symbol-grounding problem asks "How can the semantic interpretation of a formal symbol system be made intrinsic to the system, rather than just parasitic on the meanings in our heads? How can the meanings of the meaningless symbol tokens, manipulated solely on the basis of their (arbitrary) shapes, be grounded in anything but other meaningless symbols?" [23] The system developer confronts such questions in developing systems that have significant knowledge and domain semantic content. If the intent of the developer is to build a system that uses knowledge in more and less general ways, the developer must attempt to ground the conceptual structures provided by requirements analysis with both the domain terminology and the world the domain represents. While the proposed grounding model does not fully answer Harnad's questions for all cases, it does provide a framework for a practical response: agents that implement the links to the domain terminology and the domain referents provide meaning that is reasonable and acceptable to the stakeholders.

The assumption is that grounding is an essential part of rendering the meanings of the concepts and that grounding requires constraints to be provided within the conceptual structures and contacts to be made to the world that the domain presents. In relation to cognitive science this corresponds to notions of conceptual web grounding and external grounding where the external grounding is provided by both the domain terminology and references to the world presented by the domain. [24].

Grounding is one of three problems that the system developer must resolve. The other problems are bounding and concurring. The system developer must assist in establishing the bounds of the application, its knowledge and concepts. Such bounds may not be immediately obvious. For example, there may be time, language, and granularity bounds on the problem the software is to help solve. However there are also bounding problems that arise through the use of concepts and symbols in multiple ways by multiple domain experts. These conflicts must be negotiated so that all can concur about the basic meaning of the concepts. In related ways the end user (the user of the developed software) will also engage in bounding and concurring. When the system is deployed, the end user will have some issue or topic in mind and the software must therefore allow the end user to bound the content to that issue or topic. This will require some degree of concurrence by the user to use concepts in a particular way. Thus the system developer can use the conceptual structures proposed here to build tools that allow users to bound the topics or issues of interest in appropriate ways.

The system developer may also develop auxiliary tools that allow users to both actively ground and terminologically ground the models that are embedded in the software, disambiguating concepts where feasible and providing new individuals and concepts. In this way the users become part of the developmental process, expanding the content of the system using the conceptual structures presented here.

The grounding model has provided some responses to the three questions posed earlier.

- *How do different users of knowledge deal with their different terminologies, domains and purposes?* The knowledge grounding model provides explicit links to domain terminologies through actors that can mediate between the concepts as represented in the software system and the terminology used by the stakeholders.
- *How does the knowledge base ensure that the acquired knowledge is consistent (e.g., does not render previous conclusions invalid)?* Although we have not directly addressed this issue, the use of conceptual graphs provides the logical foundation upon which consistency model checking can be built.
- *How does knowledge of the changing world become captured into the knowledge base's representation?* The knowledge grounding model provides a framework in which actors can actively determine changing values in the domain.

7 Conclusion and Future Work

The follow-on to this work consists of incorporating this framework into one or more existing CG tools, as well as further elaborating the relationship between terminology and system models. A larger set of problems occurs in the process of establishing the domain's boundaries and gaining concurrence among the various stakeholders in a system to be developed. We will be pursuing both of these avenues.

References

[1] ISO/IEC, ISO/IEC 24707:2007 - Information technology - Common Logic (CL) - A framework for a family of logic-based languages, International Organization for Standardization, Geneva, Switzerland (2007), http://standards.iso.org/ittf/PubliclyAvailableStandards/c039175_ISO_IEC_24707_2007E.zip

[2] Sowa, J.F.: Conceptual Structures: Information Processing in Mind and Machine. Addison-Wesley, Reading, Mass (1984)

[3] Sowa, J. F.: Knowledge Representation: Logical, Philosophical, and Computational Foundations: Brooks/Cole (2000)

[4] Devlin, K.: Modeling real reasoning. LNCS. Springer, Heidelberg (to appear, 2007), http://www.stanford.edu/~kdevlin/ModelingReasoning.pdf
[5] Richards, J., Heuer, J.: Psychology of Intelligence Analysis: Lulu.com (1999)
[6] Delugach, H.S.: Towards Building Active Knowledge Systems With Conceptual Graphs. In: Ganter, B., de Moor, A., Lex, W. (eds.) ICCS 2003. LNCS, vol. 2746, pp. 296–308. Springer, Heidelberg (2003)
[7] Delugach, H.S.: Active Knowledge Systems. In: Hitzler, P., Schärfe, H. (eds.) Conceptual Structures in Practice. Chapman and Hall/CRC Press (2008)
[8] Delugach, H.S.: CharGer - A Conceptual Graph Editor, Univ. of Alabama in Huntsville (2003), http://projects.sourceforge.net/charger
[9] Genest, D. : CoGITaNT, LIRMM - Montpellier (2003), http://cogitant.sourceforge.net/index.html
[10] Kabbaj, A.: Development of Intelligent Systems and Multi-Agents Systems with Amine Platform. In: Øhrstrøm, P., Schärfe, H., Hitzler, P. (eds.) Conceptual Structures: Inspiration and Application, pp. 286–299. Springer, Heidelberg (2006)
[11] Delugach, H.S.: Specifying Multiple-Viewed Software Requirements With Conceptual Graphs. Jour. Systems and Software 19, 207–224 (1992)
[12] ISO, ISO 1087-1:2000 - Terminology work - Vocabulary. Part 1: Theory and application, International Organization for Standardization, Geneva, Switzerland (2000)
[13] ISO, ISO 704:2000 - Terminology work - Principles and methods, International Organization for Standardization, Geneva, Switzerland (2000)
[14] Abbott, E.A., Flatland: A Romance of Many Dimensions (1884): Penguin Classics (1998)
[15] Hodges, W.: A Shorter Model Theory. Cambridge University Press, Cambridge (1997)
[16] Raban, R., Delugach, H.S.: Animating Conceptual Graphs. In: Lukose, D., Delugach, H.S., Keeler, M., Searle, L., Sowa, J.F. (eds.) ICCS 1997. LNCS, vol. 1257, pp. 431–445. Springer, Heidelberg (1997)
[17] Peterson, J.L.: Petri Net Theory and the Modeling of Systems. Prentice Hall PTR, Upper Saddle River (1981)
[18] Kripke, S.A.: Naming and Necessity. Blackwell Publishing, Malden (2003)
[19] Guha, R.V.: Context: a formalization and some applications. In: Computer Science, Stanford University (1992)
[20] Mealling, M., Denenberg, R.: RFC 3305 - Report from the Joint W3C/IETF URI Planning Interest Group: Uniform Resource Identifiers (URIs), URLs, and Uniform Resource Names (URNs): Clarifications and Recommendations (2002)
[21] Ganter, B., Wille, R.: Formal Concept Analysis: Mathematical Foundations. Springer, Heidelberg (1999)
[22] Rochowiak, D.: A pragmatic understanding of "knowing that" and "knowing how": the pivotal role of conceptual structures. In: Lukose, D., Delugach, H.S., Keeler, M., Searle, L., Sowa, J.F. (eds.) ICCS 1997. LNCS, vol. 1257, pp. 25–40. Springer, Heidelberg (1997)
[23] Harnad, S.: The symbol grounding problem. Physica D 42, 335–346 (1990)
[24] Goldstone, R.L., Feng, Y., Rogosky, B.: Connecting concepts to the world and each other. In: Pecher, D., Zwaan, R. (eds.) Grounding cognition: the role of perception and action in memory, language and thinking, pp. 292–314. Cambridge Univ. Press, Cambridge (2005)

Scenario Argument Structure vs Individual Claim Defeasibility: What Is More Important for Validity Assessment?

Boris A. Galitsky[1] and Sergei O. Kuznetsov[2]

[1] Knowledge-Trail, Inc.
9 Charles Str Natick MA 01760
bgalitsky@searchspark.com
[2] Higher School of Economics, Moscow, Russia
skuznetsov@hse.ru

Abstract. We conduct comparative analysis of two sources of argumentation-related information to assess validity of scenarios of interaction between agents. The first source is an overall structure of a scenario, which included communicative actions in addition to attack relations and is learned from previous experience of multi-agent interactions. In our earlier studies we proposed a concept-based learning technique for this source. Scenarios are represented by directed graphs with labeled vertices (for communicative actions) and arcs (for temporal and attack relations). The second source is a traditional machinery to handle argumentative structure of a dialogue, assessing the validity of individual claims. We build a system where data for both sources are visually specified, to assess a validity of customer complaints. Evaluation of contribution of each source shows that both sources of argumentation-related information are essential for assessment of multi-agent scenarios. We conclude that concept learning of scenario structure should be augmented by defeasibility analysis of individual claims to successfully reason about scenario truthfulness.

1 Introduction

Understanding and simulating behavior of human agents, as presented in text or other medium, is an important problem to be solved in a number of decision-making and decision support tasks [3]. One class of the solutions to this problem involves learning argument structures from previous experience with these agents, from previous scenarios of interaction between similar agents [8]. Another class of the solutions for this problem, based on the assessment of quality and consistency of argumentation of agents, has been attracting attention of the behavior simulation community as well [1].

In the context of agent-based decision support systems, the study of dynamics of *argumentation* [14] has proven to be a major feature for analyzing the course of interaction between conflicting agents (e.g. in argument-based negotiation or in multiagent dialogues. The issue of argumentation semantics of communicative models has also been addressed in the literature (eg [15]). Formal models of valuables norms and procedures for rational discussion have been introduced ([12]). However, when there

P. Eklund and O. Haemmerlé (Eds.): ICCS 2008, LNAI 5113, pp. 282–296, 2008.

is a lack of background domain-dependent information, the evolution of *dialogues* ought to be taken into account in addition to the communicative actions these arguments are attached to. Rather than trying to determine the epistemic status of those arguments involved, in one of our previous studies [8] we were concerned with the emerging *structure* of such dialogues in conflict scenarios, based on inter-human interaction. The structure of these dialogues is considered in order to compare it with similar structures for other cases to mine for relevant ones for the purpose of assessing its truthfulness and exploration of a potential resolution strategy.

In our earlier studies we proposed a concept learning technique for scenario graphs, which encode information on the sequence of communicative actions, the subjects of communicative actions, the causal [4,], and argumentation attack relationships between these subjects [6,8]. Scenario knowledge representation and learning techniques were employed in such problems as predicting an outcome of international conflicts, assessment of an attitude of a security clearance candidate, mining emails for suspicious emotional profiles, and mining wireless location data for suspicious behavior [7]. A performance evaluation in these domains demonstrated an adequateness of graph-based representation in rather distinct domains and applicability in a wide range of applications involving multi-agent interactions.

In this study we perform a comparative analysis of the *two sources* of argumentation-related information mentioned above to assess validity of scenarios of interaction between agents. The *source* 1) of information on argumentation is an overall structure of a scenario, which included communicative actions in addition to attack relations and is learned from previous experience of multi-agent interactions. Scenarios are represented by directed graphs with labeled vertices (for communicative actions) and arcs (for temporal and causal relationships between these actions and their parameters) [4]. The *source* 2) is a traditional machinery to handle argumentative structure of a dialogue, assessing the validity of individual claims, which has been a subject of multiple applied and theoretical AI studies.

2 Learning Argumentation in Dialogue

We approximate an *inter-human interaction scenario* as a sequence of communicative actions (such as *inform, agree, disagree, threaten, request*), ordered in time, with *attack* relation between some of the subjects of these communicative language. Scenarios are simplified to allow for effective matching by means of *graphs*. In such graphs, communicative actions and attack relations are the most important component to capture similarities between scenarios. Each vertex in the graph will correspond to a communicative action, which is performed by an (artificial) agent. As we are modeling dialogue situations for solving a conflict, we will borrow the terms *proponent* and *opponent* from dialectical argumentation theory [14] to denote such agents. An arc (oriented edge) denotes a sequence of two actions.

In our simplified model of communication semantics [6] communicative actions will be characterized by three parameters: (1) *agent name*, (2) *subject* (information transmitted, an object described, etc.), and (3) *cause* (motivation, explanation, etc.) for this subject. When representing scenarios as graphs we take into account all these parameters. Different arc types bear information whether the subject stays the same or not. **Thick arcs** link vertices that correspond to communicative actions with the **same subject**, whereas thin arcs link vertices that correspond to communicative actions with different subjects. We will make explicit conflict situations in which the cause of one communicative action M1 "attacks" the cause or subject of another communicative action M2 via an argumentation arc A (or *argumentation link)* between the vertices for these communicative actions. This attack relationship expresses that the cause of first communicative action ("from") *defeats* the subject or cause of the second communicative action ("to"). Such defeat relationship is *defeasible*, as it may be subject to other defeats, as we will see later.

A pair of vertices for a thick or thin arc may or may not be linked by the attack relation: a subject of the first communicative action is supported by a cause for the same (respectively, different) subjects of the second communicative action. However, we are concerned with argumentation arcs which link other than consecutive vertices (communicative actions) as shown at Fig. 1.

For the sake of example, consider the text given below representing a complaint scenario in which a client is presenting a complaint against a company because he was charged with an overdraft fee which he considers to be unfair (Fig. 1). We denote both parties in this complaint scenario as ***Pro*** and ***Con*** (proponent and opponent), to make clear the dialectical setting. In this text communicative actions are shown in **bold**. Some expressions appear underline, indicating that they are defeating earlier statements. Fig. 2 shows the associated graph, where straight thick and thin arcs represent temporal sequence, and curve arcs denote defeat relationships.

- *(Pro) I* ***explained*** *that I made a deposit, and then wrote a check which bounced due to a bank error.*
- *(Con) A customer service representative* ***confirmed*** *that it usually takes a day to process the deposit.*
- *(Pro) I* ***reminded*** *that I was unfairly charged an overdraft fee a month ago in a similar situation.*
- *(Con) They* ***explained*** *that the overdraft fee was due to insufficient funds as disclosed in my account information.*
- *(Pro) I* ***disagreed*** *with their fee because I made a deposit well in advance and wanted this fee back*
- *(Con) They* ***denied*** *responsibility saying that nothing can be done at this point and that I need to look into the account rules closer.*

Fig. 1. A conflict scenario with attack relations

Note that first two sentences (and the respective subgraph comprising two vertices) are about the current transaction (*deposit*), three sentences after (and the respective subgraph comprising three vertices) address the *unfair charge*, and the last sentence is probably related to both issues above. Hence the vertices of two respective subgraphs are linked with thick arcs: *explain-confirm* and *remind-explain-disagree*. It must be remarked that the underlined expressions help identify where conflict among arguments arise. Thus, the company's claim *as disclosed in my account information* defeats the client's assertion *due to a bank error*. Similarly, the expression <u>I made a deposit well in advance</u> defeats *that it usually takes a day to process the deposit* (makes it non-applicable). The former defeat has the intuitive meaning "*existence of a rule or criterion of procedure attacks an associated claim of an error*", and the latter defeat has the meaning "*the rule of procedure is not applicable to this particular case*". It can be noticed that this complaint scenario is not sound because it seems that the complainant does not understand the procedure of *processing the deposit* nor distinguishes it from an *insufficient funds* situation (source 2). However, this scenario itself and its associated argumentation patterns do not have surface-level explicit inconsistencies, if one abstracts from the domain-specific (banking) knowledge (source 1).

Our task is to classify (for example, by determining its validity) a new complaint scenario without background knowledge, having a dataset of scenarios for each class. We intend to automate the above analysis given the formal representation of the graph (obtained from a user-company interaction in the real world, filled in by the user via a special form where communicative actions and argumentation links are specified).

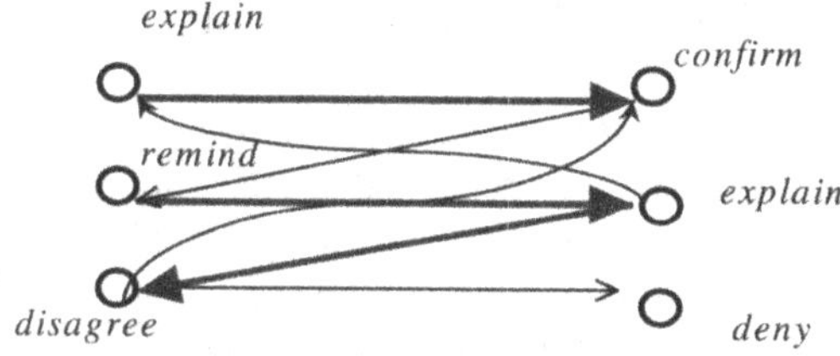

Fig. 2. The graph for approximated scenario (Fig. 1)

Let us enumerate the constraints for the scenario graph:

1) Each vertex is either assigned with the proponent (drawn on the left side of each graph in Fig. 2) or to the opponent (drawn on the right side).
2) Thin and thick arcs point from a vertex to the subsequent one in the temporal sequence (from the proponent to the opponent or vice versa);
3) Curly arcs, staying for *attack* relations, jump over several vertices in either direction.

Similarity between scenarios is defined by means of maximal common subscenarios. Since we describe scenarios by means of labeled graphs, we outline the definitions of labeled graphs and domination relation on them (see [9,11]).Given ordered set G of graphs (V,E) with vertex- and edge-labels from the sets $(\mathcal{L}_V, \preceq$ and $(\mathcal{L}_\mathcal{E}, \preceq)$. A labeled graph Γ from G is a quadruple of the form $((V,l),(E,b))$, where V is a set of vertices, E is a set of edges, $l\colon V \to \mathcal{L}_V$ is a function assigning labels to vertices, and $b\colon E \to \mathcal{L}_\mathcal{E}$ is a function assigning labels to edges.

The order is defined as follows: For two graphs $\Gamma_1 := ((V_1,l_1),(E_1,b_1))$ and $\Gamma_2 := ((V_2,l_2),(E_2,b_2))$ from G we say that Γ_1 *dominates* Γ_2 or $\Gamma_2 \leq \Gamma_1$ (or Γ_2 is a *subgraph* of Γ_1) if there exists a one-to-one mapping $\varphi\colon V_2 \to V_1$ such that it respects edges: $(v,w) \in E_2 \Rightarrow (\varphi(v), \varphi(w)) \in E_1$, and fits under labels: $l_2(v \preceq l_1(\varphi(v)), (v,w) \in E_2 \Rightarrow b_2(v,w) \preceq b_1(\varphi(v), \varphi(w))$.

This definition allows generalization ("weakening") of labels of matched vertices when passing from the "larger" graph G_1 to "smaller" graph G_2.

Now, generalization Z of a pair of scenario graphs X and Y (or their similarity), denoted by $X \sqcap Y = Z$, is the set of all inclusion-maximal common subgraphs of X and Y, each of them satisfying the following additional conditions: To be matched, two vertices from graphs X and Y must denote communicative actions of the same agent, and each common subgraph from Z contains at least one thick arc.

The following conditions hold when a scenario graph U is assigned to a class:

1) U is similar to (has a nonempty common scenario subgraph of) a positive example R^+. It is possible that the same graph has also a nonempty common scenario subgraph with a negative example R^- . This is means that the graph is similar to both positive and negative examples.

2) For any negative example R^-, if U is similar to R^- (i.e., $U \sqcap R^- \neq \emptyset$) then $U \sqcap R^- \sqsubseteq U \sqcap R^+$.

3 Assessing Defeasibility of Individual Claims

In this section we consider a realistic complaint scenario (Fig. 3), represent it as a dialogue with communicative actions (italic) and attack relations (curly arcs) on their arguments [shown in brackets] in Fig. 4. Finally, the graph with two vertices for each dialogue step (communicative actions for *receiving* and *sending*). Respective communicative action can be shown as '*; if unknown (not shown in text), Figure 5.

I have 2 loans through Huntington, both of which were automatically deducted at the appropriate times each month from my Huntington account. At the beginning of July, I began paying those loans by check from the non-Huntington account. Though I had attempted to stop Huntington from taking the funds directly from my Huntington account, they continued to do so resulting in a continuing negative balance, compounded by NSF and overdraft fees, as well as the initial debit for the loans. Calls to Huntington regarding the matter have had no effect.

I'm constantly bombarded with calls from Huntington about this so called delinquency which culminated in a threat from Huntington collections to repossess my truck and other vehicle (both loan items)

When I explained that I had been paying the loans by check AND that those checks had been debited from my other bank account, they continued to insist that no payments had been applied to either account and that Huntington was still going to repossess my vehicles. I've found corresponding checks that have posted from my primary, non-Huntington account.

It does appear, however, that one payment for $181.62 was never posted. After this, I again called Huntington and explained the situation. I was told that as long as Huntington had an open account for me, from which they'd already set up automatic withdraw, they could continue to withdraw funds for loan payment, even if the loan had already been paid by check! I was also informed that the best way to rectify the situation was to close the Huntington account.

Since getting my loan, I've had continuing trouble. The first payment was late, due to a mistake made by Huntington- which they acknowledged. Huntington told me that they'd take the late payment off my record but it appears they never did.

Fig. 3. Full complaint scenario

Asked [do not withdraw]	*Ignored* [kept withdrawing]
Requested [explanation why NSF, over-draft, etc.]	*Ignored*[] *Threatened* [delinquency]
Explained [has been paying & debited]	*Disagreed*[] *Insisted*[no payment applied] *Threatened* [repossession]
Agree [one check was never posted]	Informed [continue to withdraw funds Requested [close account]
Ask[refund late payment fee]	Promise [],
	Ignore [never did]

Fig. 4. Dialog structure of the Full complaint scenario with subjects of communicative actions and attack relation on them

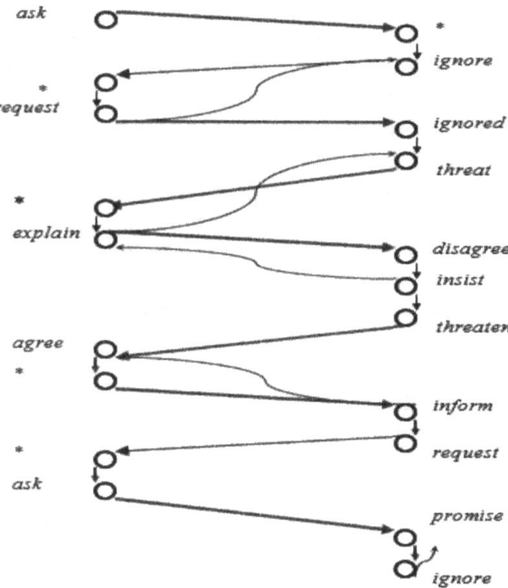

Fig. 5. Scenario graph for the above complaint

4 Interactive form for Detection of Implicit Self-attack

To verify the truthfulness of a complainant's claim, we use the special form called Interactive Argumentation Form which assists in structuring a complaint. Use of this form enforces a user to explicitly indicate all causal and argumentation links between statements which are included in a complaint (compare with [17]) . The form is used to assess whether a particular scenario has valid argumentation pattern: does it contain self-attacks (explicit for the complainant).

The form includes eight input areas where a complainant presents a component-based description of a problem (Fig. 6). At the beginning, the subject of the dispute is specified: an operation (or a sequence of operations) which are believed by a complainant to be performed by a company in a different manner to what was expected <Where company got confused>. Then the essence of the problem is described, what exactly turned out to be wrong. In the section <Company wrongdoing> the complainant sketches the way the company performed its duties, which caused the current complaint. The customer's perception of the damage is inputted in section <How it harmed me>. In the fourth section <Why I think this was wrong> the customer backs up his beliefs concerning the above two sections, <Where company got confused> and <Company wrongdoing>.

Usually, customer dissatisfaction event is followed by negotiation procedure, which is represented by two sections, <What company accepted> and <How company explained>. The acceptance section includes the circumstances which are confirmed by the company (in the complainant's opinion) to lead to the event of the customer's dissatisfaction. The latter section includes the customer's interpretation of how these issues are commented on by the company, the beliefs of its representatives on what lead to the event of the customer's dissatisfaction and the consequences. <Unclear> section includes the issues which remain misunderstood and/or unexplained by the company, in particular, problems with providing relevant information to the customer. Finally, <Systematic wrongdoing> section includes customer's conclusion about the overall business operation in similar situations, how in customer's opinion her experience can serve as a basis to judge how other customers are treated in similar situations.

Each section includes one or more sentences which provide information appropriate for this section, providing background information and/or backing up claims in this or other sections from the standpoint of the customer. Each statement which participates in (at least one) argumentation link is marked by a check box debit and check☑

All possible causal and argumentation links are shown as arrows. Arrows denote the links between the sentences in the respective sections; some arrows go one way and other both ways (only the ending portion is shown in this case). If the user does not find an arrow between two sections for a pair of inputted sentences, it means that either or both of these sentences belong to a wrong section: the data needs to be modified to obey the pre-defined structure. End of each arrow is assigned by a check-box to specify if the respective link is active for a given complaint. Bold arrows denote most important links.

Two sorts of links are specified via the Form:

1) Supporting and causal links, by which the user backs up his claims;
2) Defeating links, which are used by the user to demonstrate that certain claims of the opponent are invalid. A complainant may wish to defeat the opponents' claims. The form does not provide means to express complainant's **explicit** defeating of her own statements, because those are not expected for a "reasonable" scenario.

The list box nc is used to specify for a particular link (going either way) whether it is *supporting* or *defeating*. To specify supporting and defeating links for a number of statements for each section, multiple instances of the Interactive Argumentation Forms may be required for a given complaint.

The role of the Interactive Argumentation Form is a visual representation of argumentation, and intuitive preliminary analysis followed by the automated argumentation analysis. Since even for a typical complaint manual consideration of all argumentation links is rather hard, automated analysis of inter-connections between the complaint components is desired. We use the defeasible logic programming [10,2] approach to verify whether the complainant's claims are valid (cannot be defeated given the available data), concluding with the main claim, *Systematic wrongdoing.*

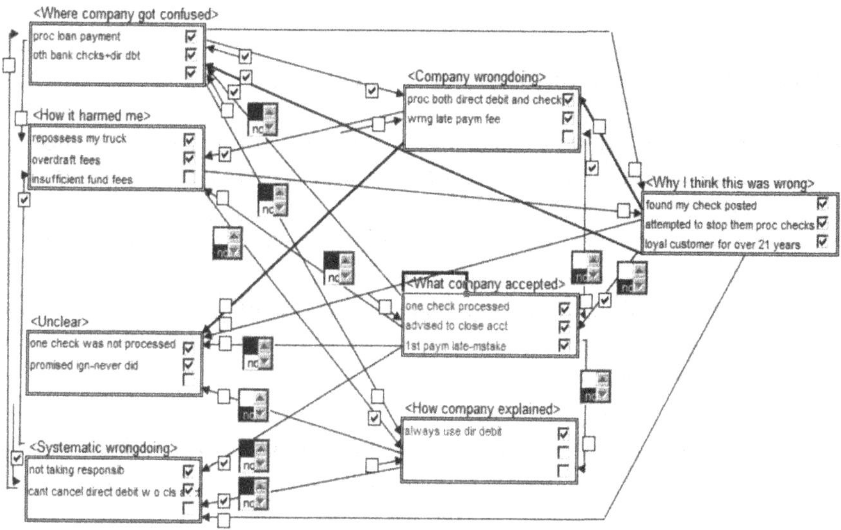

Fig. 6. Interactive Argumentation Form

The role of the Interactive Argumentation Form is a visual representation of argumentation, as well as its intuitive preliminary analysis. Since even for a typical complaint manual consideration of all argumentation links is rather hard, automated analysis of inter-connections between the complaint components is desired. We use the defeasible logic programming approach to verify whether the complainant's claims are valid (cannot be defeated given the available data).

Applying reasoning to Interactive Argumentation Form, we attempt to confirm that the statements in <Company wrongdoing> section is not defeated by any other statement, or the statement which supports <Company wrongdoing> (sections <Where company got confused>, <How it harmed me>, <Why I think this was wrong> or <What company accepted> are not defeated. It needs to be checked that the statements by the company, which defeats complainants' claims in the section <How

company explained> is in turn defeated by either complainant (in the sections <Where company got confused>,<Systematic wrongdoing>,<How it harmed me>) or other company representatives <What company accepted>.

To subject the available argumentation links to precise and accurate treatment in an automated manner, we form the defeasible logic program (de.l.p.) on the basis of these links. The defeasible logic program includes the classical clauses (including commonsense knowledge) which are always true, and the clauses which are typically true as long as they are not defeated [10]. In our approach all information specified by a complainant in the Interactive Argumentation Form may be unreliable; therefore all clauses have the form of defeasible rules where square brackets denote optional expressions: [~]*ClaimSectionTO* >- [~]*ClaimSectionFROM1(Statement11),*

ClaimSectionFROM1(Statement12),...
[[~]*ClaimSectionFROM2(Statement21),*
ClaimSectionFROM2(Statement22),...].

5 Dialectic Trees for Implicit Self-attacks

In this section we provide the definition and algorithm for building dialectic trees to discover implicit self attack in a defeasible logic program, specified by the Interactive Argumentation Form (Figure 6).

Defeasible logic program (de.l.p.) is a set of facts, strict rules Π of the form (A:-B) , and a set of defeasible rules Δ of the form (A>-B).

Let P=(Π, Δ) be a de.l.p. and L a ground literal. A *defeasible derivation* of L from P consists of a finite sequence $L_1, L_2, \ldots, L_n = L$ of ground literals, and each literal Li is in the sequence because:

(a) L_i is a fact in Π, or
(b) there exists a rule R_i in P (strict or defeasible) with head L_i and body $B_1, B_2, \ldots, B_k$, every literal of the body is an element L_j of the sequence appearing before L_j ($j < i$)

Let h be a literal, and P=(Π, Δ) a de.l.p.. We say that <A, h> is an *argument structure* for h, if A is a set of defeasible rules of Δ, such that:

1. there exists a defeasible derivation for h from =($\Pi \cup$ A)
2. the set ($\Pi \cup$ A) is non-contradictory, and
3. A is minimal: there is no proper subset A_0 of A such that A_0 satisfies conditions (1) and (2).

Hence an argument structure <A, h> is a minimal non-contradictory set of defeasible rules, obtained from a defeasible derivation for a given literal h.

We say that <A_1, h_1> *attacks* <A_2, h_2> iff there exists a sub-argument <A, h> of <A_2, h_2> ($A \subseteq A_1$) so that h and h_1 are inconsistent. Argumentation line is a sequence of argument structures where each element in a sequence attacks its predecessor. There is a number of *acceptability* requirements for argumentation lines (Garcoa and Simari 03).

We finally approach the definition of dialectic tree which gives us an algorithm to discover implicit self-attack relations in users' claims. Let $<A_0, h_0>$ be an argument structure from a program P. A *dialectical tree* for $<A_0, h_0>$ is defined as follows:

1. The root of the tree is labeled with $<A_0, h_0>$
2. Let N be a non-root vertex of the tree labeled $<A_n, h_n>$ and $\Lambda= [<A_0, h_0>, <A_1, h_1>, \ldots, <A_n, h_n>]$ the sequence of labels of the path from the root to N. Let $[<B_0, q_0>, <B_1, q_1>, \ldots, <B_k, q_k>]$ all attack $<A_n, h_n>$. For each attacker $<B_i, q_i>$ with acceptable argumentation line $[\Lambda,<B_i, q_i>]$, we have an arc between N and its *child* N_i.

In a dialectical tree every vertex (except the root) represents an attack relation to its parent, and leaves correspond to non-attacked arguments. Each path from the root to a leaf corresponds to one different acceptable argumentation line. As shown in Fig.8, the dialectical tree provides a structure for considering all the possible acceptable argumentation lines that can be generated for deciding whether an argument is defeated.

```
systematic_wrongdoing1(X) -< why_wrng1(X).
why_wrng1(X) -< how_it_harmed1(X). how_it_harmed1('reposses my track').
~ why_wrng1(X). -< how_it_harmed1(X1), company_accepted1 (X2).
company_accepted 1('one check processed').
~ why_wrng1(X) -< comp_confused1(X).  comp_confused1('proc loan payment').
~ unclear1(X)-< company_accepted2 (X1),  company_wrongdoing2(X2).
company_wrongdoing2(X) -< how_it_harmed2(X).
how_it_harmed2('overdraft fees').
~ why_wrng1(X)-<  how_it_harmed1(X1), unclear1(X2).
unclear1(X)-< company_accepted2 (X). company_accepted2 ('advised to close ac-
count').  company_accepted3 ('1st payment late - mistake').
~ unclear1(X)-<how_company_explained(X). how_company_explained('always use
direct debit'). ~ company_wrongdoing2(X) -< company_accepted3 (X).
```

Fig. 7. Defeasible logic program for a fragment of Interactive Argumentation Form on Fig. 6

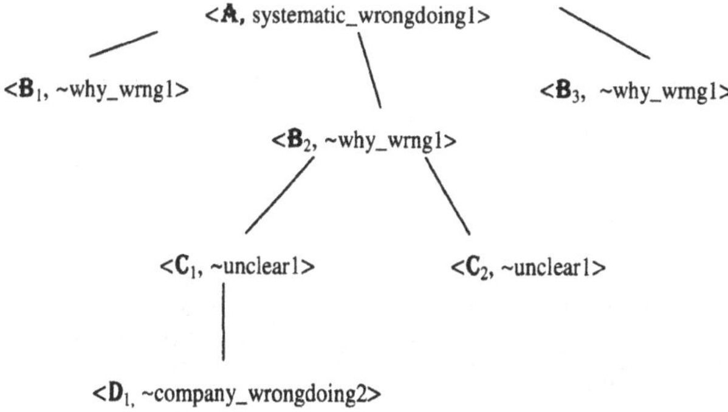

Fig. 8. Dialectic tree for the Defeasible Logic Program Figure 7

6 Evaluation

To observe the comparative contribution of argumentation data of sources 1) and 2), we used the database of textual complaints which were downloaded from the public website PlanetFeedback.com during three months starting from March 2004 and used in a number of computational studies since then. For the purpose of this evaluation, each complaint was:

1) manually assigned a validity assessment;
2) manually represented as a source 1) for concept-based learning evaluation;
3) manually represented as a source 2) for finding self-defeating claims.

This complaint preprocessing resulted in 560 complaints, divided in fourteen banks (or datasets), each of them involving 40 complaints. In each bank 20 complaints were used for training and 20 complaints for evaluation.

We performed the comparative analysis of relating scenarios to the classes of valid/invalid taking into account source 1), argument structure only; source 2), defeasibility of individual claims only, and combined sources (1-2) . Such an analysis sheds a light on the possibility to recognize a scenario (1) without background knowledge, but reusing previous assigned argument structures, and (2) with partial background knowledge, expressed as a set of attack relations between claims. Furthermore, we evaluate a cautious approach combining a) and b), where scenario is valid if a) it is similar to a valid one **or** b) it does not contain self-defeated claims, and invalid otherwise.

Classification results are shown in Tables 1 and 2. On the left, the first three columns contain bank number, and the numbers of valid/invalid complaints as manually assessed by human experts. The middle light-grayed set of columns show the classifications results based on source1: the assessment for valid/invalid scenarios (including cases which are close to neither) with false positives and false negatives (as assigned by human experts). The columns on the right show classification results based on the source 2: valid scenarios where self-defeating is not found, invalid scenarios where it is found, and the number of false positives and negatives relatively to the same assessment by human experts which was used for evaluation of source 1.

The reader can observe that classification based on the combination of sources gives substantial increase in recognition accuracy: F(Source1)= 63%, F(Source2) = 77%, and F(Source1+Source2)= 89%, which is a 26% of increase of accuracy for the source 1 and 12% increase of the accuracy for source 2.

7 Results and Discussions

In this study we observed how two sources of information on argumentation, overall argumentation pattern of a scenario and dialectic trees for individual claims, compliment each other. Comparative computational analysis of scenario classification with respect to validity showed that both sources of argumentation (the former proposed in the current study, and the latter well known in various reasoning domains) are essential to determine whether a scenario is plausible or not (contains misrepresentation or

Table 1. Results of classification based on source 1 (learning argument structure) and source 2 (dialectic trees for individual claims)

	Valid	Invalid	Cclassified as Valid	Classified as Invalid	Classified as valid but invalid	Not Classified as invalid but invalid	refusal to classify (close to neither)	Precision valid	Precision invalid	Recall valid	Recall invalid	CL as Valid (no claim is self-	CL as Invalid (found self-defeated	Classified as valid but invalid	Classified as invalid but valid	Precision valid	Precision invalid	Recall valid	Recall invalid
Bank 1	8	12	9	8	1	4	3	80%	67%	73%	35%	10	10	2	0	80%	83%	100%	83%
Bank 2	6	14	8	8	2	6	4	60%	57%	60%	31%	13	7	7	0	46%	50%	100%	50%
Bank 3	7	13	9	10	2	3	1	64%	77%	88%	42%	11	9	4	0	64%	69%	100%	69%
Bank 4	5	15	6	10	2	5	4	83%	67%	56%	34%	6	14	1	1	83%	93%	100%	88%
Bank 5	8	12	11	6	3	6	3	57%	50%	73%	29%	11	9	3	0	73%	75%	100%	75%
Bank 6	8	12	7	9	2	3	4	89%	75%	67%	36%	11	8	4	0	73%	67%	88%	67%
Bank 7	11	9	10	7	1	2	3	100%	78%	79%	37%	13	7	2	1	85%	78%	100%	70%
Bank 8	8	12	9	9	1	3	2	80%	75%	80%	39%	12	8	4	0	67%	67%	100%	67%
Bank 9	7	13	8	9	1	4	3	78%	68%	70%	36%	9	11	2	0	78%	85%	100%	85%
Bank10	9	11	10	8	3	3	2	69%	73%	82%	38%	13	7	4	0	69%	64%	100%	64%
Bank11	10	10	12	7	2	3	1	71%	70%	91%	39%	12	8	2	0	83%	80%	100%	80%
Bank12	5	15	6	11	1	4	3	71%	73%	63%	38%	10	10	6	1	50%	67%	80%	63%
Bank13	10	10	8	9	0	1	3	125%	90%	77%	41%	12	8	2	0	83%	80%	100%	80%
Bank14	8	12	7	11	0	1	2	114%	82%	80%	44%	13	7	5	0	62%	58%	100%	58%
Average	7.86	12.1	8.6	9	1.5	3.429	2.71	80%	72%	74%	37%	11.14	8.79	3.429	0.214	71%	73%	98%	71%
F-measure										77%	49%							82%	72%

Table 2. Results of the combined classification

Bank	**As assigned by experts**		**Results of classification: source1 + source2**								
	Valid	**Invalid**	**Cclassified as Valid**	**Classified as**	**Classified as valid but invalid**	**Not Classified as invalid but invalid**	**Precision valid**	**Precision invalid**	**Recall valid**	**Recall invalid**	**F-measure**
Bank 1	8	12	8	11	0	1	100%	92%	100%	92%	96%
Bank 2	6	14	7	8	3	6	70%	57%	88%	57%	69%
Bank 3	7	13	9	11	2	2	82%	85%	82%	85%	83%
Bank 4	5	15	6	14	2	1	75%	93%	86%	93%	89%
Bank 5	8	12	8	10	0	2	100%	83%	100%	83%	91%
Bank 6	8	12	7	10	2	2	78%	83%	117%	83%	97%
Bank 7	11	9	11	9	0	0	100%	100%	100%	100%	100%
Bank 8	8	12	9	9	1	3	90%	75%	90%	75%	82%
Bank 9	7	13	7	11	0	2	100%	85%	100%	85%	92%
Bank10	9	11	10	10	3	1	77%	91%	91%	91%	91%
Bank11	10	10	10	8	0	2	100%	80%	100%	80%	89%
Bank12	5	15	6	13	1	2	86%	87%	86%	87%	86%
Bank13	10	10	10	9	0	1	100%	90%	100%	90%	95%
Bank14	8	12	9	11	1	1	90%	92%	90%	92%	91%
Average	**7.86**	**12.1**	**8.4**	**10**	**1.071**	**1.857**	**89%**	**85%**	**95%**	**85%**	**89%**

self-contradiction). Hence we believe a practical argumentation management system should include scenario-oriented machine learning capability in addition to handling argumentation for individual claims.

Graph-based concept learning benefits from argumentation information in the form of dialectic tree, because a representation graph G includes more sensitive data on how claims attacking each other: G = { *Communicative_actions* + *attack_relations_on_their_subject* + *vertices of dialectic tree*}. In our previous studies (Galitsky et al 08) we verified that using attack relationship in addition to *Communicative_actions* as a way to express dialogue discourse indeed increases classification accuracy in a similar setting to the current study. Dialectic trees work well when all relevant background knowledge is available, and has been represented in a form suitable for reasoning. Since it is never the case in practical application, argumentation leverages concept learning as an additional mechanism of acquiring data for individual claims from previous experiences.

We found an adequate application domain for computing dialectic trees such as assessment of validity of customer complaints. On one hand, this domain is a good source of experimental data for evaluation of argumentation structures because of a high volume of nontrivial scenarios of multiagent interaction, yielding a wide variety of de.l.ps. On the other hand, it is an important set of long-waited features to be leveraged by customer relation management (CRM) systems.

We selected de.l.p. [10] as a most suitable approach to manage arguments in a dialogue, and employ dialectic trees to be integrated into scenario representation graph. [18] has proposed a very abstract and general argument-based framework, where he completely abstracts from the notions of argument and defeat. In contrast with approach [10] of defining an object language for representing knowledge and a concrete notion of argument and defeat, Dung's approach [18] assumes the existence of a set of arguments ordered by a binary relation of defeat. In our case the source of this order is previous experience with involved agents. [13] have developed an argumentation system for legal reasoning, that uses the language of extended logic programming. However, since they are inspired by legal reasoning, the protocol for dispute is rather different from dialectical tree of [10]. A proof of a formula takes the form of a dialogue tree, where each branch of the tree is a dialogue between a proponent and an opponent, so communicative actions are not taken into account to express strength of a claim. We have not found an argumentation study concerned with matching the dialectic trees as a source of "global" structural information about scenarios.

References

1. Chesñevar, C., Maguitman, A., Loui, R.: Logical Models of Argument. ACM Computing Surveys 32(4), 337–383 (2000)
2. Chesñevar, C., Maguitman, A.: An Argumentative Approach for Assessing Natural Language Usage based on the Web Corpus. In: Proc. Of the ECAI 2004 Conf., Valencia, Spain, pp. 581–585 (2004)
3. Fum, D., Missiera, F.D., Stoccob, A.: The cognitive modeling of human behavior: Why a model is (sometimes) better than 10,000 words. Cognitive Systems Research 8 - 3, 135–142 (2007)
4. Galitsky, B., Kuznetsov, S., Samokhin, M.: Analyzing Conflicts with Concept-Based Learning. In: ICCS 2005, Kassel, Germany (2005)
5. Galitsky, B., Kovalerchuk, B., Kuznetsov, S.O.: Learning Common Outcomes of Communicative Actions Represented by Labeled Graphs. In: ICCS 2007, pp. 387–400 (2007)
6. Galitsky, B.: Reasoning about mental attitudes of complaining customers. Knowledge-Based Systems Elsevier 19(7), 592–615 (2006)
7. Galitsky, B.: Merging deductive and inductive reasoning for processing textual descriptions of inter-human conflicts. J. Intelligent Info Systems 27(1), 21–48 (2006)
8. Galitsky, B., Gonzalez M.P., Chesnevar C.: Processing Customer Complaints Scenarios through Argument-Based Decision Making. Decision-Support Systems (in the press, 2008)
9. Ganter, B., Kuznetsov, S.: Pattern Structures and Their Projections. In: Delugach, H.S., Stumme, G. (eds.) ICCS 2001. LNCS (LNAI), vol. 2120, pp. 129–142. Springer, Heidelberg (2001)
10. García, A., Simari, G.: Defeasible Logic Programming: an argumentative approach. Theory and Practice of Logic Programming 4(1), 95–138 (2004)
11. Kuznetsov, S.O.: Learning of Simple Conceptual Graphs from Positive and Negative Examples. In: Żytkow, J.M., Rauch, J. (eds.) PKDD 1999. LNCS (LNAI), vol. 1704, pp. 384–391. Springer, Heidelberg (1999)
12. Rahwan, I., Ramchurn, S., Jennings, N., McBurney, P., Parsons, S., Sonenberg, L.: Argumentation-based negotiation. In Knowl. Eng. Rev. 18(4), 343–375 (2003)
13. Prakken, H., Sartor, G.: Argument-based logic programming with defeasible priorities. J. of Applied Non-classical logics 7, 25–75 (1997)

Griwes: Generic Model and Preliminary Specifications for a Graph-Based Knowledge Representation Toolkit

Jean-François Baget [1,2], Olivier Corby[1], Rose Dieng-Kuntz[1], Catherine Faron-Zucker[3], Fabien Gandon[1], Alain Giboin[1], Alain Gutierrez[1], Michel Leclère[2], Marie-Laure Mugnier[2], and Rallou Thomopoulos [2,4]

[1] Edelweiss, INRIA Sophia Antipolis Méditerranée
{First.Last}@inria.fr
[2] RCR, LIRMM, UMII, CNRS
{First.Last}@lirmm.fr
[3] KEWI, I3S, UNSA, CNRS
{First.Last}@unice.fr
[4] IATE Joint Research Unit, INRA Montpellier

Abstract. Griwes is an initiative to develop a common model and an open-source freeware platform shared by different graph-based frameworks. We provide an overview of its objectives, architecture and specifications. We detail some of the basic mathematical structures that are used to characterize the primitives for graph-based knowledge representation. We then propose to factorize recurrent knowledge representation primitives that can be shared across specific graph-based languages and we provide a proof of concept by showing how two languages (Simple Conceptual Graphs and RDF) can be described in this framework.

Keywords: graph-based languages, semantic web, platform.

1 Introduction

Graph-based knowledge representation formalisms are more and more common, from Conceptual graphs (CG) [19] which are historical descendants of semantic networks, to more recently proposed representations such as RDF[1], SKOS[1] or Topic Maps[2].

The web is playing an important role in the emergence of these new formalisms and in recent web architectures the RDF graph model became a core layer of the stack of standards[3]. Many knowledge representation frameworks are now used online (RDF, RDFS, SKOS, OWL, GRDDL, RDFa, µFormats, etc.)[1] allowing human and artificial agents to weave graphs describing web resources or just any entity and the relations existing between them. In a recent post[4] Tim Berners-Lee insisted on the

[1] W3C Semantic Web Activity http://www.w3.org/2001/sw/
[2] http://www.topicmaps.org/
[3] One Web http://www.w3.org/Consortium/technology
[4] Giant Global Graph, Tim Berners-Lee, http://dig.csail.mit.edu/breadcrumbs/node/215

P. Eklund and O. Haemmerlé (Eds.): ICCS 2008, LNAI 5113, pp. 297–310, 2008.

graph nature (Giant Global Graph) of the semantic web and the importance of this structure in developing and exploiting the semantic web (*i.e.* the web of data).

Reasonings on these different graph formalisms are often very similar. We could share many operations and their implementation across frameworks and even within them on different levels of their models e.g. transitive closure in RDFS class hierarchy, in SKOS concept narrower / broader links, in instances of OWL transitive properties, in CG the concept type hierarchy, etc. In fact when we compare these different languages we can find many similarities. Consider for instance the similarities between RDF/S and CG as underlined in [5] and [1]:

- both models consider assertions as positive, conjunctive and existential;
- both models represent assertions as labeled graphs;
- the class hierarchy (resp. property hierarchy) of RDFS is equivalent to the concept type (resp. relation type) hierarchy of CG;
- properties in RDF/S are first class citizens, declared outside classes just like relations are first class citizens in CG;
- subsumption in RDF/S is equivalent to projection in CG;

The reasonings on these different graph-based frameworks are sometimes also shared with other non graph-based formalisms e.g., databases. [19]

Tools designed and developed for these different graph-based frameworks are tailored to specific languages and/or scenarios and this criticism includes the tools we have been working on in the past years such as Cogitant [9] or Corese [5]. These experiences convinced us that it would be interesting to share these efforts and avoid re-designing and re-implementing the same structures and operators again and again. For this reason we started the project Griwes that stands for Graph-based Representations and Inferences for Web Semantics. The main objective of this initiative is to bootstrap an open-source platform, to share efforts on developing graph-based data structures and algorithms with anyone who wants to contribute. This also implies a proper definition of the considered graph structures shared by the different graph-based formalisms.

In the rest of this article we give an overview of the objectives and architecture of Griwes and we position it w.r.t. other contributions in the field (section 2). We then give some details of the basic mathematical structures that are used to characterize the primitives for graph-based knowledge representation (section 3). We proceed with the layer factorizing recurrent knowledge representation primitives that can be shared across specific graph-based languages (section 4). Finally we provide a proof of concept by showing how two languages (Simple Conceptual Graphs and RDF) can be described in this framework (section 5). We conclude with a discussion on several difficulties and perspectives we identified. These sections are extracted from the working draft of a more detailed research report from Griwes available online[5].

2 Griwes Initiative

This section is an introduction to the Griwes initiative to develop a common model and an open-source freeware platform shared by different graph-based frameworks.

---

[5] http://www-sop.inria.fr/acacia/project/griwes/

2.1 Objectives of the Griwes Initiative

In order to develop a common model and an open-source freeware platform shared by different graph-based frameworks, the objectives of the Griwes initiative can be divided into four kinds of tasks:

- Identification of users' and developers' profiles in the graph-based knowledge modeling communities and semantic web communities, and definition of usage scenarios for the platform;
- Definition of a common representation model shared by different graph-based formalisms and of architectural principles for the organization of the toolkit, allowing the platform to federate contributions and extensions and fostering reuse across graph-based representation models;
- Implementation of the API, interfaces and components in an open-source freeware platform.
- Bootstrapping a community of contributors for this platform (users and developers).

2.2 Architecture of the Griwes Toolkit

As summarized in figure 1, the current vision of the framework distinguishes three layers of abstraction and one transversal component for interaction:

- *Structure layer*: this layer gathers and defines the basic mathematical structures (e.g. oriented acyclic labeled graph) that are used to characterize the primitives for knowledge representation (e.g. type hierarchy)
- *Knowledge layer*: this layer factorizes recurrent knowledge representation primitives (e.g. a rule) that can be shared across specific knowledge representation languages (e.g. RDF/S, Conceptual Graphs).
- *Language & Strategy*: this layer is two-sided. One side gathers definitions specific to languages (e.g. RDF triple). The other side identifies the strategies that can be applied to these languages (e.g. validation of a knowledge base, completion of a fact by rules).

The *interaction and interfaces* aspect was deemed transversal to these layers. It gathers events (e.g. additional knowledge needed) and reporting capabilities (e.g. validity warning) needed to synchronize conceptual representations and interface representations. In Griwes, we intend to analyze the requirements of that aspect for each layer as soon as the first draft of these layers is stable.

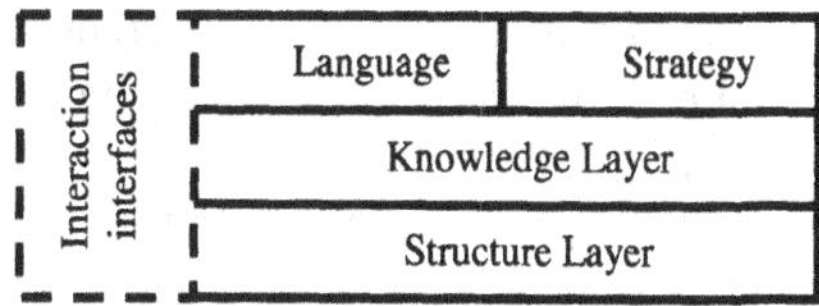

Fig. 1. The three abstraction layers of the current architecture of Griwes

Before delving into some extracts of the Knowledge and Structure layers, the next section reviews a number of contributions that prefigured, inspired and justified this initiative.

2.3 Related Work

There exist a growing number of platforms to reason on graph-based knowledge formalisms, be they in the conceptual graph families or in the RDF graph family.

On2Brocker [7] is an early ontology-based system to handle RDF annotations. Ontologies, queries and rules are expressed in Frame Logic. The query engine translates Frame Logic data into Horn Logic to answer a query. Triple [17] is a query language initially designed for RDF/S and DAML+OIL. Its core is an RDF query language based on Horn Logic extended with syntactical features supporting namespaces, resources and statements (triples). This core language is compiled into Horn Logic programs executed by a Prolog engine. The core Triple language is extended with rules for axiomatizing the semantics of RDFS; they can be used together with a Horn Logic based inference engine to derive additional knowledge from an RDF Schema specification. DAML+OIL or OWL DL cannot be mapped to Horn Logic directly and therefore Triple accesses a Description Logic classifier to handle these extensions. Triple has a layered architecture to handle different knowledge models. Both On2Brocker and Triple remain focused on logic-based engines not exploiting the graph structures of the RDF model.

Sesame [3] is a generic architecture for persistent storing of RDF(S) data into Data Based Management Systems (DBMS) and querying of RDF(S) data with the RQL language. RQL [15] is an RDF query language defined by means of a set of core queries, a set of basic filters and a way to build new queries through functional composition and iterators. When parsing an RQL query, Sesame builds an optimized query tree model from this composition which is then evaluated through a set of calls to the storage and inference modules of Sesame. Sesame supports querying at the semantic level but does not support XML Schema Datatypes, nor does it support inference rules.

DAMLJessKB [13], its successor OWLJessKB and the e-Wallet [8] are tools for reasoning with the Semantic Web and DAML or OWL-Lite. They map the RDF triples and the ontologies into facts of the CLIPS-like language of Jess[6] and apply rules implementing the semantics of RDF, RDFS, XSD and DAML or OWL-Lite. These systems can perform class instance reasoning and terminological reasoning about the relationships among classes. In addition, the e-Wallet is able to run rules to complete the knowledge base, to invoke external services to obtain new knowledge, to answer queries and to control the precision and truthfulness of answers to preserve privacy. Here again these engines remain focused on production rule reasoning not exploiting the graph structure of the RDF model and relying on their internal logic language for query expression.

Jena [4] is one of the most complete platforms offering persistence and reasoning for RDF as well as SPARQL querying. It includes a forward-chaining engine (RETE) and a backward-chaining engine to allow hybrid reasoning and to implement the

---

[6] JESS engine http://herzberg.ca.sandia.gov/

semantics of RDFS and OWL. Jena relies on a fixed database structure for large storage and on a custom data structure for main-memory storage.

WebKB [14] is an early ontology server and Web robot based on Conceptual Graphs. WebKB interprets and automatically translates into CGs chunks of knowledge statements expressed in a CG linear notation and embedded in Web documents. It also provides commands to query lexical or structural properties of HTML documents or to display specializations or generalizations of a concept or a relation or a CG. OntoSeek [10] is another early system that relies on Conceptual Graphs for ontology-driven content matching. Queries and resource annotations are lexical conceptual graphs to match one against the other. Neither WebKB nor OntoSeek handle RDF(S) data or rules. Moreover they both focus on a specific family of web applications not aiming at allowing different mapping to their graph-based representations and not providing a generic expressive query language.

With the OWL recommendation at W3C, Description Logics (DL) became especially important in the spectrum of logic-based systems on the web. Several systems exist here: Fact and its successor Fact++ [20], KAON2 [16], KAON that remains focused on RDFS, Racer [11] and Pellet [18]. These engines offer classical DL operations such as identification, classification and validation. Queries are usually limited to conjunctive queries and the graph structure of the RDF model is not exploited at the core of these engines.

To summarize, none of these contributions is offering a pivot model and an open-source platform to efficiently implement querying and reasoning on graph-based models. Most of them are tied to specific languages, logics or even applications.

Members of Griwes also developed platforms of their own over the last decade. Let us mention two of them: Cogitant [9] dedicated to conceptual graph reasoning and Corese [5] dedicated to a conceptual graph operationalisation of RDF/S.

Our own tools based upon CGs implementations and also contributions like Amine [12] relying on a combination of Prolog and CGs, suffer from their closed design preventing reuse and cross-pollination. The next section is a guided tour of some extracts of the specifications of Griwes as defined in the current working draft of its research report.

3 Structure Layer

The structure layer is the core layer of the architecture of Griwes. We extracted here some definitions of the basic mathematical structures that we chose to characterize the primitives for knowledge representation.

3.1 ERGraphs: Entity-Relation Graphs

Our core representation primitive is intended to describe a set of entities and relationships between these entities; it is called an Entity-Relation graph (ERGraph in short). An entity is anything that can be the topic of a conceptual representation. A relationship, or simply relation, might represent a property of an entity or might relate two or more entities.

The relations can have any number of arguments including zero and these arguments are totally ordered. In graph theoretical terms, an ERGraph is an *oriented hypergraph*, where nodes represent the entities and hyperarcs represent the relations on these entities. However, a hypergraph has a natural graph representation associated with it: a bipartite graph, with two kinds of nodes respectively representing entities and relations, and edges linking a relation node to the entity nodes arguments of the relation; the edges incident to a relation node are totally ordered according to the order on the arguments in the relation.

The nodes (Entities) and hyperarcs (Relations) in an ERGraph have labels. At the structure level, they are just elements of a set L that can be defined in intension or in extension. Labels obtain a meaning at the knowledge level.

Definition of an ERGraph: An ERGraph relative to a set of labels L is a 4-tuple $G=(E_G, R_G, n_G, l_G)$ where

- E_G and R_G are two disjoint finite sets respectively, of nodes called entities and of hyperarcs called relations.
- $n_G : R_G \to E_G^*$ associates to each relation a finite tuple of entities called the arguments of the relation. If $n_G(r)=(e_1,...,e_k)$ we note $n_G^i(r)=e_i$ the ith argument of r.
- $l_G: E_G \cup R_G \to L$ is a labelling function of entities and relations.

In some knowledge representation primitives and some algorithms it is useful to distinguish some entities of a graph. For this purpose we define a second core primitive, called λ–ERGraph.

Definition of a λ-ERGraph: A λ-ERGraph λ_G is a couple of an ERGraph G and a tuple of entities of G: $\lambda_G = ((e_1,...e_k), G)$, $e_i \in E_G$. We say that k is the size of λ_G and that $(e_1,...e_k)$ are distinguished in G.

Definition of an induced SubERGraph: Let $G=(E_G, R_G, n_G, l_G)$ be an ERGraph. Let $E_{G'}$ be a subset of E_G. The SubERGraph of G induced by $E_{G'}$ is the ERGraph $G'=(E_{G'}, R_{G'}, n_{G'}, l_{G'})$ defined by: (1) $R_{G'}= \{ r \in R_G \mid \forall 1 \leq i \leq card(n_G(r)), n_G^i(r) \in E_{G'} \}$ (2) $n_{G'}$ is the restriction of n_G to $R_{G'}$ (3) $l_{G'}$ is the restriction of l_G to $E_{G'} \cup R_{G'}$

Definition of a Merge: let $G=((g_1,...g_k), G')$ et $H=((h_1,...h_k), H')$ two λ-ERGraphs of same size, the merge of H in G modifies G' by adding a copy $C(H')$ of H' to G' and then for $1 \leq i \leq k$ by merging the entities $C(h_i)$ and g_i.

Note that the labels of the merged entities are obtained by applying a method defined at higher levels.

3.2 Mapping between ERGraphs

Intuitively, a Mapping associates entities of a query ERGraph to entities of an ERGraph in a knowledge base of ERGraphs. Mapping entities of graphs is a fundamental operation for comparing and reasoning with ERGraphs.

Definition of an EMapping: Let G and H be two ERGraphs, an EMapping from H to G is a partial function M from E_H to E_G *i.e.* a binary relation that associates each element of E_H with at most one element of E_G ; not every element of E_H has to be associated with an element of E_G.

Definition of an ERMapping: Let G and H be two ERGraphs, an ERMapping from H to G is an EMapping M from H to G such that: Let H' be the SubERGraph of H induced by $M^{-1}(E_G)$, $\forall r' \in R_{H'}$ $\exists r \in R_G$ such that $card(n_{H'}(r')) = card(n_G(r))$ and $\forall$ $1 \leq i \leq card(n_G(r))$, $M(n_{H'}{}^i(r')) = n_G{}^i(r)$. We call r a support of r' in M and note $r \in M(r')$

Mapping is a basic operation used in many more complex operations e.g. rule application. Let us note that by default an EMapping is partial. This enables us to manipulate and reason on EMappings *during* the process of mapping graphs. When this process is finished, the EMapping – if any – is said total: all the entities of the query graph H are mapped. In general we use specific mappings that preserve some chosen characteristics of the graphs (e.g., compatibility of labels, structural information etc.); figure 2 shows their hierarchy.

In particular an ERMapping constrains the structure of the graphs being mapped and an $\text{EMapping}_{<X>}$ constrains the labelling of entities in the graphs being mapped. An **ERMapping** is an EMapping that leads to map each relation in H to a relation in G with the same arity. An $\textbf{EMapping}_{<X>}$ is an EMapping that satisfies a compatibility relation X on entities labels. An $\textbf{ERMapping}_{<X>}$ is both an ERMapping and an $\text{EMapping}_{<X>}$. A Homomorphism is a total ERMapping. Other specializations include: injective mappings, surjective mappings, faithful mappings (preserve the absence of hyperarcs), etc.

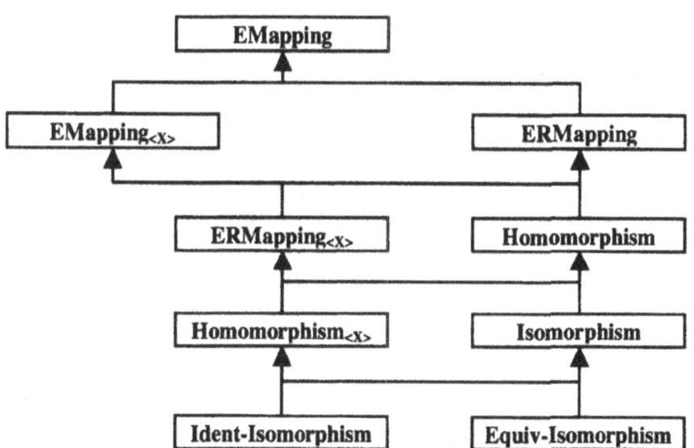

Fig. 2. EMapping specialization hierarchy

In conceptual graph projections, many systems map not only entities, but relations as well. The notion of projection as defined in conceptual graphs corresponds to a $\text{Homomorphism}_{<X>}$ that is to say a total $\text{ERMapping}_{<X>}$, where X is a preorder over the label set L.

3.3 Proofs of a Mapping

We define the proof of a mapping as a kind of "reification" of the mapping; a proof provides a static view over the dynamic operation of mapping, enabling thus to access information relative to the *state* of the mapping. Formally the proof of a mapping is the set(s) of associations detailing the exact association from each entity and relation of the query graph H to entities and relations of G.

We follow the hierarchy of mappings outlined in the previous section and associate with each kind of EMapping a notion of proof: EProof, ERProof and ERProof$_{<X>}$. For instance the proof for a homomorphism corresponds to the proof of a total ERMapping$_{<X>}$ where X is a preorder over the label set L and defined as follows:

Definition of an EProof: Let G and H be two ERGraphs, and M an EMapping from H to G. The EProof of M is a set $M_E = \{ (e_H,e_G) \in E_H \times E_G \mid e_G=M(e_H) \}$.

Definition of an ERProof: Let G and H be two ERGraphs, and M an ERMapping from H to G. Let H' be the SubERGraph of H induced by $M^{-1}(E_G)$. An ERProof of M is a couple $P=(M_E,M_R)$ where M_E is the EProof of M and $M_R= \{(r_1,r'_1),\dots (r_k,r'_k)\}$ with $\{r_1,\dots,r_k\}=R_{H'}$ and $\forall 1 \leq i \leq k\ r'_i \in M(r_i)$.

Definition of an ERProof$_{<X>}$: Let G and H be two ERGraphs, and M an EMapping from H to G. An ERProof$_{<X>}$ of M is a couple $P=(M_{EX},M_{RX})$ where M_{EX} is the EProof$_{<X>}$ of M and $M_{RX}= \{(r_1,r'_1,p_1)\dots (r_k,r'_k,p_K)\}$ where $\{(r_1,r'_1)\dots (r_k,r'_k)\}$ is the second element of an ERProof of M and $\forall 1 \leq i \leq k\ p_i$ is a proof of $(l_G(M(r)), l_H(r)) \in X$.

At this point we make no assumption on the structure of p_i and the means to obtain it. A system for comparing labels should be able to produce such proofs, e.g. a chain of subsumption relations which transitive closure confirms the comparison of two labels. Note that several different ERProofs can be associated to a same ERMapping (e.g. when there are two twin relations in G that can support a same relation of H).

3.4 Constraints System for Mappings

An EMapping constraint system is a function $\mathcal{C}$ that sets additional conditions that an EMapping must satisfy in order to be correct.It takes the form of an evaluable expression which must evaluate to true for an EMapping to satisfy the constraint system.

Definition of an EMapping Constraint System: An EMapping constraint system for an EMapping M from H to G is a function $\mathcal{C}(E)$ where E is the triple (H,P,V) called the environment, with P the proof of M and V a binary relation associating to variables v_i a unique entity or relation of H. This function can evaluate to *{true, false, unknown, error}*.

An EMapping M satisfies (resp. violates) a constraint system $\mathcal{C}$ if $\mathcal{C}(M)=true$ (resp. if $\mathcal{C}(M)=false$).

This facet of the specifications was motivated by scenarios using expressive query languages such as SPARQL [6]. For instance, let us consider the following SPARQL query and in particular its FILTER clause (line 7):

```
1. PREFIX inria: <http://www.inria.fr#>
2. SELECT ?student ?name
3. WHERE {
4.   ?student rdf:type inria:Student
5.   ?student inria:name ?name .
6.   ?student inria:age ?age .
7.   FILTER (xsd:integer(?age) > 22 && regex(?name, "A.*")) }
```

The triples of the query pattern can be seen as a graph pattern requesting students (line 4) with their name (line 5) and their age (line 6):

```
[Student]-
    (name)->[?name]
    (age)->[?age].
```

Line 7 however is an additional constraint pattern that has to be satisfied in order for the matching to be correct; it specifies that the integer value of the age has to be greater than 22 and that the name should start with an "A".

These kinds of constraints motivated the definition of constraint systems in our specifications but constraint systems are also envisaged to provide efficient access means to indexes of graphs, for instance to retrieve all the arcs of a graph satisfying a given constraint system.

4 Knowledge Layer

In our architecture, a knowledge base B is defined by a vocabulary, one or several bases of facts, optionally a base of rules and a base of queries. *B= (Vocabulary, Fact Base* $^+$*, Rule Base*, Query Base*).*

A vocabulary is a set of none necessarily disjoint named sets of elements called vocabulary subsets together with preorders on the union of these sets:

Definition of a Vocabulary: A Vocabulary V is a tuple $V = \left(U = \bigcup_{1 \le i \le k} V_i, (\le_1, \ldots, \le_q) \right)$ where V_i are k sets of elements and $\le_i$ are q preorders on U.

Definition of a Fact: A Fact is an ERGraph.

Definition of a Base of Facts: A Base of Facts is a set of Facts.

Let us note that every ERGraph G in a base of facts respects $l_G : E_G \cup R_G \rightarrow L$ where L is constructed from the set U of elements of the vocabulary V of the knowledge base.

Definition of a Query: A Query is a couple $Q=(q, \mathcal{C})$ of a λ-ERGraph $q=((e_1,\ldots e_k), G)$ and a Constraint system $\mathcal{C}$.

The answers to a query depend on the kind of EMapping used to query the base. In the next definitions, the letter X stands for a type of EMapping;

X-Answer to a Query: Let $Q=(((e_1,\ldots e_k), G), \mathcal{C})$ be a query and F be a Fact. $A=(a_1,\ldots a_k)$ is an X-Answer to Q in F iff there exists an EMapping M of type X from G to F satisfying $\mathcal{C}$ such that $M(e_i)=a_i$.

Note: the proof of an X-Answer is the proof of the EMapping associated to that X-Answer.

Definition of a Base of Queries: A Base of Queries is a set of Queries.

Definition of a Rule: A Rule is a couple $R=(H,C)$ of a Query $H=(G, \mathcal{C})$ and a λ-ERGraph C of the same size as G. H is the hypothesis of the rule, and C is its conclusion.

***X*-applicable Rule:** A rule $R=(H,C)$ is X-applicable to a fact F iff there exists an X-Answer to H in F.

X-applying a Rule: Let $R=(H,C)$ be a rule X-applicable to a fact F, and A be an X-Answer to H in F. The X-Application of R on F with respect to A merges C in (A,F).

Definition of a Base of Rules: A Base of Rules is a set of Rules.

Definition of an ERFunction: An ERFunction F is a function associating to an ERProof P a label or an error.

Definition of a functional ERGraph: A functional ERGraph is an ERGraph where some entities or relations are labelled with ERFunctions.

Evaluation of a functional ERGraph: The evaluation of a functional ERGraph G with respect to an EProof P and an environment E is a copy G' of G where every functional label is replaced by the evaluation of the function against P. If any of the evaluations returns an error then $G'=\emptyset$.

Definition of a Functional Rule: A functional rule is a rule $R=(H,C)$ where C is a functional λ-ERGraph.

X-applying a Functional Rule: let $R=(H,C)$ be a functional rule X-applicable to a fact F, and A be an X-Answer to H in F and P be a proof of that X-Answer. The X-functional-Application of R on F with respect to P merges the evaluation of C with respect to P in (A,F).

Definition of Co-Reference: A Co-Reference relation R is an equivalence relation over the set of entities of G..

Definition of a Normal Form: let G be an ERGraph with a co-reference relation R and a function $fusion(E_1,E_2,..., E_n)$ that returns a new entity from a set of entities, the normal form of G is the graph $NF(G)$ obtained by merging every entities of a same equivalence class defined by R as a new entity calculated by calling *fusion* on the entities of this class.

Co-reference and fusion are abstract functions which must be specified at the language level.

5 Validating Against Two Languages: Simple Graphs and RDF

This article focuses on the structure layer and the knowledge layer of Griwes and does not include a description of the language and strategy layer still under discussions. However this section shows how the primitives of the pivot model defined in Griwes can be used to represent the semantics of two languages: Simple (conceptual) Graphs and RDF. This practice would, ultimately, be the objective of the language layer.

5.1 Representing Simple Graphs in the Griwes Model

Non-surprisingly, the SG [2] graphs map smoothly to the core model of Griwes since this model was inspired by the conceptual graphs formalism.

Primitive SG	Griwes translation
Primitive concept type	Member of a specific finite vocabulary sub-set *TC* defined in extension. This finite vocabulary sub-set has a partial order $\leq_{TC}$.
Primitive relation type	Member of a specific finite vocabulary sub-set *TR* defined in extension and providing a label *l*, an arity *k*, and a signature $s \in (TCC)^k$. This finite vocabulary sub-set has a partial order $\leq_{TR}$ defined only for labels with the same arity.
Conjunctive concept type	Member of a specific vocabulary sub-set *TCC* defined in intension; sub-set of power set of Primitive concept types. This finite vocabulary sub-set has a partial order $\leq_{TCC}$ derived from $\leq_{TC}$. NB: $TC \subset TCC$
Individual marker and Generic marker *	Member of a specific finite vocabulary sub-set $M=I\cup\{*\}$ defined in intension. This finite vocabulary sub-set has a partial order $\leq_M$ such that $\forall i \in M\ i \leq_M *$.
Concept	An entity where the label is a couple (t, m) with $t \in TCC$ and $m \in M$. We define $\leq_C$ a partial order on these labels such that $(t_1, m_1) \leq_C (t_2, m_2)$ iff $t_1 \leq_{TCC} t_2$ and $m_1 \leq_M m_2$.
Relation	a relation where the label is a type $t \in TR$
Fact	*A Fact.*
Simple Graph	An ERGraph respecting labelling functions.
Query	A query $Q=(q, C)$ with $\mathcal{C}=\emptyset$.
Rule	A rule R=(H, C) with $\mathcal{C}=\emptyset$.
Banned concept type	Member of a specific vocabulary sub-set *BT* sub set of power set of primitive concept types; members of this sub-set should never be used in other sets of the vocabulary, in facts, in queries or rules.
Support	the vocabulary *V*.
Graph specialization	Let $\leq$ be the partial order defined by $\leq_C$ when applied to two entities, by $\leq_{TR}$ when applied to two relations, and not holding for any other case. A graph *G* specializes a graph *H* if there exists a homomorphism$_\leq$ from *H* to *G*.
Graph deduction	*H* is deduced from *G* iff the normal form *NF(G)* specializes *H* or *G* is inconsistent; *NF(G)* is defined by $coref_{SG}$ and $fusion_{SG}$.

5.2 Representing RDF in the Griwes Model

This section shows how the RDF graph model can be mapped to the core model of Griwes. Mappings given in the following table rely on the following preorder.

Definition: let $\leq_{RDF}$ be a preorder over *V* such that

- $x \leq_{RDF} y$ if $y \in$ Blanks
- $x \leq_{RDF} y$ if $x, y \in Literals^2$ and value(x)=value(y)
- $x \leq_{RDF} y$ if x=y

Primitive RDF	Griwes translation
Blank	Member of a specific vocabulary sub-set defined in intension.
Literal	Member of a specific vocabulary sub-set defined in intension.
Literal ^^datatype	Member of a specific vocabulary sub-set defined in intension.
Literal @lang	Member of a specific vocabulary sub-set defined in intension.
URI ref	Member of a specific vocabulary sub-set defined in intension.

Triple: subject, predicate, object (x p y)	a relation in an ERGraph ; it would naturally be binary but additional coding information may be added with n-ary relations e.g. quadratic relations specifying the source and the property reified. The ERGraph *G* includes the relation R_p such that $n_G(R_p)=(e_x,e_p,e_y)$
RDF graph G (i.e. a **set** of triples on a given vocabulary)	An ERGraph such that for each distinct term *t* appearing in a triple of *G* the ERGraph *E* associated to *G* contains a distinct entity *e(t)* and for each triple *(s,p,o)* of *G*, *E* contains a relation *r* such that $n_E(r)=(e(s),e(p),e(o))$. Remark : a well-formed RDF ERGraph: - has no isolated entity; - first element of relations must not be a Literal; - a property name is only a URI ref; One may have to work on non-well-formed RDF ERGraph.
RDF nodes	Entities appearing in position 1 and 3 of a relation.
Vocabulary (set of names)	Vocabulary.
RDF Vocabulary (rdf:Property, rdf:type)	a specific vocabulary sub-set defined in extension for RDF.
Simple RDF entailment	H entails G iff there exists a Homomorphism$_{\leq RDF}$ from G to the normal form NF(H) defined by *coref*$_{RDF}$ and *fusion*$_{RDF}$.
RDF axioms	the ERGraph representation of the triples of the axiomatic triples of RDF are asserted in every base of facts.
x rdf:type t	as any other triple. (NB: *t* can be integrated in the label of the entity representing *x*)
RULE 1 **IF** *x p y* in RDF graph *G* **THEN** *p rdf:type rdf:Property*	*R=(H,C)* where *H=((e(y)),H')* with *H'* is the graph associated with *{(x,y,z)}* where *x*, *y* and *z* are blanks and *C=((e(u)),C')* with *C'* the graph associated with *{(u, rdf:type, rdf:Property)}* where *u* is a blank and *rdf:type* and *rdf:Property* are URI refs of the RDF vocabulary.
RULE 2 **IF** *x p y^^d* in RDF graph *G* and *y^^d* well-typed **THEN** *y^^d rdf:type d*	R=(Q,D) a functional rule, where *Q=(H,C)* with *H=((e(z)),H')* with *H'* is the graph associated with *{(x,y,z)}* where *x*, *y* and *z* are blanks, C is satisfied iff e(z) is labelled by a well-typed datatype literal. *D=((e(a)),D')* is the lambda functional ERGraph associated with *{(a, rdf:type, fun:getType(im(e(z)))), (x, fun:id(im(r(y)), fun:getNormalForm(im(e(z)))) , (fun:getNormalForm(im(e(z))), rdf:type, fun:getType(im(e(z)))) }* where *a* is a blank and *rdf:type* is a URI ref of the RDF vocabulary and *fun:getType()* is a function extracting the type from a literal.

6 Discussion

In this article we presented an initiative to design a common model and specify a platform to share state-of-the-art structures and algorithms across several graph-based knowledge representation frameworks such as RDF/S, Conceptual Graphs, Topic Maps or SKOS. This article is extracted from the working draft of a more detailed research report from Griwes available online[7].

We identified a number of limitations and problems that we intend to address in a near future:

- **Generalization of lambdas to relation labels:** we may have to consider two tuples in lambda graphs, a tuple of entities and a tuple of relations (or a tuple of entities and relations) in order to use variables on relations as allowed in SPARQL.

[7] http://www-sop.inria.fr/acacia/project/griwes/

- **Structure of proofs:** at this point of the design, we made no assumptions on the structure of the proofs and the means to obtain them; this may have to be detailed in the future and extended to reasoning in general.
- **Index of graphs:** in order to wrap different efficient accesses to graphs and also heterogeneous arc producers (e.g. database wrappers) we are currently working on introducing indexes as companion structures of graphs that provide constrained listing of the components of a graph to support efficient access mechanisms.
- **Relations with different arities:** in the ERMapping, we may have to generalize the constraint on arity and matching for instance to map relations with different arities or different orders in their arguments.
- **Complex modifiers in queries:** a query language like SPARQL introduces constructors for representing optional parts in queries, disjunctive parts, constraints with complex scopes, constraints between different answers to a query, etc. These extensions will require additional work.
- **Architectural choices:** for instance there is an ongoing discussion on the status of queries and the fact they should or should not be linked to knowledge base.
- **Subtleties in domains of interpretations:** the distinction between terms and values in SPARQL-RDF is full of complex cases that require us to find the right compromise between efficiency and size of data.

To illustrate these questions, let us just detail this last example to consider the options one could have in representing datatyped literals and their value. Currently RULE 2 of the RDF mapping presented here does not cover coreference between a Literal Entity and its datatyped value representation. We identified three solutions to this problem:

- Explicitly indicate coreference between these entities and handle them in the algorithms;
- Consider composite labels representing sets of literals and modify preorders on labels and normalization so as to indicate original destinations of arcs on the arcs themselves;
- Use hyperarcs containing the literal representation, its type and its value and modify ERMappings to handle a variable number of arguments in the arc.

The current work in Griwes includes discussing these options and finding the right compromise between efficiency, generality and feasibility.

To summarize, we now have a first draft of three layers of our architecture. We intend to refine and extend this architecture and, even more importantly, to start the open-source design of the corresponding APIs and their implementations.

Acknowledgments. We are grateful to the COLOR funding program of INRIA.

References

1. Baget, J.B.: RDF entailment as a graph homomorphism. In: Proceedings of the 4th International Semantic Web Conference. LNCS, pp. 82–96. Springer, Heidelberg (2005)
2. Baget, J.-F., Mugnier, M.-L.: The SG Family: Extensions of Simple Conceptual Graphs. In: IJCAI, pp. 205–212 (2001)

3. Broekstra, J., Kampman, A., van Harmelen, F.: Sesame: A Generic Architecture for Storing and Querying RDF and RDF Schema. In: Horrocks, I., Hendler, J. (eds.) ISWC 2002. LNCS, vol. 2342, pp. 54–68. Springer, Heidelberg (2002)
4. Carroll, J. J., Dickinson, I., Dollin, C., Reynolds, D., Seaborne, A., Wilkinson, K.: Jena: Implementing the semantic web recommendations. Technical Report HP Lab (2003)
5. Corby, O., Dieng-Kuntz, R., Faron-Zucker, C.: Querying the Semantic Web with the Corese Search Engine. In: ECAI, pp. 705–709. IOS Press, Amsterdam (2004)
6. Corby, O., Faron-Zucker, C.: Implementation of SPARQL Query Language based on Graph Homomorphism. In: Proc. 15th International Conference on Conceptual Structures (2007)
7. Fensel, D., Angele, J., Decker, S., Erdmann, M., Schnurr, H.-P., Staab, S., Studer, R., Witt, A.: On2broker: Semantic-based access to information sources at the WWW. In: World Conference on the WWW and Internet (1999)
8. Gandon, F., Sadeh, N.: Semantic Web Technologies to Reconcile Privacy and Context Awareness. Web Semantics: Science, Services and Agents on the World Wide Web 1(3), 241–260 (2004)
9. Genest, D., Salvat, E.: A Platform Allowing Typed Nested Graphs: How CoGITo Became CoGITaNT. In: Mugnier, M.-L., Chein, M. (eds.) ICCS 1998. LNCS (LNAI), vol. 1453, pp. 154–161. Springer, Heidelberg (1998)
10. Guarino, N., Masolo, C., Vetere, G.: Ontoseek: Content-based access to the Web. IEEE Intelligent, Systems 14(3), 70–80 (1999)
11. Haarslev, V., Möller, R.: Racer: An OWL Reasoning Agent for the Semantic Web. In: Proceedings of the International Workshop on Applications, Products and Services of Web-based Support Systems, Halifax, Canada, October 13, pp. 91–95 (2003)
12. Kabbaj, A., Bouzoubaa, K., ElHachimi, K., Ourdani, N.: Ontology in Amine Platform: Structures and Processes. In: 14th Proc. Int. Conf. Conceptual Structures, ICCS 2006, Aalborg, Denmark (2006)
13. Kopena, J., Regli, W.: DAMLJessKB: A Tool for Reasoning with the Semantic Web. IEEE Intelligent Systems 18(3), 74–77 (2003)
14. Martin, P., Eklund, P.: Knowledge Retrieval and the World Wide Web. IEEE Intelligent, Systems 15(3), 18–25 (2000)
15. Miller, L., Seaborne, A., Reggiori, A.: Three Implementations of SquishQL, a Simple RDF Query Language. In: Horrocks, I., Hendler, J. (eds.) ISWC 2002. LNCS, vol. 2342, pp. 423–435. Springer, Heidelberg (2002)
16. Motik, Sattler, U.: KAON2, A Comparison of Reasoning Techniques for Querying Large Description Logic Aboxes. In: Hermann, M., Voronkov, A. (eds.) LPAR 2006. LNCS (LNAI), vol. 4246, pp. 227–241. Springer, Heidelberg (2006)
17. Sintek, M., Decker, S.: Triple: A Query, Inference and Transformation Language for the Semantic Web. In: Horrocks, I., Hendler, J. (eds.) ISWC 2002. LNCS, vol. 2342, pp. 364–378. Springer, Heidelberg (2002)
18. Sirin, E., Parsia, B., Grau, B.C., Kalyanpur, A., Katz, Y.: Pellet: A practical OWL-DL reasoner. Journal of Web Semantics 5(2) (2007)
19. Sowa, J.F.: Conceptual graphs for a database interface. IBM Journal of Research and Development 20(4), 336–357 (1976)
20. Tsarkov, D., Horrocks, I.: FaCT++ Description Logic Reasoner: System Description. LNCS, vol. 4130. Springer, Heidelberg (2006)

Author Index

Lecture Notes in Artificial Intelligence (LNAI)

Vol. 5113: P. Eklund, O. Haemmerlé (Eds.), Conceptual Structures: Knowledge Visualization and Reasoning. X, 311 pages. 2008.

Vol. 5110: W. Hodges, R. de Queiroz (Eds.), Logic, Language, Information and Computation. VIII, 313 pages. 2008.

Vol. 5108: P. Perner, O. Salvetti (Eds.), Advances in Mass Data Analysis of Images and Signals in Medicine, Biotechnology, Chemistry and Food Industry. VI, 173 pages. 2008.

Vol. 5097: L. Rutkowski, R. Tadeusiewicz, L.A. Zadeh, J.M. Zurada (Eds.), Artificial Intelligence and Soft Computing – ICAISC 2008. XVI, 1269 pages. 2008.

Vol. 5078: E. André, L. Dybkjær, W. Minker, H. Neumann, R. Pieraccini, M. Weber (Eds.), Perception in Multimodal Dialogue Systems. X, 311 pages. 2008.

Vol. 5077: P. Perner (Ed.), Advances in Data Mining. XI, 428 pages. 2008.

Vol. 5076: R. van der Meyden, L. van der Torre (Eds.), Deontic Logic in Computer Science. X, 279 pages. 2008.

Vol. 5064: L. Prevost, S. Marinai, F. Schwenker (Eds.), Artificial Neural Networks in Pattern Recognition. IX, 318 pages. 2008.

Vol. 5040: M. Asada, J.C.T. Hallam, J.-A. Meyer, J. Tani (Eds.), From Animals to Animats 10. XIII, 530 pages. 2008.

Vol. 5032: S. Bergler (Ed.), Advances in Artificial Intelligence. XI, 382 pages. 2008.

Vol. 5027: N.T. Nguyen, L. Borzemski, A. Grzech, M. Ali (Eds.), New Frontiers in Applied Artificial Intelligence. XVIII, 879 pages. 2008.

Vol. 5012: T. Washio, E. Suzuki, K.M. Ting, A. Inokuchi (Eds.), Advances in Knowledge Discovery and Data Mining. XXIV, 1102 pages. 2008.

Vol. 5009: G. Wang, T. Li, J.W. Grzymala-Busse, D. Miao, A. Skowron, Y. Yao (Eds.), Rough Sets and Knowledge Technology. XVIII, 765 pages. 2008.

Vol. 5001: U. Visser, F. Ribeiro, T. Ohashi, F. Dellaert (Eds.), RoboCup 2007: Robot Soccer World Cup XI. XIV, 566 pages. 2008.

Vol. 4994: A. An, S. Matwin, Z.W. Raś, D. Ślęzak (Eds.), Foundations of Intelligent Systems. XVII, 653 pages. 2008.

Vol. 4953: N.T. Nguyen, G.S. Jo, R.J. Howlett, L.C. Jain (Eds.), Agent and Multi-Agent Systems: Technologies and Applications. XX, 909 pages. 2008.

Vol. 4946: I. Rahwan, S. Parsons, C. Reed (Eds.), Argumentation in Multi-Agent Systems. X, 235 pages. 2008.

Vol. 4944: Z.W. Raś, S. Tsumoto, D.A. Zighed (Eds.), Mining Complex Data. X, 265 pages. 2008.

Vol. 4938: T. Tokunaga, A. Ortega (Eds.), Large-Scale Knowledge Resources. IX, 367 pages. 2008.

Vol. 4933: R. Medina, S. Obiedkov (Eds.), Formal Concept Analysis. XII, 325 pages. 2008.

Vol. 4930: I. Wachsmuth, G. Knoblich (Eds.), Modeling Communication with Robots and Virtual Humans. X, 337 pages. 2008.

Vol. 4929: M. Helmert, Understanding Planning Tasks. XIV, 270 pages. 2008.

Vol. 4924: D. Riaño (Ed.), Knowledge Management for Health Care Procedures. X, 161 pages. 2008.

Vol. 4923: S.B. Yahia, E.M. Nguifo, R. Belohlavek (Eds.), Concept Lattices and Their Applications. XII, 283 pages. 2008.

Vol. 4914: K. Satoh, A. Inokuchi, K. Nagao, T. Kawamura (Eds.), New Frontiers in Artificial Intelligence. X, 404 pages. 2008.

Vol. 4911: L. De Raedt, P. Frasconi, K. Kersting, S. Muggleton (Eds.), Probabilistic Inductive Logic Programming. VIII, 341 pages. 2008.

Vol. 4908: M. Dastani, A. El Fallah Seghrouchni, A. Ricci, M. Winikoff (Eds.), Programming Multi-Agent Systems. XII, 267 pages. 2008.

Vol. 4898: M. Kolp, B. Henderson-Sellers, H. Mouratidis, A. Garcia, A.K. Ghose, P. Bresciani (Eds.), Agent-Oriented Information Systems IV. X, 292 pages. 2008.

Vol. 4897: M. Baldoni, T.C. Son, M.B. van Riemsdijk, M. Winikoff (Eds.), Declarative Agent Languages and Technologies V. X, 245 pages. 2008.

Vol. 4894: H. Blockeel, J. Ramon, J. Shavlik, P. Tadepalli (Eds.), Inductive Logic Programming. XI, 307 pages. 2008.

Vol. 4885: M. Chetouani, A. Hussain, B. Gas, M. Milgram, J.-L. Zarader (Eds.), Advances in Nonlinear Speech Processing. XI, 284 pages. 2007.

Vol. 4874: J. Neves, M.F. Santos, J.M. Machado (Eds.), Progress in Artificial Intelligence. XVIII, 704 pages. 2007.

Vol. 4870: J.S. Sichman, J. Padget, S. Ossowski, P. Noriega (Eds.), Coordination, Organizations, Institutions, and Norms in Agent Systems III. XII, 331 pages. 2008.

Vol. 4869: F. Botana, T. Recio (Eds.), Automated Deduction in Geometry. X, 213 pages. 2007.

Vol. 4865: K. Tuyls, A. Nowe, Z. Guessoum, D. Kudenko (Eds.), Adaptive Agents and Multi-Agent Systems III. VIII, 255 pages. 2008.

Vol. 4850: M. Lungarella, F. Iida, J.C. Bongard, R. Pfeifer (Eds.), 50 Years of Artificial Intelligence. X, 399 pages. 2007.

Vol. 4845: N. Zhong, J. Liu, Y. Yao, J. Wu, S. Lu, K. Li (Eds.), Web Intelligence Meets Brain Informatics. XI, 516 pages. 2007.

Vol. 4840: L. Paletta, E. Rome (Eds.), Attention in Cognitive Systems. XI, 497 pages. 2007.

Vol. 4830: M.A. Orgun, J. Thornton (Eds.), AI 2007: Advances in Artificial Intelligence. XIX, 841 pages. 2007.

Vol. 4828: M. Randall, H.A. Abbass, J. Wiles (Eds.), Progress in Artificial Life. XII, 402 pages. 2007.

Vol. 4827: A. Gelbukh, Á.F. Kuri Morales (Eds.), MICAI 2007: Advances in Artificial Intelligence. XXIV, 1234 pages. 2007.

Vol. 4826: P. Perner, O. Salvetti (Eds.), Advances in Mass Data Analysis of Signals and Images in Medicine, Biotechnology and Chemistry. X, 183 pages. 2007.

Vol. 4819: T. Washio, Z.-H. Zhou, J.Z. Huang, X. Hu, J. Li, C. Xie, J. He, D. Zou, K.-C. Li, M.M. Freire (Eds.), Emerging Technologies in Knowledge Discovery and Data Mining. XIV, 675 pages. 2007.

Vol. 4811: O. Nasraoui, M. Spiliopoulou, J. Srivastava, B. Mobasher, B. Masand (Eds.), Advances in Web Mining and Web Usage Analysis. XII, 247 pages. 2007.

Vol. 4798: Z. Zhang, J.H. Siekmann (Eds.), Knowledge Science, Engineering and Management. XVI, 669 pages. 2007.

Vol. 4795: F. Schilder, G. Katz, J. Pustejovsky (Eds.), Annotating, Extracting and Reasoning about Time and Events. VII, 141 pages. 2007.

Vol. 4790: N. Dershowitz, A. Voronkov (Eds.), Logic for Programming, Artificial Intelligence, and Reasoning. XIII, 562 pages. 2007.

Vol. 4788: D. Borrajo, L. Castillo, J.M. Corchado (Eds.), Current Topics in Artificial Intelligence. XI, 280 pages. 2007.

Vol. 4775: A. Esposito, M. Faundez-Zanuy, E. Keller, M. Marinaro (Eds.), Verbal and Nonverbal Communication Behaviours. XII, 325 pages. 2007.

Vol. 4772: H. Prade, V.S. Subrahmanian (Eds.), Scalable Uncertainty Management. X, 277 pages. 2007.

Vol. 4766: N. Maudet, S. Parsons, I. Rahwan (Eds.), Argumentation in Multi-Agent Systems. XII, 211 pages. 2007.

Vol. 4760: E. Rome, J. Hertzberg, G. Dorffner (Eds.), Towards Affordance-Based Robot Control. IX, 211 pages. 2008.

Vol. 4755: V. Corruble, M. Takeda, E. Suzuki (Eds.), Discovery Science. XI, 298 pages. 2007.

Vol. 4754: M. Hutter, R.A. Servedio, E. Takimoto (Eds.), Algorithmic Learning Theory. XI, 403 pages. 2007.

Vol. 4737: B. Berendt, A. Hotho, D. Mladenic, G. Semeraro (Eds.), From Web to Social Web: Discovering and Deploying User and Content Profiles. XI, 161 pages. 2007.

Vol. 4733: R. Basili, M.T. Pazienza (Eds.), AI*IA 2007: Artificial Intelligence and Human-Oriented Computing. XVII, 858 pages. 2007.

Vol. 4724: K. Mellouli (Ed.), Symbolic and Quantitative Approaches to Reasoning with Uncertainty. XV, 914 pages. 2007.

Vol. 4722: C. Pelachaud, J.-C. Martin, E. André, G. Chollet, K. Karpouzis, D. Pelé (Eds.), Intelligent Virtual Agents. XV, 425 pages. 2007.

Vol. 4720: B. Konev, F. Wolter (Eds.), Frontiers of Combining Systems. X, 283 pages. 2007.

Vol. 4702: J.N. Kok, J. Koronacki, R. Lopez de Mantaras, S. Matwin, D. Mladenič, A. Skowron (Eds.), Knowledge Discovery in Databases: PKDD 2007. XXIV, 640 pages. 2007.

Vol. 4701: J.N. Kok, J. Koronacki, R. Lopez de Mantaras, S. Matwin, D. Mladenič, A. Skowron (Eds.), Machine Learning: ECML 2007. XXII, 809 pages. 2007.

Vol. 4696: H.-D. Burkhard, G. Lindemann, R. Verbrugge, L.Z. Varga (Eds.), Multi-Agent Systems and Applications V. XIII, 350 pages. 2007.

Vol. 4694: B. Apolloni, R.J. Howlett, L. Jain (Eds.), Knowledge-Based Intelligent Information and Engineering Systems, Part III. XXIX, 1126 pages. 2007.

Vol. 4693: B. Apolloni, R.J. Howlett, L. Jain (Eds.), Knowledge-Based Intelligent Information and Engineering Systems, Part II. XXXII, 1380 pages. 2007.

Vol. 4692: B. Apolloni, R.J. Howlett, L. Jain (Eds.), Knowledge-Based Intelligent Information and Engineering Systems, Part I. LV, 882 pages. 2007.

Vol. 4687: P. Petta, J.P. Müller, M. Klusch, M. Georgeff (Eds.), Multiagent System Technologies. X, 207 pages. 2007.

Vol. 4682: D.-S. Huang, L. Heutte, M. Loog (Eds.), Advanced Intelligent Computing Theories and Applications. XXVII, 1373 pages. 2007.

Vol. 4676: M. Klusch, K.V. Hindriks, M.P. Papazoglou, L. Sterling (Eds.), Cooperative Information Agents XI. XI, 361 pages. 2007.

Vol. 4667: J. Hertzberg, M. Beetz, R. Englert (Eds.), KI 2007: Advances in Artificial Intelligence. IX, 516 pages. 2007.

Vol. 4660: S. Džeroski, L. Todorovski (Eds.), Computational Discovery of Scientific Knowledge. X, 327 pages. 2007.

Vol. 4659: V. Mařík, V. Vyatkin, A.W. Colombo (Eds.), Holonic and Multi-Agent Systems for Manufacturing. VIII, 456 pages. 2007.

Vol. 4651: F. Azevedo, P. Barahona, F. Fages, F. Rossi (Eds.), Recent Advances in Constraints. VIII, 185 pages. 2007.

Vol. 4648: F. Almeida e Costa, L.M. Rocha, E. Costa, I. Harvey, A. Coutinho (Eds.), Advances in Artificial Life. XVIII, 1215 pages. 2007.

Vol. 4635: B. Kokinov, D.C. Richardson, T.R. Roth-Berghofer, L. Vieu (Eds.), Modeling and Using Context. XIV, 574 pages. 2007.

Vol. 4632: R. Alhajj, H. Gao, X. Li, J. Li, O.R. Zaïane (Eds.), Advanced Data Mining and Applications. XV, 634 pages. 2007.

Vol. 4629: V. Matoušek, P. Mautner (Eds.), Text, Speech and Dialogue. XVII, 663 pages. 2007.